Lecture Notes in Mathematics 1940

Editors:
J.-M. Morel, Cachan
F. Takens, Groningen
B. Teissier, Paris

FONDAZIONE
CIME
ROBERTO CONTI

CENTRO INTERNAZIONALE MATEMATICO ESTIVO
INTERNATIONAL MATHEMATICAL SUMMER CENTER

C.I.M.E. means Centro Internazionale Matematico Estivo, that is, International Mathematical Summer Center. Conceived in the early fifties, it was born in 1954 and made welcome by the world mathematical community where it remains in good health and spirit. Many mathematicians from all over the world have been involved in a way or another in C.I.M.E.'s activities during the past years.

So they already know what the C.I.M.E. is all about. For the benefit of future potential users and co-operators the main purposes and the functioning of the Centre may be summarized as follows: every year, during the summer, Sessions (three or four as a rule) on different themes from pure and applied mathematics are offered by application to mathematicians from all countries. Each session is generally based on three or four main courses (24−30 hours over a period of 6-8 working days) held from specialists of international renown, plus a certain number of seminars.

A C.I.M.E. Session, therefore, is neither a Symposium, nor just a School, but maybe a blend of both. The aim is that of bringing to the attention of younger researchers the origins, later developments, and perspectives of some branch of live mathematics.

The topics of the courses are generally of international resonance and the participation of the courses cover the expertise of different countries and continents. Such combination, gave an excellent opportunity to young participants to be acquainted with the most advance research in the topics of the courses and the possibility of an interchange with the world famous specialists. The full immersion atmosphere of the courses and the daily exchange among participants are a first building brick in the edifice of international collaboration in mathematical research.

C.I.M.E. Director
Pietro ZECCA
Dipartimento di Energetica "S. Stecco"
Università di Firenze
Via S. Marta, 3
50139 Florence
Italy
e-mail: zecca@unifi.it

C.I.M.E. Secretary
Elvira MASCOLO
Dipartimento di Matematica
Università di Firenze
viale G.B. Morgagni 67/A
50134 Florence
Italy
e-mail: mascolo@math.unifi.it

For more information see CIME's homepage: http://www.cime.unifi.it

CIME's activity is supported by:

− Istituto Nazionale di Alta Mathematica "F. Severi"
− Ministero dell'Istruzione, dell'Università e della Ricerca

Jacek Banasiak · Vincenzo Capasso
Mark A.J. Chaplain · Mirosław Lachowicz
Jacek Miękisz

Multiscale Problems in the Life Sciences

From Microscopic to Macroscopic

Lectures given at the Banach Center
and C.I.M.E. Joint Summer School
held in Będlewo, Poland
September 4–9, 2006

Editors:
Vincenzo Capasso
Mirosław Lachowicz

 Springer

Authors and Editors

Jacek Banasiak
School of Mathematical Sciences
University of Kwa-Zulu-Natal (Westville)
Private Bag X54001
Durban 4000
South Africa
banasiak@ukzn.ac.za

Vincenzo Capasso
Department of Mathematics &
ADAMSS (Centre for Advanced Applied
  Mathematical and Statistical Sciences)
University of Milan
Via Saldini 50
20133 Milano
Italy
vincenzo.capasso@unimi.it

Mark A.J. Chaplain
The SIMBIOS Centre
Division of Mathematics
University of Dundee
Dundee DD1 4HN
Scotland
chaplain@maths.dundee.ac.uk

Mirosław Lachowicz
Jacek Miękisz
Institute of Applied Mathematics
  and Mechanics
Faculty of Mathematics, Informatics
  and Mechanics
University of Warsaw
ul. Banacha, 2
02-097 Warsaw
Poland
lachowic@mimuw.edu.pl
miekisz@mimuw.edu.pl

ISBN: 978-3-540-78360-2   e-ISBN: 978-3-540-78362-6
DOI: 10.1663/978-3-540-78362-6

Lecture Notes in Mathematics ISSN print edition: 0075-8434
                ISSN electronic edition: 1617-9692

Library of Congress Control Number: 2008921566

Mathematics Subject Classification (2000): 35K57, 47BXX, 60HXX, 60FXX, 65-XX, 82CXX, 91AXX, 92CXX, 92DXX

Cover design: WMXDesign GmbH

Printed on acid-free paper

9 8 7 6 5 4 3 2 1

springer.com

Preface

"... It also happens...
that branches which were thought
to be completely disparate
are suddenly seen to be related ..."

Michiel Hazewinkel, 1977

During September 4–9, 2006 the Stefan Banach International Center (BC) of the Institute of Mathematics of the Polish Academy of Sciences and the (Italian) International Summer Institute for Mathematics (CIME) jointly organized at the Mathematical Research and Conference Center (MRCC), Będlewo, Poland, the School *From a Microscopic to Macroscopic Description of Complex Systems*. In addition to the main speakers, whose contributions are included in this volume, there were a significant number of participants from 11 countries. In parallel with the School, a workshop on *Modelling Cellular Systems with Applications to Tumor Growth* was organized in the framework of the activity of the EU MCRTN "MRTN-CT-2004-503661". The courses were targeted at Ph.D. students and young researchers and have had an educational character, whereas the workshop offered presentations on particular applications to modelling tumour growth phenomena.

The aim of the School has been to offer a broad presentation of updated methods suitable to provide a mathematical framework for the development of a hierarchy of models of complex systems in the natural sciences, with special attention to biology and medicine. The mastering of complexity implies the sharing of different tools which require a much higher level of communication between different mathematical and scientific schools, for solving classes of problems of the same nature. Nowadays, more than ever, one of the most important challenges derives from the bridging of parts of a system evolving at different time and space scales, especially with respect to computational affordability. Therefore, the courses have had a rather general character and method; the main role is played here by stochastic processes, positive semi-groups, asymptotic analysis, continuum theory and game theory.

For many biological systems only non-negative states or solutions make sense. The theory of Banach lattices and positive operators, developed in the series of lectures *Positivity in the Natural Sciences"* by Jacek Banasiak, provides a mathematical framework to address such problems. The lectures show how the interplay of positivity and compactness yields very strong results in many fields ranging from well-posedness of the problem at hand, through long time behaviour of solutions (including emergence of chaos), to asymptotic analysis of systems displaying multiple scale phenomena. Theoretical results are applied to a variety of specific problems occurring in natural sciences, including birth-and-death type models that describe the development of drug resistance in cancer cells, blood cells' evolution equation, singularly perturbed models of sole migration, or diffusion approximation of the Fokker–Planck equation.

As a paradigmatic microcosm for all of biology, i.e. as an observable system where mutation and evolution take place, in the course on *Cancer* by Mark Chaplain, the different aspects of the growth phases of a tumour are described(raise query?) solid tumours (the most frequent of all cancers) progress through several key stages of growth from a single transformed/mutated cell, to a multicell spheroid (avascular growth), vascular growth in connection with blood vessels, and finally invasive growth of the local tissue and metastasis to distant sites where secondary tumours occur. Modelling these growth phases involves a mixture of continuum models (ordinary, delay and partial differential equations – reaction–diffusion–taxis equations) and individual-based models (cellular automata, discrete modelling techniques). The lecture notes present a range of mathematical techniques to examine a family of models (qualitative analysis of DEs, asymptotic analysis, numerical analysis and computation) and provide a general framework for developing quantitative and predictive models.

The mathematical framework for searching for links between solutions related to equation modelling at the microscopic, mesoscopic and macroscopic levels is the topic of the course *Links between microscopic and macroscopic description* by Mirosław Lachowicz. Usually, the description of biological populations is carried out on a macroscopic level of interacting sub-populations. The mathematical structures are deterministic reaction–diffusion equations. They describe the (deterministic) evolution of densities of subpopulations rather than the interactions between their individual entities. However, in many cases the description on a micro-scale of interacting entities (e.g. cells) seems to be more appropriate. The problem of relationships between the various scales of description seems to be one of the most important problems of the mathematical modelling of complex systems, e.g. in the modeling of tumour growth. The following strategy can be applied. One starts with the deterministic macroscopic model for which the identification of parameters by an experiment is easier. Then one provides the theoretical framework for modelling at the

microscopic scale in such a way that the corresponding models at the macro-
and micro-scales are asymptotically equivalent, i.e. the solutions are close to
each other in a properly chosen norm. Then, if the microscopic model is chosen suitably, one may hope that it covers not only the macroscopic behaviour
of the system in question, but also some of its microscopic features. The microscopic model by its nature is richer and it may describe a larger variety
of phenomena. In mathematical terms, we are interested in the links between
the following mathematical structures: at the micro-scale of stochastically
interacting entities (cells, individuals,...), in terms of continuous stochastic
semigroups, the meso-scale of statistical entities, in terms of continuous nonlinear semigroups related to the solutions of Boltzmann-type nonlocal kinetic
equations, and the macroscale of densities of interacting entities in terms of
dynamical systems related to reaction–diffusion equations.

The notes on *Rescaling Stochastic Processes: Asymptotics*, by Vincenzo
Capasso and Daniela Morale, investigate the links among different scales,
from a more probabilistic point of view. As already mentioned, particular
attention is being paid to the mathematical modelling of the social behaviour of interacting individuals in a biological population, on the one hand
because there is an intrinsic interest in the dynamics of population herding,
and on the other hand, as agent-based models are being used in complex optimization problems. Among other interesting features, these systems lead to
self-organization phenomena, which exhibit interesting spatial patterns. Here,
we show how properties on the macroscopic level depend on interactions at
the microscopic level; in particular, suitable laws of large numbers are shown
to imply the convergence of the evolution equations for empirical spatial distributions of interacting individuals to nonlinear reaction–diffusion equations
for a so-called mean field, as the total number of individuals becomes sufficiently large. As a working example, an interacting particle system modelling
social behaviour has been proposed, based on a system of stochastic differential equations, driven by both aggregating/repelling and external forces. To
support a rigorous derivation of the asymptotic nonlinear integro-differential
equation, compactness criteria for convergence in metric spaces of measures,
and problems of existence of a weak/entropic solution have been analyzed.
Further the temporal asymptotic behaviour of the stochastic system of a fixed
number of interacting particles has been discussed. This leads to the problem
of the existence of nontrivial invariant probability measures.

These microscopic interactions between individuals can often be described
within game-theoretic models. This theme has been discussed in the notes on
Evolutionary Game Theory and Population Dynamics by Jacek Miȩkisz. In
such evolutionary models, individuals adapt to a changing environment and
are subject to selection pressure and mutations. We will present deterministic and stochastic models of adaptive dynamics and discuss the stability of
equilibria in appropriate dynamical systems such as time-delay equations and
Markov chains.

As Co-Directors, we are pleased to thank both institutions, BC and CIME, in particular the former Director of the Institute of Mathematics and founder of MRCC, Professor Bogdan Bojarski, and the Director of CIME Professor Pietro Zecca, for letting us organize the joint school, and for their continuous support. It has had a special meaning at the time of extension of Europe towards the East; in this way a concrete occasion has been offered to the young participants to contribute actively in building a common European Research Area. They have had the chance of experiencing directly that different Schools in Europe may actively contribute to making the European Union a highly competitive scientific community.

Our special thanks are due to our Colleagues, Professors Banasiak, Chaplain and Miękisz, for their careful preparation and stimulating presentation of the material, both at the school and in these Lecture Notes. We thank Dr Gabriela Lorelai Litcanu for her work in preparation of the workshop organized in connection with the courses. All the participants contributed to the creation of an exceptionally friendly atmosphere which also characterized the various social events organized in the beautiful environment of the Będlewo Palace. We thank the Director and the whole staff of MRCC in Będlewo for their warm and efficient hospitality. Thanks are due to Dr Daniela Morale for her editorial assistance during the preparation of this volume.

Milan, Warsaw,   *Vincenzo Capasso*
March 2007   *Mirosław Lachowicz*

Contents

Positivity in Natural Sciences

Jacek Banasiak

School of Mathematical Sciences, University of KwaZulu-Natal,
Durban, South Africa
banasiak@ukzn.ac.za

Summary. Many problems in natural sciences require a notion of positivity: only non-negative densities, population sizes or probability make sense in real life. This imposes certain constrains in the modelling but also, coupled with metric structure of underlying spaces, offers new powerful tools for analyzing complex systems. The presented lectures describe such an intertwining of topological and order structures in analysis of linear dynamical systems, mainly arising in mathematical biology and kinetic theory. Among covered topics the reader will find a survey of classical topics such as well-posedness and detailed analysis of long-time and asymptotic behaviour of infinite dimensional linear systems as well as recent results on emergence of chaos and phase transitions in such systems. The lectures are concluded by multiple scale asymptotic analysis of singularly perturbed kinetic problems leading to various diffusion equations which preserve the coarse structure of the original models.

1 Introduction

Laws of physics and, increasingly, also those of other sciences are in many cases expressed in terms of differential or integro–differential equations. If one models systems evolving with time, then the variable describing time plays a special role, as the equations are built by balancing the change of the system in time against its 'spatial' behaviour. In mathematics such equations are called *evolution equations.* Such equations usually are formulated pointwise; that is, all the operations, such as differentiation and integration, are understood in the classical (calculus) sense and the equation itself is supposed to be satisfied for all values of the independent variables in the relevant domain:

$$\frac{\partial}{\partial t} u(t,x) = [\mathcal{A}u(t,\cdot)](x), \quad x \in \Omega$$

$$u(t,0) = \overset{\circ}{u}, \tag{1}$$

where $\mathcal{A}$ is a certain expression, differential, integral, or functional, that can be evaluated at any point $x \in \Omega$ for all functions from a certain subset S.

However, when we are trying to solve (1), we change its meaning by imposing various *a priori* restrictions on the solution to make (1) amenable to particular techniques. Quite often (1) does not provide a complete description of the dynamics even if it looks complete from the modelling point of view. Then the obtained solution maybe be not what we have been looking for. This becomes particularly important if we cannot get our hands on the actual solution but use 'soft analysis' to find relevant properties of it. These lecture notes predominantly are devoted to one particular way of looking at the evolution of a system in which we describe time changes as transitions from one state to another; that is, the evolution is described by a family of operators $(G(t))_{t\geq0}$, parameterized by time, that map an initial state of the system onto all subsequent states in the evolution; that is, the solutions are represented as

$$u(t) = G(t)u_0. \tag{2}$$

The family $(G(t))_{t\geq0}$ is called a *semigroup* and u_0 is an initial state.

In this approach we place the process in some abstract space X which is chosen partially for the relevance to the problem and partially for mathematical convenience. For example, if (1) describes the evolution of an ensemble of particles, then u is the particle density function and the natural space seems to be $L_1(\Omega)$ since in this case the norm of a nonnegative u, that is, the integral of u over Ω, gives the total number of particles in the ensemble. It is important to note that this choice is not unique but rather is a mathematical intervention into the model, which could change it in a quite dramatic way. For instance, staying with this case, we could choose the space of measures on Ω with the same interpretation of the norm. On the other hand, if we are interested in controlling maximum concentration of the particles, a more proper choice would be some reasonable space with a supremum norm, such as, for example, the space of bounded continuous functions on Ω, $C_b(\Omega)$.

Once we select our space, the right-hand side can be interpreted as an operator $A : D(A) \to X$, defined on some subset $D(A)$ of X (not necessarily equal to X), such that $x \to [Au](x) \in X$. With this, (1) can be written as an ordinary differential equation in X:

$$u_t = Au, \qquad t > 0,$$
$$u(0) = u_0 \in X. \tag{3}$$

The domain $D(A)$ is also not uniquely defined by the model. Clearly, we would like to choose it in such a way that the solutions originating from $D(A)$ could be differentiated and that they belong to $D(A)$ for all t so that both sides of the equation would make sense. As we shall see, semigroup theory in some sense forces $D(A)$ upon us, although it is not necessarily the optimal choice from the modelling point of view. Although throughout the lectures we assume that the underlying space is given, finding good $D(A)$ on which we define the realisation A of the expression $\mathcal{A}$, is a more complicated thing and has major implications as to whether we are getting from the model what we bargained for.

Though we also discuss a general theory, our main focus is on models preserving some notion of *positivity*: non-negative inputs should give non-negative outputs (in a suitable sense of the word).

1.1 What can go Wrong?

In the discussion above we have alluded to the fact that the mathematical model not always has all desirable properties. The following 'pathologies' occur quite often.

Dishonesty. Models are based on certain laws coming from the applied sciences and we expect the solutions to equations of these models to return these laws. However, this is not always true: we will see models built on the basis of population conservation principles, solutions of which, for certain classes of parameters, do not preserve populations. Such models are called *dishonest*. Dishonesty could be a sign of a phase transition happening in the model, or simply indicate limits of validity of the model.

Multiple solutions. Even if all side conditions relevant to the modelled process seem to have been built into the model, we may find that the model does not provide full description of the dynamics; while for some classes of parameters the model gives uniquely determined solutions, for others there exist multiple solutions.

We will see that methods based on positivity techniques provide a comprehensive explanation of these two 'pathological' phenomena.

1.2 And if Everything Seems to be Fine?

If we make sure that the abstract model (3) gives a reasonable description of the phenomena at hand, we can analyse its further properties. One of the questions most often asked by practitioners pertains to stability and long time behaviour of solutions. In particular, in population theory an important problem is the existence of dominating long time pattern of evolution. More precisely, we can pose the following questions, [5]:

1. Does there exists a special solution u^* of (3) of the form $u^*(t) = e^{\lambda^* t} u_0^*$ for some real λ^* and an element $u_0^* \in X$ such that for any other solution there is a constant C such that

$$u(t) = C e^{\lambda^* t} u_0^* + O(\exp{(\lambda^* - \epsilon)t}) \qquad (4)$$

for some $\epsilon > 0$ (independent of u)? An added bonus would be if u_0^* could be selected positive.
2. If this is impossible, may be there is a finite-dimensional projection P which commutes with the semigroup $G(t)$ and such that

$$e^{-\lambda^* t} G(t) - P \to 0, \quad \text{exponentially fast.} \qquad (5)$$

3. More generally, we may ask whether there exists a finite dimensional projection P which commutes with the semigroup $G(t)$ and such that $G(t)|_{PX}$ can be extended to a group of operators of the form e^{tM} with all eigenvalues of M satisfying $\Re\lambda = \lambda^*$ and, for some $\epsilon > 0$,

$$(I - P)G(t) = O(e^{-(\lambda^*-\epsilon)t}), \quad \text{as } t \to \infty. \tag{6}$$

Those familiar with the finite dimensional population theory will recognize that in the first case we have primitive irreducible transition matrix while in the second and third the matrix is only irreducible with different properties of the largest eigenvalue.

We say that the semigroup $(G(t))_{t\geq0}$ has *asynchronous exponential growth* (AEG) if (4) is verified (and *positive* AEG if u^* is positive). If only (5) is satisfied, then we say that $(G(t))_{t\geq0}$ has *multiple asynchronous growth* (MAEG) and, finally, if (6) holds, then we say that $(G(t))_{t\geq0}$ has *extended asynchronous growth* (EAEG).

The name 'asynchronous exponential growth' comes precisely from the population biology where it is observed that in many cases initially synchronized populations lose synchrony after just a few generations. It reflects the fact that whatever distribution was observed at the initial time, the population evolves towards an asymptotic distribution, where the proportion of individuals in a given stage is constant.

Here the interplay of compactness and positivity techniques can produce, in infinite dimensional case, results which are very close to the classical Frobenius-Perron theory. However, unlike for the finite-dimensional case, some models which behave perfectly well for some classes of parameters, can exhibit phase transitions or degenerate into chaotic behaviour for others. It is worthwhile to note that phase transitions and chaos usually are associated with nonlinear phenomena. Here we will see that they can occur in linear systems but for this these systems must be infinite dimensional.

We shall demonstrate this on two examples. One is related to the classical birth-and-death type problem, the other is a variant of the age structured population model.

2 Spectral Properties of Operators

The considerations below will be carried in an arbitrary Banach space. However, in the present lectures most applications are restricted to the Banach spaces, which are commonly used in the population theory due to their natural interpretation. It is worthwhile to understand that, in applications, working in a particular Banach space simply means that the functions we are working must satisfy a numerical restriction which is important in the modelling process. In population theory usually we are interested in the evolution of an ensemble of elements the state of which is described by a function $n(t, x)$

representing either a number of elements in a given state (if the number of states is finite or countable) or the density of particles in the state x, if x is a continuous variable. In many cases we are interested in tracking the total number of elements of the population which, for a time t, is given by

$$\sum_{x \in \Omega} n(t, x),$$

where Ω is the state space, if Ω is countable and

$$\int_{x \in \Omega} n(t, x) dx,$$

if Ω is a continuum. To make full use of the tools of functional analysis, we must allow entries of arbitrary sign, so it is not surprising that using this point of view we will be working either in

$$l_1 := \{(n_i)_{i \in \mathbb{N}}; \sum_{i=1}^{\infty} |n_i| < \infty\}$$

or

$$L_1(\Omega) := \{x \to n(x); \int_{\Omega} |n(x)| dx < \infty\},$$

where in the first case we took $\Omega = \mathbb{N}$. We noted in Introduction that if we are more interested in maximal concentration of elements, then we should work in spaces of functions with supremum norm, such as $C(\Omega)$ (the space of continuous function on a compact set Ω).

If uncomfortable with abstract notions, one can substitute one of the spaces described above for the abstract Banach space X to get a better understanding of the main ideas of the lectures.

2.1 Operators

Let X, Y be real or complex Banach spaces. The norm in X will be denoted by $\| \cdot \|_X$ or, if no misunderstanding can occur, by $\| \cdot \|$.

An *operator* from X to Y is a linear rule $A : D(A) \to Y$, where $D(A)$ is a linear subspace of X, called the *domain* of A. We use the notation $(A, D(A))$ to denote the operator A with domain $D(A)$. By $\mathcal{L}(X, Y)$ we denote the space of all bounded operators between X and Y; that is, the operators for which

$$\|A\| := \sup_{\|x\| \leq 1} \|Ax\| = \sup_{\|x\| = 1} \|Ax\| < +\infty. \tag{7}$$

The space $\mathcal{L}(X, X)$ is abbreviated as $\mathcal{L}(X)$. For $A, B \in \mathcal{L}(X, Y)$, we write $A \subset B$ if $D(A) \subset D(B)$ and $B|_{D(A)} = A$. We define the *image* of A by

$$Im\ A = \{y \in Y; \ y = Ax \text{ for some } x \in D(A)\}$$

and the *kernel* of A by

$$Ker\ A = \{x \in D(A);\ Ax = 0\}.$$

Furthermore, the *graph* of A is defined as

$$G(A) = \{(x, y) \in X \times Y;\ x \in D(A), y = Ax\}. \tag{8}$$

We say that the operator A is *closed* if $G(A)$ is a closed subspace of $X \times Y$. Equivalently, A is closed if and only if for any sequence $(x_n)_{n \in \mathbb{N}} \subset D(A)$, if $\lim_{n \to \infty} x_n = x$ in X and $\lim_{n \to \infty} Ax_n = y$ in Y, then $x \in D(A)$ and $y = Ax$.

An operator A in X is *closable* if the closure of its graph $\overline{G(A)}$ is itself a graph of an operator, that is, if $(0, y) \in \overline{G(A)}$ implies $y = 0$. Equivalently, A is closable if and only if for any sequence $(x_n)_{n \in \mathbb{N}} \subset D(A)$, if $\lim_{n \to \infty} x_n = 0$ in X and $\lim_{n \to \infty} Ax_n = y$ in Y, then $y = 0$. In such a case the operator whose graph is $\overline{G(A)}$ is called the *closure* of A and denoted by $\overline{A}$.

Compact Operators

An operator $K \in \mathcal{L}(X, Y)$ is called *compact* (resp. *weakly compact*) if the image of the unit ball in X is a relatively compact (resp. weakly compact) subset of Y. Most relevant properties of compact operators are preserved if the operator $K \in X$ is power compact; that is, if K^m is compact for some $m \in \mathbb{N}$. Importance of power compact operators stems, in particular, from the fact that in certain spaces ($C(\Omega)$, $L_1(\Omega)$) the square of a weakly compact operator is compact.

In applications it is often needed that AK be power compact for any $A \in \mathcal{L}(X)$. Such operators are called *strictly power compact*. Since the space of weakly compact (and also compact) operators is a two sided ideal in $\mathcal{L}(X)$, weakly compact operators in $C(\Omega), L_1(\Omega)$ are strictly power compact.

Example 1. Consider the integral operator given formally by

$$Tf(x) = \int_\Omega k(x, y) f(y) dy,$$

where $\Omega \subseteq \mathbb{R}^n$. The operator T is compact from $L_p(\Omega)$ to $L_p(\Omega)$ if $k \in L_{p,q}(\Omega \times \Omega)$, where $1/p + 1/q = 1$, provided $p > 1$. For $p = 1$, the corresponding assumption

$$k \in L_{1,\infty}(\Omega \times \Omega) \tag{9}$$

is not sufficient for compactness. For such T to be compact, we require e.g. $k \in C(\Omega, L_\infty(\Omega))$ (see [40, p.53]). However, under assumption (9), the operator T is weakly compact and thus strictly power compact ([27]).

2.2 Spectral Properties of a Single Operator

Let A be any operator in X. The *resolvent set* of A is defined as

$$\rho(A) = \{\lambda \in \mathbb{C}; \ \lambda I - A : D(A) \to X \text{ is invertible}\}. \tag{10}$$

We call $(\lambda I - A)^{-1}$ the resolvent of A and denote it by $R(\lambda, A) = (\lambda I - A)^{-1}$, for $\lambda \in \rho(A)$. The complement of $\rho(A)$ in $\mathbb{C}$ is called the *spectrum* of A and denoted by $\sigma(A)$. In general, it is possible that either $\rho(A)$ or $\sigma(A)$ is empty. The spectrum is usually subdivided into several subsets. We follow the approach of [30, 42] which, though being not the most common, is very suitable for the description of long time behaviour of semigroups. First,

- *Point spectrum* $\sigma_p(A)$ is the set of $\lambda \in \sigma(A)$ for which the operator $\lambda I - A$ is not one-to-one. In other words, $\sigma_p(A)$ is the set of all eigenvalues of A.

A generalization of the point spectrum, which will play an important role later, is the approximate spectrum:

- *Approximate spectrum* $\sigma_a(A)$ is the set of $\lambda \in \sigma(A)$ for which either the operator $\lambda I - A$ is not one-to-one or the range $Im\ A$ is not closed.

The name approximate spectrum comes from the following property which is often used to as a definition [30, Lemma IV.1.9].

Lemma 1. *If $(A, D(A))$ is a closed operator in X, then $\lambda \in \sigma_a(A)$ if and only if there is a sequence $(x_n)_{n \in \mathbb{N}} \subset D(A)$ such that $\|x_n\| = 1$, $n \in \mathbb{N}$, and $\lim_{n \to \infty} \|Ax_n - \lambda x_n\| = 0$.*

The last part of the spectrum is

- *Residual spectrum* $\sigma_r(A)$ is the set of $\lambda \in \sigma(A)$ for which $Im\ (\lambda I - A)$ is not dense in X.

Clearly, the σ_p, σ_a and σ_r are not disjoint (in particular, $\sigma_p \subset \sigma_a$) but

$$\sigma(A) = \sigma_a(A) \cup \sigma_r(A).$$

Moreover, $\sigma_r(A) = \sigma_p(A^*)$ (A^* denotes the adjoint of A) and the topological boundary of $\sigma(A)$ satisfies, [30, Proposition 1.10]

$$\partial\sigma(A) \subset \sigma_a(A). \tag{11}$$

Remark 1. Typically, $\sigma(A)$ is divided into $\sigma_p(A)$ (defined as above), the continuous spectrum $\sigma_c(A)$ which is the set of $\lambda \in \sigma(A)$ for which the operator $\lambda I - A$ is one-to-one and its range is dense in, but not equal to, X and the residual spectrum is defined as the set of $\lambda \in \sigma(A)$ for which the operator $\lambda I - A$ is one-to-one and its range is not dense in X. Clearly, $\sigma_c(A) \subset \sigma_a(A)$ but we shall not explore further relations between these two definitions. However, the continuous spectrum will come in handy in e.g. Theorem 28.

The resolvent of any operator A satisfies the *resolvent identity*

$$R(\lambda, A) - R(\mu, A) = (\mu - \lambda)R(\lambda, A)R(\mu, A), \qquad \lambda, \mu \in \rho(A), \qquad (12)$$

from which it follows, in particular, that $R(\lambda, A)$ and $R(\mu, A)$ commute. It follows that $\rho(A)$ is an open set and $R(\lambda, A)$ is an analytic function of $\lambda \in \rho(A)$ which can be written as the power series

$$R(\lambda, A) = \sum_{n=0}^{\infty} (\mu - \lambda)^n R(\mu, A)^{n+1} \qquad (13)$$

for $|\mu - \lambda| < \|R(\mu, A)\|^{-1}$. For any bounded operator the spectrum is a compact subset of $\mathbb{C}$ so that $\rho(A) \neq \emptyset$. If A is bounded, then the limit

$$r(A) = \lim_{n \to \infty} \sqrt[n]{\|A^n\|} \qquad (14)$$

exists and is called *the spectral radius*. Clearly, $r(A) \leq \|A\|$.

Theorem 1. *[49] We have*

$$R(\lambda, A) = \sum_{n=0}^{\infty} \lambda^{-(n+1)} A^n, \qquad (15)$$

where the series converges in the operator norm for $|\lambda| > r(A)$ and diverges for $|\lambda| < r(A)$. Moreover

$$r(A) = \sup_{\lambda \in \sigma(A)} |\lambda|. \qquad (16)$$

To show that $\lambda \in \mathbb{C}$ belongs to the spectrum we often use the following result.

Theorem 2. *Let A be a closed operator. If $\lambda \in \rho(A)$, then $dist(\lambda, \sigma(A)) = 1/r(R(\lambda, A)) \geq 1/\|R(\lambda, A)\|$. In particular, if $\lambda_n \to \lambda$, $\lambda_n \in \rho(A)$, then $\lambda \in \sigma(A)$ if and only if $\{\|R(\lambda_n, A)\|\}_{n \in \mathbb{N}}$ is unbounded.*

The *peripheral spectrum* of a bounded operator A is the set

$$\sigma_{per,r(A)} = \{\lambda \in \sigma(A); |\lambda| = r(A)\}. \qquad (17)$$

Clearly, $\sigma_{per,r(A)}(A)$ is compact and, by (16), non-empty.

For an unbounded operator A the role of the spectral radius often is played by the *spectral bound* $s(A)$ defined as

$$s(A) = \sup\{\Re\lambda; \lambda \in \sigma(A)\}, \qquad (18)$$

and the peripheral spectrum of A in this case is accordingly defined as

$$\sigma_{per,s(A)} = \{\lambda \in \sigma(A); \Re\lambda = s(A)\}. \qquad (19)$$

Suppose $A \in \mathcal{L}(X)$ and $f(z) = \sum_{n=0}^{\infty} a_n z^n$ is an analytic function in a disc containing $\sigma(A)$. Then we can define a function $f(A)$ by

$$f(A) = \sum_{i=0}^{\infty} a_n A^n,$$

where the series is convergent as $\sigma(A)$ is contained in a circle with radius $r(A)$. An alternative definition can be obtained via the Dunford integral, [27–29].

Spectra of A and $f(A)$ are related by the Spectral Mapping Formula

$$\sigma(f(A)) = f(\sigma(A)). \tag{20}$$

Decomposition of the Spectrum

Let A be a closed operator. An important case occurs if $\sigma(A)$ can be decomposed into two disjoint parts, one of which is compact and the other closed. We shall focus on the case when the compact part consists of an isolated point λ_0. This means that the resolvent can be expanded into a Laurent series

$$R(\lambda, A) = \sum_{n=-\infty}^{\infty} (\lambda - \lambda_0)^n B_n \tag{21}$$

for $0 < |\lambda - \lambda_0| < \delta$ for sufficiently small δ. The coefficients B_n are bounded operators given by the formula

$$B_n = \frac{1}{2\pi i} \int_{\gamma} (\lambda - \lambda_0)^{-n-1} R(\lambda, A) d\lambda, \quad n \in \mathbb{Z} \tag{22}$$

where γ is a positively oriented simple curve in $\rho(A)$, which surrounds λ_0. Application of the Cauchy integral formula gives

$$B_{-n} B_{-k} = B_{-n-k+1}, \quad n, k \in \mathbb{N} \tag{23}$$

The coefficient $P = B_{-1}$ is called the *residue* of A. If there exists k such that $B_{-k} \neq 0$ while B_{-n}, $n > k$, then λ_0 is called the *pole* of $R(\lambda, A)$ *of order* k. We have

$$B_k = \lim_{\lambda \to \lambda_0} (\lambda - \lambda_0)^k R(\lambda, A).$$

The following properties can be found in e.g. [49].

Theorem 3. *1. The operator B_{-1} is a projection on X with $Im\ B_{-1}$ and $Im\ (I - B_{-1})$ closed.*
2. The restriction of A to $Im\ B_{-1}$ is bounded and has spectrum $\{\lambda_0\}$.
3. If $dim\ Im\ B_{-1} < \infty$, then λ_0 is a pole of $R(\lambda, A)$.

4. If λ_0 is a pole of $R(\lambda, A)$ of order k, then it is an eigenvalue of A and

$$Im \; B_{-1} = Ker \; (\lambda_0 I - A)^k = Ker \; (\lambda_0 I - A)^{k+j},$$
$$Im \; (I - B_{-1}) = Im \; (\lambda_0 I - A)^k = Im \; (\lambda_0 I - A)^{k+j}, \qquad (24)$$

for $j \geq 0$, with $X = Ker \; (\lambda_0 I - A)^k \oplus Im \; (\lambda_0 I - A)^k$.

Let us prove the first part of (4), which frequently occurs in applications. Multiplying (21) by $(\lambda I - A)$ we obtain

$$I = \sum_{n=-\infty}^{\infty} (\lambda - \lambda_0)^{n+1} B_n + \sum_{n=-\infty}^{\infty} (\lambda - \lambda_0)^n (\lambda_0 I - A) B_n$$

so that

$$(\lambda_0 I - A) B_{-n} = -B_{-(n+1)}.$$

Since n is the order of the pole, $B_{-(n+1)} = 0$. However, since $B_{-n} \neq 0$, there is f such that $x = B_{-n} f \neq 0$ is an eigenvector corresponding to λ_0.   □

We define

$$Ker_\infty (\lambda_0 I - A) = \bigcup_{k \geq 0} Ker \; (\lambda_0 I - A)^k;$$

$Ker_\infty (\lambda_0 I - A)$ is called the generalized eigenspace of A corresponding to the eigenvalue λ_0. The number $dim \; Im \; B_{-1}$ is called the *algebraic multiplicity* of λ_0, denoted m_a, while $m_g = dim \; Ker \; (\lambda_0 I - A)$ is called the *geometric multiplicity*. If $m_a = 1$, then λ_0 is called an *algebraically simple* pole. If k is the order of the pole ($k := \infty$, if λ_0 is an essential singularity), then

$$m_g + k - 1 \leq m_a \leq m_g k$$

$(0 \cdot \infty := \infty)$. Thus, $m_a < \infty$ if and only if λ_0 is a pole with $m_g < \infty$.

If A is closed with $\rho(A) \neq \emptyset$, then λ_0 is an isolated point of $\sigma(A)$ if and only if $(\lambda - \lambda_0)^{-1}$ is isolated in $\sigma(R(\lambda, A))$ and the residues and orders of the respective poles coincide. In particular, if A has compact resolvent, then $\sigma(A)$ consists only of poles of finite algebraic multiplicity.

Turning Approximate Eigenvalues into Eigenvalues

There is a very useful construction extending a given Banach space, called the *ultrapower* of X ([1]) or *F-product* ([42]). Here we shall discuss it in a restricted setting. Let $l_\infty(X)$ (resp. $c_0(X)$) be the vector space of bounded (resp. converging to 0) sequences $(x_n)_{n \in \mathbb{N}} \subset X$. We denote

$$\hat{X} = l_\infty(X)/c_0(X)$$

with the classes of equivalence denoted by $\hat{x} = (x_n)_{n \in \mathbb{N}} + c_0(X)$. The space $\hat{X}$ becomes a Banach space under the norm $\|\hat{x}\| = \limsup_{n \to \infty} \|x_n\|$. There is a natural embedding $X \ni x \to (x, x, \ldots) + c_0(X) \in \hat{X}$ so that X can be identified with a closed subspace of $\hat{X}$.

Bounded operators on X give rise to bounded operators on $\hat{X}$: for $A \in \mathcal{L}(X)$ and $\hat{x} = (x_n)_{n \in \mathbb{N}} + c_0(X)$ we have

$$\hat{A}\hat{x} = \widehat{(Ax_1, Ax_2, \ldots)}$$

and it can be proved that $\|A\| = \|\hat{A}\|$.

If $(x_n)_{n \in \mathbb{N}}$ is an approximate eigenvector of A with approximate eigenvalue λ, then $\|Ax_n - \lambda x_n\| \to 0$ as $n \to \infty$. But this is the same as saying that $\hat{x} = (x_1, x_2, \ldots) + c_0(X)$ is an eigenvector of $\hat{A}$ with the same eigenvalue. Actually, even more is true.

Theorem 4. *[24, p.290] Let $A \in \mathcal{L}(X)$. Then*

1. $\sigma(A) = \sigma(\hat{A})$;
2. $\sigma_a(A) = \sigma_a(\hat{A}) = \sigma_p(\hat{A})$;
3. $\widehat{R(\lambda, A)} = R(\lambda, \hat{A})$ *for* $\lambda \in \rho(A) = \rho(\hat{A})$;
4. $\lambda_0 \in \sigma(A)$ *is a pole of* $R(\lambda, A)$ *of order p if and only if $\lambda_0 \in \sigma(\hat{A})$ is a pole of* $R(\lambda, \hat{A})$ *of order p.*

Unfortunately, for unbounded operators and semigroups the situation becomes more complicated and we shall return to this topic later.

Spectrum of Compact and Power Compact Operators

The main results, summarizing the spectral properties of compact and power compact operators, are given in the following theorem.

Theorem 5. *If K is compact (or power compact), then*

(i) $\sigma(K)$ is at most countable and, if $\dim X = \infty$, then it contains $\{0\}$;
(ii) If $\sigma(K)$ is infinite and $\{\lambda_1, \lambda_2, \cdots\}$ is any enumeration of it, then $\lambda_n \to 0$ as $n \to \infty$.
(iii) If $0 \neq \lambda \in \sigma(K)$, then λ is a pole of the resolvent and thus is an eigenvalue.

Proof. For compact operators this result is known as the Fisher-Riesz theory, see e.g. [1, Section 7.1]. Extension to power compact operators is possible due to the Spectral Mapping Theorem which gives $\sigma(K^n) = \{\lambda^n; \ \lambda \in \sigma(K)\}$ and proves the assertions about the spectrum. To prove (iii), let $X_\lambda := B_{-1}X$ for $0 \neq \lambda \in \sigma(K)$ and the restriction K_{X_λ}. Then X_λ is invariant with respect to K and thus $K|_{X_\lambda}$ is power compact with $\sigma(K|_{X_\lambda}) = \{\lambda\}$ by Theorem 3(2). If $\dim X_\lambda = \infty$, then this contradicts point (i) of the present theorem. Thus $\dim X_\lambda < \infty$ and the result follows by Theorem 3 (4). $\square$

Essential Spectrum

As we have seen above, it is important to separate 'good' points of spectrum from the 'bad' ones. The concept of *essential spectrum* have been introduced with this idea in mind.

Definition 1. *[24] The essential spectrum of A, denoted by $\sigma_e(A)$ is the set of $\lambda \in \sigma(A)$ which satisfy at least one of the following conditions*

(i) $Im\,(\lambda I - A)$ is not closed;
(ii) $dim\,K_\infty(\lambda I - A) = \infty$;
(iii) λ is an accumulation point of $\sigma(A)$.

Essential spectrum is closely related to the concept of Fredholm points of A. We say that λ is a *Fredholm point* of A, and write $\lambda \in \rho_\Phi(A)$, if $Ker\,(\lambda I - A)$ is finite dimensional and $Im\,(\lambda I - A)$ is closed of finite codimension. The *Fredholm spectrum* of A, denoted $\sigma_\Phi(A)$, is the set of $\lambda \in \mathbb{C}$ which are not Fredholm points of A. Clearly,

$$\sigma_\phi(A) \subset \sigma_e(A),$$

but, in general, these sets are different (for instance, there may exist non-isolated Fredholm points of A.

Remark 2. Several authors (see e.g. [30, 42]) define the essential spectrum as the Fredholm spectrum. It has the additional advantage that it coincides with the normal spectrum of the canonical image of A in the quotient space $\mathcal{L}(X)/\mathcal{K}(X)$, where $\mathcal{K}(X)$ is the ideal of compact operators in X. This allows to define the Fredholm norm of A as

$$\|A\|_\Phi = dist(A, \mathcal{K}(X)) = \inf\{\|A - K\|,\ K \in \mathcal{K}(X)\}. \tag{25}$$

As we shall see later, for the purpose of these lectures, the difference between both definitions is not significant.

We have the following result [21, 24].

Theorem 6. *Suppose $\lambda_0 \in \sigma(A)$ and $dim\,Ker\,(\lambda_0 I - A) < +\infty$. Then $\lambda_0 \in \sigma(A) \setminus \sigma_e(A)$ if and only if $R(\lambda, A)$ is analytic in a neighbourhood of λ_0 and has a pole at λ_0.*

Without assumption that $dim\,Ker\,(\lambda_0 I - A) < +\infty$ we can prove only that if $\lambda_0 \in \sigma(A) \setminus \sigma_e(A)$, then λ_0 is a pole of $R(\lambda, A)$. In particular, if λ_0 is a non-essential point of $\sigma(A)$, then $Im\,(\lambda_0 I - A)$ is of finite codimension (see (24)) and thus $\lambda_0 \in \rho_\Phi(A)$.

We note some properties of the interior and the boundary of the spectrum, [5]:

(a) $int\,\sigma \subset \sigma_e$;
(b) $\partial\sigma_e \subset \sigma_\Phi$.

We can use the characterization (16) of the spectral radius to define analogous concepts related to the essential and Fredholm spectra of A:

$$r_e(A) = \sup_{\lambda \in \sigma_e(A)} |\lambda|, \qquad r_\Phi(A) = \sup_{\lambda \in \sigma_\Phi(A)} |\lambda|. \tag{26}$$

Clearly, we have $r_\Phi(A) \leq r_e(A)$. On the other hand, since $\sigma_e(A)$ is a compact set (for A bounded), there is $\lambda \in \sigma_e(A)$ with $|\lambda| = r_e(A)$. Such λ is in $\partial\sigma_e(A)$, hence, by (b) above, it is in $\sigma_\Phi(A)$. Therefore $r_\Phi(A) \geq r_e(A)$ and

$$r_\Phi(A) = r_e(A). \tag{27}$$

Using the above discussion, the essential radius can be characterized as follows

$r_e(A)$ is the smallest $r \in \mathbb{R}_+$ such that every $\lambda \in \sigma(A)$ satisfying $|\lambda| > r$ is an isolated pole of finite algebraic multiplicity. For any $r > r_e(A)$, the set $\{\lambda \in \sigma(A); |\lambda| \geq r\}$ is finite.

The last statement follows from the fact that the spectrum of a bounded operator is compact and any accumulation point of $\sigma(A)$ belongs to $\sigma_e(A)$.

3 Banach Lattices and Positive Operators

In many processes in the natural sciences only nonnegative solutions are meaningful. This is the case when the solution is a probability, a density function, the absolute temperature, and so on. Thus, mathematical models of such processes should have the property that nonnegative data yield nonnegative solutions. If we work in concrete spaces of functions, then the notion of positivity is natural: either pointwise for continuous functions or almost everywhere in the spaces of measurable functions. However, in a general setting we have to find an abstract notion generalizing the pointwise concepts of positivity.

3.1 Defining Order

In a given vector space X an order can be introduced either geometrically, by defining the so-called *positive cone* (in other words, what it means to be a *positive element* of X), or through an axiomatic definition. We follow the second approach.

Definition 2. *Let X be an arbitrary set. A partial order (or simply, an order) on X is a binary relation, denoted here by '$\geq$', which is reflexive, transitive, and antisymmetric, that is,*

(1) $x \geq x$ for each $x \in X$;
(2) $x \geq y$ and $y \geq x$ imply $x = y$ for any $x, y \in X$;
(3) $x \geq y$ and $y \geq z$ imply $x \geq z$ for any $x, y, z \in X$.

We need a number of related conventions and definitions. The notation $x \leq y$ means $y \geq x$. $x > y$ means $x \geq y$ and $x \neq y$. An *upper bound* for a set $S \subset X$ is an element $x \in X$ satisfying $x \geq y$ for all $y \in S$. An element $x \in S$ is said to be *maximal* if there is no $S \ni y \neq x$ for which $y \geq x$. A *lower bound* for S and a *minimal element* are defined analogously. A *greatest element* (resp. a *least element*) of S is an $x \in S$ satisfying $x \geq y$ (resp. $x \leq y$) for all $y \in S$.

The *supremum* of a set is its least upper bound and the *infimum* is the greatest lower bound. The supremum and infimum of a set need not exist. It is worthwhile to emphasize that an element s, which is an upper bound of S, is a supremum of the set S if, for any upper bound y of S, we have $s \leq y$.

Let $x, y \in X$ and $x \leq y$. The *order interval* $[x, y]$ is defined by

$$[x, y] := \{z \in X; \; x \leq z \leq y\}.$$

For a two-point set $\{x, y\}$ we write $x \wedge y$ or $\inf\{x, y\}$ to denote its infimum and $x \vee y$ or $\sup\{x, y\}$ to denote supremum. We say that X is a *lattice* if every pair of elements (and so every finite collection of them) has both supremum and infimum. From now on, unless stated otherwise, X is a real vector space.

Definition 3. *An ordered vector space is a vector space X equipped with partial order which is compatible with its vector structure in the sense that*

(4) $x \geq y$ implies $x + z \geq y + z$ for all $x, y, z \in X$;
(5) $x \geq y$ implies $\alpha x \geq \alpha y$ for any $x, y \in X$ and $\alpha \geq 0$.

The set $X_+ = \{x \in X; \; x \geq 0\}$ is referred to as the positive cone of X.

If the ordered vector space X is also a lattice, then it is called a vector lattice or a Riesz space.

Typical examples of Riesz spaces are provided by function spaces with order defined pointwise or pointwise almost everywhere.

We only consider Archimedean spaces; that is, spaces havig the property that if, for any $x \in X_+$ we have $\inf_{n \in \mathbb{N}}\{n^{-1}x\} = 0$, then $x = 0$.

The operations of taking supremum or infimum in a Riesz space have several useful properties which make them similar to the numerical case. In particular, we can define the positive and negative part of $x \in X$, and its absolute value, respectively, by

$$x_+ = \sup\{x, 0\}, \quad x_- = \sup\{-x, 0\}, \quad |x| = \sup\{x, -x\}.$$

The functions $(x, y) \rightarrow \sup\{x, y\}, (x, y) \rightarrow \inf\{x, y\}, x \rightarrow x_\pm$ and $x \rightarrow |x|$ are collectively referred to as the *lattice operations* of a Riesz space. They are related by

$$x = x_+ - x_-, \qquad |x| = x_+ + x_-. \tag{28}$$

Also the absolute value has a number of useful properties that are reminiscent of the properties of the scalar absolute value; that is, for example, $|x| = 0$ if and only if $x = 0$, $|\alpha x| = |\alpha||x|$ for any $x \in X$ and any scalar α.

The existence of suprema or infima of finite sets, ensured by the definition of a Riesz space, does not extend to infinite sets. This warrants introducing a more restrictive class of spaces.

Definition 4. *We say that a Riesz space X is Dedekind (or order) complete if every nonempty and bounded from above subset of X has a least upper bound in X. X is said to be σ-Dedekind or (σ-order) complete, if every bounded from above nonempty countable subset of X has a least upper bound X.*

Example 2. The space $C([0,1])$ is not σ-order complete (and thus also not order complete). To see this, consider the sequence of functions given by

$$f_n(x) = \begin{cases} 1 & \text{for } 0 \leq x \leq \frac{1}{2} - \frac{1}{n}, \\ n\left(\frac{1}{2} - x\right) & \text{for } \frac{1}{2} - \frac{1}{n} < x \leq \frac{1}{2}, \\ 0 & \text{for } \frac{1}{2} < x < 1. \end{cases}$$

This is clearly an increasing sequence bounded from above by $g(x) \equiv 1$. However, it converges pointwise to a discontinuous function $f(x) = 1$ for $x \in [0, 1/2)$ and $f(x) = 0$ for $x \in [1/2, 0]$. In general, spaces $C(\Omega)$ are not σ-order complete unless Ω consists of isolated points.

On the other hand, the spaces l_p, $1 \leq p \leq \infty$, are clearly order complete, as taking the coordinatewise suprema of sequences bounded from above by an l_p sequence produces a sequence which is in l_p.

The spaces $L_p(\Omega), p \in \{0\} \cup [1, \infty]$ are also order complete but the proof is much more delicate, see [8, Example 2.52].

3.2 Banach Lattices

As the next step, we investigate the relation between the lattice structure and the norm when X is both a normed and an ordered vector space.

Definition 5. *A norm on a vector lattice X is called a lattice norm if*

$$|x| \leq |y| \quad \text{implies} \quad \|x\| \leq \|y\|. \tag{29}$$

A Riesz space X complete under the lattice norm is called a Banach lattice.

Property (29) gives the important identity:

$$\|x\| = \|\,|x|\,\|, \qquad x \in X. \tag{30}$$

If X is a normed lattice, then all lattice operations are uniformly continuous in the norm of X with respect to all variables involved.

Positive operators will be discussed in more detail below. However, we need some terminology related to operators at this instance. An operator A defined on X is said to be positive if $Ax \geq 0$ for $x \geq 0$. A positive operator A is said to be a *lattice homomorphism* if $A(x \vee y) = Ax \vee Ay$. It can be proved, [1, Theorem 1.34], that this is equivalent to A preserving all other lattice operations (e.g. $|Ax| = |x|$, $(Ax)^+ = Ax^+$, etc). If A is a one-to-one lattice homomorphism, it is called a *lattice isomorphism* and if, additionally, A is an isometry, then it is called a *lattice isometry.*

Bounded positive functionals form a convex cone in X^* and thus define a natural ordering of X^*. It can be proved, [3, Theorem 12.1], that the normed dual of a normed Riesz space is a Banach lattice under this order. In addition, the evaluation map $X \to X^{**}$ is a lattice isometry so that X becomes a Riesz subspace of X^{**}.

Sublattices, Ideals, Bands, etc.

A vector subspace X_0 of a vector lattice X, which is ordered by the order inherited from X, may fail to be a vector sublattice of X in the sense that X_0 may be not closed under lattice operations. For instance, the subspace

$$X_0 := \{f \in L_1(\mathbb{R}); \int\limits_{-\infty}^{\infty} f(t)dt = 0\}$$

does not contain any nontrivial nonnegative function, and thus it is not closed under the operations of taking $f_\pm$ or $|f|$. Accordingly, we call X_0 a *vector sublattice* or a *Riesz subspace* if X_0 is closed under lattice operations.

A subset S of a vector lattice is called *solid* if for any $x, y \in X$ from $y \in S$ and $|x| \le |y|$ it follows that $x \in S$. A solid linear subspace is called *ideal*; ideals are automatically Riesz subspaces. A *band* in X is an ideal that contains suprema of all its subsets. Any subset $S \subset X$ uniquely determines the smallest (in the inclusion sense) Riesz subspace (resp. ideal, band) in X containing S, called the *Riesz subspace* (resp. *ideal, band*) generated by S.

Example 3. Closed ideals can be used to construct new useful Banach lattices by taking quotients. Let X be a Banach lattice and E a closed ideal in X. Then the quotient space X/E is a Banach space. We can define an order in X/E through the following relation. For $X/E \ni \tilde{x}, \tilde{y}$ we say that $\tilde{x} \le \tilde{y}$ if there are $x_1 \in \tilde{x}$ and $y_1 \in \tilde{y}$ such that $x_1 \le y_1$ in X and one can prove that X/E with this order and the canonical quotient norm is a Banach lattice.

Consider, in particular, the F-product discussed in Subsection 2.2. If X is a Banach lattice, then the absolute value on $l_\infty(X)$ is given by

$$|(x_n)_{n \in \mathbb{N}}| = (|x_n|)_{n \in \mathbb{N}}.$$

Since $c_0(X)$ is a closed ideal in $l_\infty(X)$, then $\hat{X}$ is a lattice with the canonical injection becoming a lattice homomorphism.

Example 4. Closed ideals in $X = L_p(\Omega)$, which are not equal to X, are precisely the sets of the form

$$I = \{f \in X; \exists_{\Omega' \subset \Omega, \mu(\Omega')} f|_{\Omega'} = 0 \text{ a.e.}\}.$$

The set I clearly is a closed ideal. On the other hand, let $f(x) > 0$ a.e. on Ω and $g \in X_+$. Consider sets $\Omega_n = \{x \in \Omega; f(x) \geq 1/n\}$ and define $g_n(x) = 0$ on Ω_n, $g_n = \min\{g, n\}$ otherwise. We have $0 \leq g_n \leq n^2 f$, hence $g_n \in X$ and, clearly $g_n \to g$ in $L_p(\Omega)$ since $\mu(\Omega_n) \to 0$ as $n \to \infty$.

In the theory developed an important role is played by *principal ideals*, which are ideals generated by a single point, say, x; such an ideal is given by

$$E_x = \{y \in X; \text{ there exists } \lambda \geq 0 \text{ such that } |y| \leq \lambda|x|\}.$$

If for some vector $e \in X$ we have $E_e = X$, then e is called an *order unit*.

A *principal band* generated by $x \in X$ is given by

$$B_x = \{y \in X; \sup_{n \in \mathbb{N}}\{|y| \wedge n|x|\} = |y|\}.$$

An element $e \in X$ is said to be a *weak unit* if $B_e = X$. It follows that, in a vector lattice X, $e > 0$ is a weak unit if and only if, for any $x \in X$, $|x| \wedge e = 0$ implies $x = 0$. Every order unit is a weak unit. If $X = C(\Omega)$, where Ω is compact, then any strictly positive function is an order unit. On the other hand, L_p and l_1 spaces, $1 \leq p < +\infty$, will not typically have order units (L_p include functions that could be unbounded, for l_p one can always find a sequence converging to 0 at a slower rate than a given one). However, any strictly positive a.e. L_p function is a weak order unit.

An intermediate notion between order unit and weak order unit is played by *quasi-interior points*. We say that $0 \neq u \in X_+$ is a quasi-interior point of X if $\overline{E_u} = X$. We have

Lemma 2. *[1, Lemma 4.15] For $0 \neq u \in X_+$ the following are equivalent.*

(a) u is a quasi-interior point of X;
(b) For each $x \in X_+$ we have $\lim_{n \to \infty} \|x \wedge nu - x\| = 0$;
(b) If $0 < x^* \in X_+^*$, then $<x^*, u> > 0$.

AM- and AL-spaces

Two important classes of Banach lattices, which will play a significant role later, are provided by *AL-* and *AM-* spaces.

Definition 6. *We say that a Banach lattice X is*

(i) an   AL-space if $\|x + y\| = \|x\| + \|y\|$ for all $x, y \in X_+$,
(ii) an   AM-space if $\|x \vee y\| = \max\{\|x\|, \|y\|\}$ for all $x, y \in X_+$.

Example 5. Standard examples of AM-spaces are offered by the spaces $C(\overline{\Omega})$, where $\overline{\Omega}$ is either a bounded subset of $\mathbb{R}^n$, or in general, a compact topological space. Also the space $L_\infty(\Omega)$ is an AM-space. On the other hand, most known examples of AL-spaces are the spaces $L_1(\Omega)$. We observe later that these examples exhaust all (up to a lattice isometry) cases of AM- and AL-spaces. However, particular representations of these spaces can be very different.

It can be proved, [3, Theorem 12.22] and [1, Theorem 3.3], that a Banach lattice X is an AL-space (respectively, AM-space) if and only if its dual X^* is an AM-space (respectively, AL-space). Moreover, if X is an AL-space, then X^* is a Dedekind complete AM-space with unit e^* defined by

$$X^* \ni e^*(x) = \|x_+\| - \|x_-\|$$

for $x \in X$ (thus e^* is the norm of x on the positive cone). Moreover, if X is an AM-space with unit e, then X^{**} is also an AM-space with unit e.

Any AM-space X with unit e can be equivalently normed by

$$\|x\|_\infty = \inf\{\lambda > 0; \ |x| \le \lambda e\}$$

(see, e.g., [3, p. 188]). In this norm the unit ball of X coincides with the order interval $[-e, e]$. On the other hand, any Banach lattice contains AM-spaces with unit. Precisely speaking, [3, Theorem 12.20], the principal ideal generated by any element $u \in X$ with the norm

$$\|x\|_\infty = \inf\{\lambda > 0; \ |x| \le \lambda|u|\}, \tag{31}$$

becomes an AM-space with unit $|u|$, whose closed unit ball coincides with the order interval $[-|u|, |u|]$.

The following results give the full characterisation of AL- and AM- spaces.

Theorem 7. *[3, Theorem 12.26] A Banach lattice is an AL-space if and only if it is lattice isometric to an $L_1(\Omega)$ space.*

Theorem 8. *[3, Theorem 12.28] A Banach lattice X is an AM-space with unit if and only if it is lattice isometric to some $C(\Omega)$ for a unique (up to a homeomorphism) compact Hausdorff space Ω. In particular, X is an AM-space if and only if it is lattice isometric to a closed vector sublattice of a $C(\Omega)$ space.*

Proof. The compact space Ω turns out to be

$$\begin{aligned} \Omega &= \{x^* \in B_{1,+}^*; \ x^* \text{ extr. p. of } B_1^* \text{ with } \|x^*\| = \|x^*(e)\| = 1\} \\ &= \{x^* \in B_{1,+}^*; \ x^* \text{ lat. hom. with } \|x^*\| = \|x^*(e)\| = 1\}. \end{aligned}$$

Here, B_1^* is the unit ball in the dual space and extreme points of a set are understood as points which do not belong to any proper segment with endpoints in this set. Establishing this equality is a difficult part of the proof. It follows that Ω is non-empty (by Krein-Milman theorem) and weakly* compact. Thus, Ω equipped with the weak* topology will be our compact topological space. For $x \in X$ we define the mapping

$$(Tx)(x^*) = <x^*, x>, \quad x^* \in \Omega.$$

It can be proved that T is a norm preserving lattice isomorphism from X into $C(\Omega)$. Since $(Te)(x^*) = x^*(e) = 1$ for all $x^* \in \Omega$, $T(E)$ is closed and separates points of Ω, the Stone-Weierstrass theorem yields $T(E) = C(\Omega)$. $\square$

Using the last theorem, we see that 'locally' each Banach lattice is lattice isomorphic to $C(\Omega)$. More precisely, given $0 < u \in X$ we take the principal ideal E_u which can be converted into an AM-space normed by (31). This norm is not equivalent to the norm in X. However, if we have a bounded operator defined on X, then the transferred operator on $C(\Omega)$ is a positive everywhere defined operator and thus bounded (by Theorem 10). Conversely, operators specific to $C(\Omega)$, such as multiplication or composition operators, can be transferred to bounded operators on X_u. If u is a quasi-internal point and the given operator happens to be bounded in the norm of X, then it can be extended by density to the whole Banach lattice. We shall use this construction later to define the modulus of an element of a complex Banach lattice and the signum operator.

3.3 Positive Operators

Definition 7. *A linear operator A from a Banach lattice X into a Banach lattice Y is called positive, denoted by $A \geq 0$, if $Ax \geq 0$ for any $x \geq 0$.*

An operator A is positive if and only if $|Ax| \leq A|x|$. This follows easily from $-|x| \leq x \leq |x|$ so, if A is positive, then $-A|x| \leq Ax \leq A|x|$. Conversely, taking $x \geq 0$, we obtain $0 \leq |Ax| \leq A|x| = Ax$.

Positive operators are fully determined by their behaviour on the positive cone. Precisely speaking, [8, Theorem 2.64]),

Theorem 9. *If $A : X_+ \to Y_+$ is additive, then A extends uniquely to a positive linear operator from X to Y. Keeping the notation A for the extension, we have, for each $x \in X$,*

$$Ax = Ax_+ - Ax_-. \tag{32}$$

Next we give another frequently used property of positive operators.

Theorem 10. *If A is an everywhere defined positive operator from a Banach lattice to a normed Riesz space, then A is bounded.*

Proof. If A were not bounded, then we would have a sequence $(x_n)_{n \in \mathbb{N}}$ satisfying $\|x_n\| = 1$ and $\|Ax_n\| \geq n^3$, $n \in \mathbb{N}$. Because X is a Banach space, $x := \sum_{n=1}^{\infty} n^{-2}|x_n| \in X$. Because $0 \leq |x_n|/n^2 \leq x$, we have $\infty > \|Ax\| \geq \|A(|x_n|/n^2)\| \geq \|A(x_n/n^2)\| \geq n$ for all n, which is a contradiction. $\square$

The norm of a positive operator can be evaluated by

$$\|A\| = \sup_{x \geq 0, \|x\| \leq 1} \|Ax\|.$$

As a consequence, we note that if $0 \leq A \leq B$, then $\|A\| \leq \|B\|$. Moreover, it is worthwhile to emphasize that if there exists K such that $\|Ax\| \leq K\|x\|$ for $x \geq 0$, then this inequality holds for any $x \in X$.

Irreducible Operators

An important class of positive operators are *irreducible operators*. We say that an operator A on a Banach lattice X is irreducible if $\{0\}$ and X are the only invariant ideals under A. We say that A is *strongly irreducible* if Au is quasi-interior point for any $u > 0$. Strongly irreducible operators are irreducible. Indeed, any closed ideal $E \neq \{0\}$ contains a positive point u so that $Au \in AE \subset E$ provided E is invariant. Since Au is quasi-interior, this implies $E = X$ We shall return to this concept in Subsection 6.2.

Signum Operator

An important role in the following considerations is played by the multiplication by the *signum operator*. In function spaces the definition is obvious: given $u \neq 0$ and $f \in C(\Omega)$, we define $S_u f = u|u|^{-1}f$. Clearly, in this setting S_u is a linear isometry satisfying $|S_u f| = |f|$; its inverse is $S_{\bar{u}}$, where $\bar{u}$ is the complex conjugate of u.

More generally, we consider u such that $|u|$ is a quasi-interior point of X. In this case we define this operator on $E_{|u|}$ by passing to the representation $C(\Omega)$ and transferring back the signum operator defined above to X. We note that in this setting S_h is still invertible and has the same properties as in $C(\Omega)$. By $|S_u f| = |f|$ we can extend S_u by density to $X = \overline{E_{|u|}}$ preserving invertibility.

It is possible to extend this definition to the case when $|u|$ is not a quasi-interior point but it will not be needed in what follows (see e.g. [42, p. 245]).

3.4 Relation Between Order and Norm

Existence of an order in some set X allows us to introduce in a natural way the notion of convergence. However, in general, sequences are not sufficient to properly describe all related phenomena and thus we have to resort to nets.

We say that an ordered set Δ is *directed* if any pair of elements has an upper bound. Then, by a *net* $(x_\alpha)_{\alpha \in \Delta}$ in a set X, we understand a function from the *index set* Δ into X.

A net $(x_\alpha)_{\alpha \in \Delta}$ in a normed space X converges to some point $x \in X$ if for any $\epsilon > 0$ there is $\alpha_0 \in \Delta$ such that for any $\alpha \geq \alpha_0$ we have $\|x_\alpha - x\| \leq \epsilon$.

A net $(x_\alpha)_{\alpha \in \Delta}$ in an ordered set X is said to be *decreasing* (in symbols $x_\alpha \downarrow$) if for any $\alpha_1, \alpha_2 \in \Delta$ with $\alpha_1 \geq \alpha_2$ we have $x_{\alpha_1} \leq x_{\alpha_2}$. The notation $x_\alpha \downarrow x$ means that $x_\alpha \downarrow$ and $\inf\{x_\alpha; \ \alpha \in \Delta\} = x$. Furthermore, we write $x_\alpha \downarrow \geq x$ if the net is decreasing and $x_\alpha \geq x$ for all $\alpha \in \Delta$. Symbols $x_\alpha \uparrow$, $x_\alpha \uparrow x$, and $x_\alpha \uparrow \leq x$ have analogous meaning.

One of the basic results relating order and convergence in norm is

Proposition 1. *Let X be a normed lattice. Then:*

(1) The positive cone X_+ is closed.

(2) If $X \ni x_\alpha \uparrow$ and $\lim_{\alpha \in \Delta} x_\alpha = x$ in X, then $x = \sup\{x_\alpha; \ \alpha \in \Delta\}$.
(3) If $X \ni x_\alpha \downarrow$ and $\lim_{\alpha \in \Delta} x_\alpha = x$ in X, then $x = \inf\{x_\alpha; \ \alpha \in \Delta\}$.

Proof. (1) Because $X_+ = \{x \in X; \ x_- = 0\}$ and lattice operation $X \ni x \to x_- \in X$ is continuous we see that X_+ is closed.

(2) For any fixed $\alpha \in \Delta$ we have

$$\lim_{\beta \in \Delta} (x_\beta - x_\alpha) = x - x_\alpha$$

in norm and $x_\beta - x_\alpha \in X_+$ for $\beta \geq \alpha$ so that $x - x_\alpha \in X_+$ for any $\alpha \in \Delta$ by (1). Thus x is an upper bound for the net $\{x_\alpha\}_{\alpha \in \Delta}$. On the other hand, if $x_\alpha \leq y$ for all α, then $0 \leq y - x_\alpha \xrightarrow{n} y - x$ so that, again by (1), we have $y \geq x$ and hence $x = \sup\{x_\alpha; \ \alpha \in \Delta\}$.

The proof of (3) is analogous. $\square$

Example 6. The converse of Proposition 1(2) is false; that is, we may have $x_\alpha \uparrow x$ but $(x_\alpha)_{\alpha \in \Delta}$ does not converge in norm. Indeed, consider $\mathbf{x}_n = (1, 1, 1 \ldots, 1, 0, 0, \ldots) \in l_\infty$, where 1 occupies only the n first positions. Clearly, $\sup_{n \in \mathbb{N}} \mathbf{x}_n = \mathbf{x} := (1, 1, \ldots, 1, \ldots)$ but $\|\mathbf{x}_n - \mathbf{x}\|_\infty = 1$.

This example justifies introducing a special class of Banach lattices.

Definition 8. *We say that a Banach lattice X has order continuous norm if for any net $(x_\alpha)_{\alpha \in \Delta}$, $x_\alpha \downarrow 0$ implies $\|x_\alpha\| \downarrow 0$.*

Theorem 11. *[3, Theorem 12.9] For a Banach lattice X, the statements below are equivalent.*

(1) X has order continuous norm;
(2) If $0 \leq x_n \uparrow \leq x$ holds in X, then $(x_n)_{n \in \mathbb{N}}$ is a Cauchy sequence;
(3) X is σ-order complete and $x_n \downarrow 0$ implies $\|x_n\| \to 0$.

Moreover, every Banach lattice with order continuous norm is order complete.

Example 7. For $1 \leq p < \infty$, the Banach lattice $L_p(\Omega)$ has order continuous norm. Indeed, let $f_n \downarrow 0$ almost everywhere. Then $\|f_n\|^p = \int_\Omega f_n^p d\mu \to 0$ from the dominated convergence theorem and the statement follows from Theorem 11(3) as $L_p(\Omega)$ is σ-order complete by Example 2.

On the other hand, $L_\infty(\Omega)$ is order complete by Example 2 but its norm is not order continuous. To see this, consider the σ-algebra Σ of measurable subsets of Ω and let Δ be the subset of Σ containing the sets which differ from Ω by sets of positive measure, directed by the relation of inclusion. Finally, take the net $(\chi_\alpha)_{\alpha \in \Delta}$ of characteristic functions of sets from Δ. Then $\chi_\Omega - \chi_\alpha \downarrow 0$ but $\|\chi_\Omega - \chi_\alpha\| = 1$ for all $\alpha \in \Delta$.

The importance of Banach lattices with order continuous norm stems mainly from property 2 of Theorem 11 which states that increasing sequences dominated in the order sense must necessarily converge in norm. There is an important subset of this class with a stronger property that increasing and norm bounded sequences are norm convergent.

Definition 9. *We say that a Banach lattice X is a KB-space (Kantorovič–Banach space) if every increasing norm bounded sequence of elements of X_+ converges in norm in X.*

Example 8. We observe that if $x_n \uparrow x$, then $\|x_n\| \leq \|x\|$ for all $n \in \mathbb{N}$ and thus any KB-space has order continuous norm by Theorem 11. Hence, spaces which do not have order continuous norm cannot be KB-spaces. This rules out the spaces of continuous functions, l_∞ and $L_\infty(\Omega)$.

To see that the KB-class is indeed strictly smaller, let us consider the space c_0. First we prove that it has order continuous norm. It is clearly σ-order complete. Let the sequence $(\mathbf{x}_n)_{n \in \mathbb{N}}$, given by $\mathbf{x}_n = (x_k^n)_{k \in \mathbb{N}}$, satisfy $\mathbf{x}_n \downarrow 0$. For a given $\epsilon > 0$, we find k_0 such that $|x_k^1| < \epsilon$ for all $k \geq k_0$. Because $(\mathbf{x}_n)_{n \in \mathbb{N}}$ is decreasing, we also have $|x_k^n| < \epsilon$ for all $k \geq k_0$ and $n \geq 1$. Then, we find n_0 such that $|x_k^n| < \epsilon$ for all $n \geq n_0$ and $1 \leq k \leq k_0$ and combining these estimates we see that $\|\mathbf{x}_n\| < \epsilon$ for all $n \geq n_0$ so $\|\mathbf{x}_n\| \to 0$.

On the other hand, let us again take the sequence $\mathbf{x}_n = (1, 1, \ldots, 1, 0, 0, \ldots)$ where 1 occupies n first positions. It is clearly norm bounded and increasing, but it does not converge in norm to any element of c_0. Hence, the c_0 norm is not order continuous.

Next we characterize the KB-spaces which often appear in applications.

Theorem 12. *[8, Theorem 2.82] Assume that X is a weakly sequentially complete Banach lattice. If $(x_n)_{n \in \mathbb{N}}$ is increasing and $(\|x_n\|)_{n \in \mathbb{N}}$ is bounded, then there is $x \in X$ such that $\lim_{n \to \infty} x_n = x$ in X.*

The next result shows the same property for AL-spaces.

Theorem 13. *Any AL-space is a KB-space.*

Proof. If $(x_n)_{n \in \mathbb{N}}$ is an increasing and norm bounded sequence, then for $0 \leq x_n \leq x_m$, we have

$$\|x_m\| = \|x_m - x_n\| + \|x_n\|$$

as $x_m - x_n \geq 0$ so that

$$\|x_m - x_n\| = \|x_m\| - \|x_n\| = |\|x_m\| - \|x_n\||.$$

By assumption, $(\|x_n\|)_{n \in \mathbb{N}}$ is monotonic and bounded, and hence convergent, we see that $(x_n)_{n \in \mathbb{N}}$ is Cauchy. □

3.5 Complexification

Our main interest is in real operators on real Banach spaces. However, when we want to use spectral theory, we need to move the problem to a complex space. This is done by the procedure called *complexification*.

Definition 10. *Let X be a real vector lattice. The complexification X_C of X is the set of pairs $(x, y) \in X \times X$ where, following the scalar convention, we write $(x, y) = x + iy$. Vector operations are defined as in scalar case The partial order in X_C is defined by*

$$x_0 + iy_0 \leq x_1 + iy_1 \quad \text{if and only if} \quad x_0 \leq x_1 \text{ and } y_0 = y_1. \qquad (33)$$

The operations of the complex adjoint, real part, and imaginary part of $z = x + iy$ are defined through:

$$\bar{z} = \overline{x + iy} = x - iy, \quad \Re z = \frac{z + \bar{z}}{2} = x, \quad \Im z = \frac{z - \bar{z}}{2i} = y.$$

Remark 3. Note, that from the definition, $x \geq 0$ in X_C is equivalent to $x \in X$ and $x \geq 0$ in X. In particular, X_C with partial order (33) is not a lattice.

It is a more complicated task to introduce a norm on X_C because standard product norms, in general, fail to preserve the homogeneity of the norm.

First we introduce the modulus on X_C. In the scalar case we have

$$\sup_{\theta \in [0, 2\pi]} (\alpha \cos \theta + \beta \sin \theta) = |\alpha + i\beta|. \qquad (34)$$

Mimicking this, for $x + iy \in X_C$ we define

$$|x + iy| = \sup_{\theta \in [0, 2\pi]} (x \cos \theta + y \sin \theta).$$

It can be proved that this element exists. This follows because elements over which we take the supremum belong to the principal ideal generated by $|x| + |y|$ and, as we noted when discussing AM-spaces, such an ideal is an AM-space with unit $|x| + |y|$ and thus it is lattice isometric to some $C(\Omega)$. For $C(\Omega)$ the existence of $|x + iy|$ is proved pointwise by the argument leading to (34).

Such a defined modulus has all standard properties of the scalar complex modulus, [2, Problem 3.2.2] and thus one can define a norm on the complexification X_C by

$$\|z\|_c = \|x + iy\|_c = \||x + iy|\|. \qquad (35)$$

As the norm $\|\cdot\|$ is a lattice norm, we have $\|z_1\|_c \leq \|z_2\|_c$ whenever $|z_1| \leq |z_2|$, hence $\|\cdot\|_c$ becomes a lattice norm on X_C.

Definition 11. *A complex Banach lattice is an ordered complex Banach space X_C that arises as the complexification of a real Banach lattice X, according to Definition 10, equipped with the norm (35).*

We extend A to X_C according to the formula $A_C(x+iy) = Ax+iAy$, and observe that if A is a positive operator between real Banach lattices X and Y then, for $z = x+iy \in X_C$, we have $(Ax)\cos\theta+(Ay)\sin\theta = A(x\cos\theta+y\sin\theta) \leq A|z|$, therefore $|A_C z| \leq A|z|$. Hence, for positive operators,

$$\|A_C\|_c = \|A\|. \tag{36}$$

There are examples, where $\|A\| < \|A_C\|_c$.

Note that the standard $L_p(\Omega)$ and $C(\Omega)$ norms are of the type (35). These spaces have a nice property of preserving the operator norm even for operators which are not necessarily positive, see [8, p. 63].

Example 9. Any positive linear operator A on X_C is a real operator; that is, $A : X \to X$. In fact, let $X_C \ni x = x_+ - x_-$. By definition, $Ax_+ \geq 0$ and $Ax_- \geq 0$ so $Ax_+, Ax_- \in X$ and thus $Ax = Ax_+ - Ax_- \in X$.

Remark 4. If for a linear operator A we prove that it generates a semigroup of say, contractions, in X, then this semigroup will be also a semigroup of contractions on X_C, hence, in particular, A is a dissipative operator in the complex setting. Due to this observation we confine ourselves to real operators in real spaces.

3.6 Spectral Radius of Positive Operators

Let $A \in \mathcal{L}(X)$. First we note that that the peripheral spectrum $\sigma_{per,r(A)}$, see (17), is non-empty. Also, $r(A) \in \{|\lambda|; \ \lambda \in \sigma(A)\}$. This follows from the compactness of $\sigma(A)$.

As a more serious application of the theory of Banach lattices, we prove the abstract version of the Perron-Frobenius theorem.

First we note that we can carry the considerations in the complexification of X, if necessary. Since all operators are positive, the operator norms in the real lattice and its complexification are equal, see (36), and we shall not distinguish them in the proofs.

Theorem 14. *Let $r(A)$ be the spectral radius of a positive operator A on a Banach lattice X. Then $r(A) \in \sigma(A)$.*

Proof. Let $\lambda_n = r(A) + 1/n$, then $\lambda_n \in \rho(A)$ for any n. Since $\lambda_n \to r(A)$, $r(A) \in \sigma(A)$ will follow, by Theorem 2, if $\lim_{n\to\infty} \|R(\lambda_n, A)\| = \infty$.

Since the peripheral spectrum is non-empty, let $\alpha \in \sigma(A)$ with $|\alpha| = r(A)$ and define $\mu_n = \alpha\lambda_n/|\alpha|$. We have $\mu_n \in \rho(A)$ and $\mu_n \to \alpha$ so that, invoking Theorem 2 again, $\lim_{n\to\infty} \|R(\mu_n, A)\| = \infty$. Next, for each n we pick a unit vector z_n satisfying

$$\|R(\mu_n, A)z_n\| \geq \frac{1}{2}\|R(\mu_n, A)\|$$

Using the series representation of the resolvent (15) we easily infer

$$|R(\lambda, A)z| \leq R(|\lambda|, A)|z|$$

so that $|R(\mu_n, A)z_n| \leq R(\lambda_n, A)|z_n|$ and consequently

$$\|R(\lambda_n, A)\| \geq \|R(\lambda_n, A)|z_n|\| \geq \|R(\mu_n, A)z_n\| \geq \frac{1}{2}\|R(\mu_n, A)\|$$

which proves the thesis. □

Theorem 15. *If $A : X \to X$ is a compact positive operator on a Banach lattice X with $r(A) > 0$, then $r(A)$ is an eigenvalue with positive eigenvector.*

Proof. Since $r(A) > 0$, by Theorems 14 and 5 it is an eigenvalue. As above, we put $\lambda_n = r(A) + 1/n$ so that $\lambda_n \downarrow r(A)$ and $\|R(\lambda_n, A)\| \to \infty$ as $n \to \infty$. Furthermore, for each n there is z_n with $\|z_n\| = 1$ satisfying

$$\|R(\lambda_n, A)z_n\| \geq \frac{1}{2}\|R(\lambda_n, A)\|.$$

We define $x_n = R(\lambda_n, A)z_n/\|R(\lambda_n, A)z_n\|$ and note that x_n is a positive unit vector. From

$$Ax_n - r(A)x_n = (\lambda_n - r(A))x_n + Ax_n - \lambda_n x_n = \frac{x_n}{n} - \frac{z_n}{\|R(\lambda_n, A)z_n\|}$$

we obtain

$$\|Ax_n - r(A)x_n\| \to 0, \quad n \to \infty.$$

Since A is compact, the sequence $(Ax_n)_{n \in \mathbb{N}}$ has a convergent subsequence which we denote by $(Ax_n)_{n \in \mathbb{N}}$ again. Since $r(A) > 0$ and $\|x_n\| = 1$, the above implies that $\lim_{n \to \infty} x_n = x > 0$ satisfying $Ax = r(A)x$. □

Corollary 1. *The thesis of Theorem 15 remains valid if the positive operator A only is power compact.*

Proof. If $r = r(A) > 0$ and A is power compact, then from the Spectral Mapping Theorem we have $A^k x = r^k x$ for some $x > 0$. The element $y = \sum_{i=0}^{k-1} r^i A^{k-1-i} x > 0$ (from positivity of A, x and r), hence

$$Ay - ry = A^k x - r^k x = 0.$$

□

4 First Semigroups

The semigroup theory is concerned with methods of finding solutions of the Cauchy problem.

Definition 12. *Given a complex Banach space and a linear operator $\mathcal{A}$ with $D(\mathcal{A})$, $Im\mathcal{A} \subset X$ and given $u_0 \in X$, find a function $u(t) = u(t, u_0)$ such that*

1. $u \in C^0([0, \infty), X) \cap C^1((0, \infty), X)$,
2. for each $t > 0$, $u(t) \in D(\mathcal{A})$ and

$$u'(t) = \mathcal{A}u(t), \quad t > 0, \tag{37}$$

3.

$$\lim_{t \to 0+} u(t) = u_0 \quad \text{in } X. \tag{38}$$

A function satisfying all conditions above is called the classical (or strict) solution of (37), (38).

If the solution to (37), (38) is unique, then we can introduce a family of operators $(G(t))_{t \geq 0}$ such that $u(t, u_0) = G(t)u_0$. Ideally, $G(t)$ should be defined on the whole space for each $t > 0$, and the function $t \to G(t)u_0$ should be continuous for each $u_0 \in X$, leading to well-posedness of (37), (38). Moreover, uniqueness and linearity of $\mathcal{A}$ imply that $G(t)$ are linear operators. A fine-tuning of these requirements leads to the following definition.

Definition 13. *A family $(G(t))_{t \geq 0}$ of bounded linear operators on X is called a C_0-semigroup, or a strongly continuous semigroup, if*

(i) $G(0) = I$;
(ii) $G(t + s) = G(t)G(s)$ for all $t, s \geq 0$;
(iii) $\lim_{t \to 0+} G(t)x = x$ for any $x \in X$.

A linear operator A is called the (infinitesimal) generator of $(G(t))_{t \geq 0}$ if

$$Ax = \lim_{h \to 0+} \frac{G(h)x - x}{h}, \tag{39}$$

with $D(A)$ defined as the set of all $x \in X$ for which this limit exists. Typically the semigroup generated by A is denoted by $(G_A(t))_{t \geq 0}$.

If $(G(t))_{t \geq 0}$ is a C_0-semigroup, then the local boundedness and (ii) lead to the existence of constants $M > 0$ and ω such that for all $t \geq 0$

$$\|G(t)\|_X \leq Me^{\omega t}. \tag{40}$$

We say that $A \in \mathcal{G}(M, \omega)$ if it generates $(G(t))_{t \geq 0}$ satisfying (40). The *type*, or *uniform growth bound*, $\omega_0(G)$ of $(G(t))_{t \geq 0}$ is defined as

$$\omega_0(G) = \inf\{\omega; \text{ there is } M \text{ such that (40) holds}\}. \tag{41}$$

From (39) and the condition (iii) of Definition 13 we see that if A is the generator of $(G(t))_{t \geq 0}$, then for $x \in D(A)$ the function $t \to G(t)x$ is a classical solution of the following Cauchy problem,

$$\partial_t u(t) = A(u(t)), \quad t > 0, \tag{42}$$

$$\lim_{t \to 0^+} u(t) = x. \tag{43}$$

We note that, ideally, the generator A should coincide with $\mathcal{A}$ but in reality very often it is not so. In fact, a large part of the theory discussed here is concerned with finding a relation between $\mathcal{A}$ and its realisation A which generates a semigroup. However, such problems are addressed later and for most of this section we are concerned with solvability of (42), (43); that is, with the case when $\mathcal{A}$ of (37) is the generator of a semigroup.

As we noted, for $x \in D(A)$ the function $u(t) = G(t)x$ is a classical solution to (42), (43). For $x \in X \setminus D(A)$, however, the function $u(t) = G(t)x$ is continuous but, in general, neither differentiable, nor $D(A)$-valued. Therefore, it is not a classical solution. Nevertheless, the integral $v(t) = \int_0^t u(s)ds \in D(A)$ and it is a strict solution of the integrated version of (42), (43):

$$u(t) = A \int_0^t u(s)ds + x. \tag{44}$$

We say that a function u satisfying (44) is a *mild solution* or *integral solution* of (42), (43).

Proposition 2. *Let $(G(t))_{t \geq 0}$ be the semigroup generated by $(A, D(A))$. Then $t \to G(t)x$, $x \in D(A)$, is the only solution of (42), (43) taking values in $D(A)$. Similarly, for $x \in X$, the function $t \to G(t)x$ is the only mild solution.*

Thus, if we have a semigroup, we can identify the Cauchy problem which it is a solution of. Usually, however, we are interested in the reverse question, that is, in finding the semigroup for a given equation. The answer is given by the Hille–Yoshida theorem.

4.1 Around the Hille–Yosida Theorem

Theorem 16. $A \in \mathcal{G}(M, \omega)$ *if and only if A is closed and densely defined and there exist $M > 0, \omega \in \mathbb{R}$ such that $(\omega, \infty) \subset \rho(A)$ and for all $n \geq 1, \lambda > \omega$,*

$$\|(\lambda I - A)^{-n}\| \leq \frac{M}{(\lambda - \omega)^n}. \tag{45}$$

If A is the generator of $(G(t))_{t \geq 0}$, then properties (i) and (ii) follow from the formula relating $(G(t))_{t \geq 0}$ with $R(\lambda, A)$: for $\lambda > \omega_0(G)$, where $\omega_0(G)$ is defined by (41), then $\lambda \in \rho(A)$ and

$$R(\lambda, A)x = \int_0^\infty e^{-\lambda t} G(t)x \, dt \tag{46}$$

is valid for all $x \in X$.

Another widely used formula relating A with $(G(t))_{t \geq 0}$ is:

$$G(t)x = \lim_{n \to \infty} \left(I - \frac{t}{n}A\right)^{-n} x = \lim_{n \to \infty} \left(\frac{n}{t}R\left(\frac{n}{t},A\right)\right)^n x \qquad (47)$$

for any $x \in X$, and the limit is uniform in t on bounded intervals.

As we noticed earlier, a given operator $(A, D(A))$ can generate at most one C_0-semigroup. Using the Hille–Yosida theorem we can prove a stronger result which is useful later.

Proposition 3. *Assume that the closure $(\overline{A}, D(\overline{A}))$ of an operator (A, D) generates a C_0-semigroup in X. If $(B, D(B))$ is also a generator such that $B|_D = A$, then $(B, D(B)) = (\overline{A}, D(\overline{A}))$.*

Without the assumption that the closure of A is a generator there may be infinitely many extensions of a given operator which generate a semigroup: consider the semigroups generated by the realizations of the Laplacian subject to Dirichlet, Neumann, or mixed boundary conditions – all the generators coincide if restricted to the space of C_0^∞ functions.

4.2 Dissipative Operators

Let X be a Banach space (real or complex) and X^* be its dual. By the Hahn–Banach theorem, the *duality set*

$$\mathcal{J}(x) = \{x^* \in X^*; \; <x^*, x> = \|x\|^2 = \|x^*\|^2\} \qquad (48)$$

is nonempty for every $x \in X$.

Definition 14. *We say that an operator $(A, D(A))$ is* dissipative *if for every $x \in D(A)$ there is $x^* \in \mathcal{J}(x)$ such that*

$$\Re <x^*, Ax> \leq 0. \qquad (49)$$

An important equivalent characterisation of dissipative operators, [44, Theorem 1.4.2], is that A is dissipative if and only if for all $\lambda > 0$ and $x \in D(A)$,

$$\|(\lambda I - A)x\| \geq \lambda \|x\|. \qquad (50)$$

We note some important properties of dissipative operators.

Proposition 4. *[30] If $(A, D(A))$ is dissipative, then*

(i) *$Im(\lambda I - A) = X$ for some $\lambda > 0$ if and only if $Im(\lambda I - A) = X$ for all $\lambda > 0$.*

(ii) *A is closed if and only if $Im(\lambda I - A)$ is closed for some (and hence all) $\lambda > 0$.*

(iii) *If A is densely defined, then A is closable and $\overline{A}$ is dissipative. Moreover, $\overline{Im(\lambda I - A)} = Im(\lambda I - \overline{A})$.*

Combination of the Hille–Yosida theorem with the above properties gives a generation theorem for dissipative operators, known as the Lumer–Phillips theorem ([44, Theorem 1.43] or [30, Theorem II.3.15]).

Theorem 17. *For a densely defined dissipative operator $(A, D(A))$ on a Banach space X, the following statements are equivalent.*

(a) The closure $\overline{A}$ generates a semigroup of contractions.
(b) $\overline{Im(\lambda I - A)} = X$ for some (and hence all) $\lambda > 0$.

If either condition is satisfied, then A satisfies (49) for any $x^* \in \mathcal{J}(x)$.

In particular, if we know that A is closed then the density of $Im(\lambda I - A)$ is sufficient for A to be a generator. On the other hand, if we do not know a priori that A is closed then $Im(\lambda I - A) = X$ yields A being closed and consequently that it is the generator.

Example 10. If $(A, D(A))$ is a densely defined operator in X and both A and its adjoint A^* are dissipative, then $\overline{A}$ generates a semigroup of contractions in X. In fact, because $\overline{A}$ is dissipative and closed, $Im(I - \overline{A})$ is closed. If $Im(I - \overline{A}) \neq X$, then for some $0 \neq x^* \in X^*$ we have

$$0 = <x^*, x - \overline{A}x> = <x^* - \overline{A}^* x^*, x>$$

for all $x \in D(\overline{A})$. Because $\overline{A}$ is densely defined, $x^* - \overline{A}^* x^* = 0$ and because $\overline{A}^*$ is dissipative, $x^* = 0$. Hence $Im(I - \overline{A}) = X$ and $\overline{A}$ is the generator of a dissipative semigroup by Theorem 17. In particular, dissipative self-adjoint operators on Hilbert spaces are always generators.

4.3 Long Time Behaviour of Semigroups

It is important to note that the Hille–Yosida theorem is valid in both real and complex Banach spaces with the same formulation. Thus if A is an operator in a real Banach space X, generating a semigroup $(G(t))_{t \geq 0}$, then its complexification will generate a complex semigroup of the same type in the complexification X_C of X. This allows us to extend (46) to complex values of λ. Precisely, the integral in (46) is absolutely convergent for $\Re\lambda > \omega_0(G)$. Moreover, iterations of the resolvent give the following formula,

$$R(\lambda, A)^n x = \frac{(-1)^{n-1}}{(n-1)!} \frac{d^{n-1}}{d\lambda^{n-1}} R(\lambda, A) = \frac{1}{(n-1)!} \int_0^\infty t^{n-1} e^{-\lambda t} G(t) x \, dt, \quad (51)$$

valid for all $x \in X$.

Story of Four Numbers

Formula (51) yields the estimate

$$\|R(\lambda, A)^n\| \leq \frac{M}{(\Re\lambda - \omega_0(G))^n}, \qquad \Re\lambda > \omega_0(G). \tag{52}$$

An immediate consequence of the above considerations is that the spectrum of a semigroup generator is always contained in the left half-plane given by the spectral bound

$$s(A) = \sup\{\Re\lambda;\ \lambda \in \sigma(A)\}, \tag{53}$$

defined in (18). For semigroups generated by bounded operators and, in particular, by matrices, the Lyapunov theorem, see e.g. [30, Theorem I.2.10], states that the type $\omega_0(G)$ of the semigroup is equal to $s(A)$. This is no longer true for strongly continuous semigroups in general; see for example, [44, Example 4.4.2] or [42, Example A-III.1.3], where it is shown that the translation semigroup $[G(t)f](s) = f(t+s)$ on the space $X = L_p(\mathbb{R}_+) \cap E$, where E is the weighted space $E := \{f \in L_p(\mathbb{R}_+), e^s ds\}$, whose generator A is the differentiation operator, satisfies $\omega_0(G) = 0$ and $s(A) = -1$.

That the type $\omega_0(G)$ might be a rather crude estimate of $s(A)$ can be expected because the former is determined by the absolute convergence of the Laplace integral and the Laplace integral may converge as an improper integral in a possibly larger half-plane $\Re\lambda > abs(G)$, where by $abs(G)$ we denoted the abscissa of convergence (of the Laplace integral treated as an improper integral). That, $abs(G) = \inf\{\lambda \in \mathbb{C}\}$ for which

$$B_\lambda x := \lim_{\tau \to \infty} \int_0^\tau e^{-\lambda t} G(t) x \, dt \tag{54}$$

exists for all $x \in X$. Moreover, any such λ satisfies $\lambda \in \rho(A)$ and $B_\lambda x = R(\lambda, A)x$ for all $x \in X$, see e.g. [8, Proposition 3.15].

Thus at this moment we only have the obvious estimate

$$s(A) \leq \omega_0(G) < +\infty. \tag{55}$$

We can prove, however, that $abs(G)$ controls the growth of classical solutions of (42), (43), that is, of the solutions emanating from $x \in D(A)$. To make this concept precise, we define the *growth bound* $\omega_1(G)$ by

$$\omega_1(G) = \inf\{\omega; \text{ there is } M \text{ such that } \|G(t)x\| \leq Me^{\omega t}\|x\|_{D(A)}, x \in D(A), t \geq 0\}.$$

Clearly, $\omega_1(G) \leq \omega_0(G)$. The following result is true.

Proposition 5. *[8, Proposition 3.16] For a semigroup $(G(t))_{t\geq 0}$ we have*

$$\omega_1(G) = abs(G). \tag{56}$$

Fine structure of $\sigma(A)$ and the Long Time Behaviour of $(G(t))_{t\geq0}$

One of the most important questions in the theory of strongly continuous semigroups is to determine the long time behaviour of a semigroup through the spectral properties of its generator.

Spectral Mapping Theorem for semigroups

If $(G(t))_{t\geq0}$ is generated by a bounded operator A, then $G(t) = \exp tA$ and the the the Spectral Mapping Theorem (20) gives

$$\sigma(G(t)) = e^{t\sigma(A)}. \tag{57}$$

Hence

$$e^{t\omega_0(G)} = r(G(t)) = e^{ts(A)}$$

and thus (57) yields the Lyapunov theorem for dynamical systems generated by bounded operators. However, we have seen that for C_0-semigroups the spectrum of the generator does not fully determine the spectrum of the semigroup; that is, the Spectral Mapping Theorem (20) fails in this case.

Note that while the number zero can be in the spectrum of a semigroup $(G(t))_{t\geq0}$ (e.g. for eventually compact semigroups), it cannot be obtained from any finite spectral value of A through (57). Thus, we shall restrict our considerations to $\sigma(G(t)) \setminus \{0\}$. Furthermore, validity of (57) for a given $\lambda \in \sigma(G(t))$ means that there exist $k \in \mathbb{Z}$ such that

$$\mu + 2k\pi i/t \in \sigma(A) \quad \text{with} \quad \lambda = e^{t\mu}. \tag{58}$$

We note the following general result, [30, Theorems 6.2 and 6.3]

Theorem 18. *Let $(G(t))_{t\geq0}$ be the C_0-semigroup generated by A. Then*

1. $e^{t\sigma(A)} \subset \sigma(G(t))$;
2. $e^{t\sigma_p(A)} = \sigma_p(G(t)) \setminus \{0\}$;
3. $e^{t\sigma_r(A)} = \sigma_r(G(t)) \setminus \{0\}$;
4. $e^{t\sigma_a(A)} \subset \sigma_a(G(t))$

Ultrapowers in the Context of Semigroups

Theorem 18 tells us that the main obstacle for the validity of the Spectral Mapping Theorem is caused by the approximate spectrum. We have introduced a method of converting the approximate spectrum into the point spectrum in Paragraph 2.2 and it is natural to ask whether it can be used to alleviate the encountered problems. Let $(G(t))_{t\geq0}$ be a strongly continuous semigroup. As noted in Paragraph 2.2, for each $t \geq 0$, the bounded operator $G(t)$ extends to $\widehat{G(t)}$ on X preserving the norms, spectra etc. Unfortunately, the family $(\widehat{G(t)})_{t\geq0}$ is strongly continuous if and only if the generator A of $(G(t))_{t\geq0}$

is bounded. The problem is created at the first step of construction as the extension of $(G(t))_{t \geq 0}$ to $l_\infty(X)$, denoted by $(\tilde{G}(t))_{t \geq 0}$,

$$\tilde{G}(t)[(x_n)_{n \in \mathbb{N}}] := (G(t)x_n)_{n \in \mathbb{N}}$$

is not strongly continuous.

To get around this difficulty, we proceed as in the definition of the *sun-dual* (see e.g., [30, Section II.2.6]) and first define the subspace of $l_\infty(X)$ by

$$l_\infty^G(X) := \{(x_n)_{n \in \mathbb{N}} \in l_\infty(X);\ \lim_{t \to 0^+} \|G(t)x_n - x_n\| = 0,\ \text{uniformly in } n\}.$$

Clearly $l_\infty^G(X)$ is $(\tilde{G}(t))_{t \geq 0}$−invariant and it turns out that the restriction of $(\tilde{G}(t))_{t \geq 0}$ to this subspace is strongly continuous. Moreover, since a strongly continuous semigroup is uniformly continuous on compact subsets, we see that $c_0(X) \subset l_\infty^G(X)$. Then, instead of $\hat{X}$, we consider the quotient space

$$\hat{X}^G = l_\infty^G(X)/c_0(X) \tag{59}$$

and define $(\hat{G}(t)(t))_{t \geq 0}$ as the canonical projection of $(\tilde{G}(t)(t))_{t \geq 0}$ to $\hat{X}^G$:

$$\hat{G}(t)[(x_n)_{n \in \mathbb{N}} + c_0(X)] := (G(t)x_n)_{n \in \mathbb{N}} + c_0(X) \tag{60}$$

for $(x_n)_{n \in \mathbb{N}} \in l_\infty^G(X)$. Again, with the canonical injection $X \ni x \to (x, x, \ldots) \in \hat{X}^G$ the operators $\hat{G}(t)$ become extensions of $G(t)$ for any $t \geq 0$ and restrictions of $\widehat{G(t)}$ defined on $\hat{X}$. Using standard results for quotient semigroups, we find that the generator $\hat{A}$ of $(\hat{G}(t)(t))_{t \geq 0}$ on $\hat{X}^G$ is given by

$$\hat{A}[(x_n)_{n \in \mathbb{N}} + c_0(X)] = (Ax_n)_{n \in \mathbb{N}} + c_0(X) \quad \text{on}$$
$$D(\hat{A}) = \{(x_n)_{n \in \mathbb{N}} + c_0(X);\ (x_n)_{n \in \mathbb{N}} \in D(A), (x_n)_{n \in \mathbb{N}}, (Ax_n)_{n \in \mathbb{N}} \in \hat{X}^G\}$$

Unfortunately, there is a price to pay: in general it is not true that $\sigma(G(t)) = \sigma(\hat{G}(t))$. This apparent contradiction with Theorem 4 is explained by the observation that the later theorem would refer to $\widehat{G(t)}$ and not to $\hat{G}(t)$. For instance, an approximate eigenvector for $G(t)$ may fail to satisfy the condition defining $l_\infty^G(X)$ and thus fail to be an approximate eigenvector of $\hat{G}(t)$. Of course, if an approximate eigenvector $(x_n)_{n \in \mathbb{N}}$ satisfies $(x_n)_{n \in \mathbb{N}} \subset l_\infty^G(X)$, then $\hat{x} = (x_n)_{n \in \mathbb{N}} + c_0(X)$ is an eigenvector of $\hat{G}(t)$. We will see one way of getting around this difficulty below.

Eventually Uniformly Continuous Semigroups

If a semigroup $(G(t))_{t \geq 0}$ is continuous in the uniform operator topology for $t \geq 0$, then its generator is bounded and we can use classical Lyapunov theorem. However, if $(G(t))_{t \geq 0}$ is uniformly continuous for $t > 0$ (*immediately uniformly continuous*) or even for $t \geq t_0$ for some $t_0 > 0$ (*eventually uniformly continuous*), then the situation becomes non-trivial. We note that analytic or eventually compact semigroups are eventually uniformly continuous.

To prove the latter statement assume that $T(t_0)$ is compact and let $t, s \geq t_0$. Since $t \to G(t)x$ is uniformly continuous for x in compact sets (Banach-Steinhaus theorem) and $G(t_0)B_1$ is relatively compact (B_1 is the unit ball),

$$\lim_{t \to s}(G(t)x - G(s)x) = \lim_{t \to s}(G(t - t_0) - G(s - t_0))G(t_0)x = 0$$

uniformly on $x \in B_1$ giving uniform continuity of $(G(t))_{t \geq 0}$ for $t \geq t_0$.

Theorem 19. *If $(G(t))_{t \geq 0}$ is an eventually uniformly continuous semigroup with generator A, then $\sigma(G(t)) \setminus \{0\} = e^{t\sigma(A)}$.*

Proof. For the proof it suffices to show that $\sigma_a(G(t)) \setminus \{0\} \subset e^{t\sigma_a(A)}$. Furthermore, it is enough to consider $1 \in \sigma_a(G(t_1))$ for some $t_1 > 0$. In fact, any other λ and t_2 can be reduced to this situation by considering the rescaled semigroup

$$(S(t)(t))_{t \geq 0} = (e^{-t \ln \lambda / t_1} G(t t_2 / t_1))_{t \geq 0}$$

with generator $B = (t_2 A - \ln \lambda)/t_1$. The spectral properties of 1 for $S(t_1)$ are the same as of λ for $G(t_2)$ sinc $G(t_2) - \lambda I = \lambda(S(t_1) - I)$. Take $(f_n)_{n \in \mathbb{N}} \in \sigma_a(G(t_1))$; that is $\|f\| = 1$ with $\lim_{n \to \infty} \|G(t_1)f_n - f_n\| = 0$. Let $(G(t))_{t \geq 0}$ be uniformly continuous for $t \geq t_0$. We choose $k \in \mathbb{N}$ such that $kt_1 > t_0$ and define $g_n = G(kt_1)f_n$. Then we have

$$\lim_{n \to \infty} \|g_n\| = \lim_{n \to \infty} \|[G(t_1)]^n f_n\| = \lim_{n \to \infty} \|f_n\| = 1$$

as well as

$$\lim_{n \to \infty} \|G(t_1)g_n - g_n\| \leq \lim_{n \to \infty} \|G(t_1)\|^k \|G(t_1)f_n - f_n\| = 0,$$

so $(g_n)_{n \in \mathbb{N}}$ also is an approximate eigenvector with approximate eigenvalue 1. However, $(G(t))_{t \geq 0}$ is uniformly continuous on sets of the form $G(t_0)U$ where U is a bounded set. In particular, $(G(t))_{t \geq 0}$ is uniformly continuous on $(g_n)_{n \in \mathbb{N}}$ and hence $\hat{g} = (g_n)_{n \in \mathbb{N}}$ is an element in the ultrapower $\hat{X}^G$.

By comments at the end of Example 4.3, $\hat{g}$ is an eigenvector of $(\hat{G}(t))_{t \geq 0}$ with eigenvalue 1 hence, by the Spectral Mapping Theorem for the point spectrum, there is an eigenvalue $2\pi i n/t_1$ of $\hat{A}$ for some $n \in \mathbb{Z}$. Since $\sigma(A) = \sigma(\hat{A})$, we obtain the thesis. $\square$

Bad Spectrum: Chaos

Though our main interest lies with linear dynamical systems, the general framework discussed here applies to a much larger class of dynamical systems.

Let the space (X, d) be a complete metric space and $(G(t))_{t \geq 0}$ be a continuous dynamical system on X with generator A. By $O(p) = \{G(t)p\}_{t \geq 0}$ we denote the *orbit* of $(G(t))_{t \geq 0}$ originating from p.

We say that $(G(t))_{t\geq0}$ is *topologically transitive* if for any two non-empty open sets $U, V \subset X$ there is $t_0 \geq 0$ such that $G(t_0)U \cap V \neq \emptyset$. A *periodic point* of $(G(t))_{t\geq0}$ is any point $p \in X$ satisfying $G(\tau)p = p$ for some $\tau > 0$.

Definition 15. *[26] Let X be a metric space. A dynamical system $(G(t))_{t\geq0}$ in X is said to be (topologically) chaotic in X if it is transitive and its set of periodic points is dense in X.*

Devaney's definition is related to the property called hypercyclicity: a dynamical system $(G(t))_{t\geq0}$ is called *hypercyclic* if for some $x \in X$ the orbit $\{G(t)x\}_{t\geq0}$ is dense in X.

Hypercyclicity is equivalent to topological transitivity, [14, 32]. Thus, Devaney's definition means that $(G(t))_{t\geq0}$ is chaotic if it has an orbit dense in X and its set of periodic points is dense.

Positive Criteria

The classical criterion for chaoticity of linear semigroups is given in the following theorem. Recall that $\sigma_p(A)$ denotes the point spectrum of A.

Theorem 20. *[25] Let X be a separable Banach space and let A be the generator of a semigroup $(G(t))_{t\geq0}$ on X. $(G(t))_{t\geq0}$ is chaotic if*

1. *$\sigma_p(A)$, contains an open connected set U such that $U \cap i\mathbb{R} \neq \emptyset$;*
2. *There exists an analytic selection $U \ni \lambda \to x(\lambda)$ of eigenvectors of A;*
3. *$\overline{Span\{x(\lambda), \lambda \in U\}} = X$.*

Proof. The proof uses the observation that $(G(t))_{t\geq0}$ is hypercyclic if

$$X_0 = \{x \in X; \lim_{t\to\infty} G(t)x = 0\}$$
$$X_\infty = \{w \in X; \forall_{\epsilon>0}\exists_{x\in X,t>0}\|x\| < \epsilon \text{ and } \|G(t)x - w\| < \epsilon\}$$

are dense in X. Thus, if also the set of periodic points X_p is dense, then $(G(t))_{t\geq0}$ is chaotic. Condition 3. is used through the following argument. If $U' \subset U$ with an accumulation point in U and $\Phi \in X^*$ satisfy $< \Phi, x(\lambda) >= 0$ for $\lambda \in U'$, then from the principle of isolated zeros $F_\Phi(\lambda) =< \Phi, x(\lambda) >= 0$ in U which by Condition 3. is possible only if $\Phi = 0$. This in turn shows that $\overline{Span\{x(\lambda), \lambda \in U'\}} = X$. Now, it is easy to see that the sets $U_- = U \cap \{\lambda, \Re\lambda < 0\}, U_+ = U \cap \{\lambda, \Re\lambda > 0\}, U_0 = U \cap \{\lambda, \Re\lambda = 0, \Im\lambda \text{ is rational}\}$ have accumulation points in U. Moreover $Span\{x(\lambda), \lambda \in U_-\} \subset X_0$, by $x(\lambda) = G(t)e^{-\lambda t}x(\lambda)$ we see that $Span\{x(\lambda), \lambda \in U_+\} \subset X_\infty$ and $Span\{x(\lambda), \lambda \in U_0\} \subset X_p$ so that if Condition 3 is satisfied, X_0, X_∞ and X_p are dense in X and therefore $(G(t))_{t\geq0}$ is chaotic. $\qquad\square$

A reflection on the proof allows to generalize this result in a significant way. First, we observe:

Proposition 6. *If A is a closed operator in X, $x(\lambda)$ is analytic in an open connected set U and satisfies there $Ax(\lambda) = \lambda x(\lambda)$ then, denoting by a_{n,λ_0} the n-th coefficient of Taylor's expansion of $x(\lambda)$ at $\lambda_0 \in U$, we have that*

$$Z = Z_{\lambda_0} = \overline{Span\{a_{n,\lambda_0}, n \in \mathbb{N}_0\}}$$

is independent of λ_0. Moreover, for any $U' \subset U$ having an accumulation point in U we have

$$Z = \overline{Span\{x(\lambda), \lambda \in U'\}} = \overline{Span\{x(\lambda), \lambda \in U\}}.$$

Proof. The proof is an essay about the identity

$$0 = <\Phi, x(\lambda)> = \sum_{n=0}^{\infty} <\Phi, a_{n,\lambda_0}> (\lambda - \lambda_0)^n$$

$\lambda, \lambda_0 \in U'$, $\Phi \in X^*$, and the principle of isolated zeros. $\qquad\square$

Theorem 21. [15] *Suppose that conditions 1. and 2 of Theorem 20 are satisfied. Then there exists an infinite dimensional closed subspace $Y \subseteq X$ that is invariant for $(G(t))_{t \geq 0}$ such that $(G|_Y(t))_{t \geq 0}$ is chaotic.*

Proof. The proof of this result uses the previous proposition to show that closed linear spans of eigenvectors with $\Re\lambda > 0, \Re\lambda < 0$ and $\Re\lambda = 0$ are the same. Since a closed linear span of eigenvectors of the generator is invariant w.r.t. the semigroup, the theorem follows. $\qquad\square$

Semigroups that are hypercyclic in a subspace are called *sub-hypercyclic*; similarly, semigroups which are chaotic in a subspace are called *sub-chaotic*.

Negative Criteria

It is important to distinguish cases when the dynamical system cannot be chaotic, even in a subspace.

For a set $M \subset X$, denote

$$M^{\perp} = \{f \in X^*; \; <f, x> = 0, \forall x \in M\}$$

Theorem 22. *Let $(G(t))_{t \geq 0}$ be a continuous linear dynamical generated by A in a Banach space X, having an orbit dense in some subspace $X_{ch} \subset X$. Then the adjoint A^* of A and the dual dynamical system $(G^*(t))_{t \geq 0}$ have the following properties:*

(i) Let $0 \neq \phi \in X^*$. If $\{G^*(t)\phi\}_{t \geq 0}$ is bounded, then $\phi \in X_{ch}^{\perp}$,

(ii) If ϕ is an eigenvector of A^*, then $\phi \in X_{ch}^{\perp}$. In particular, $(G(t))_{t \geq 0}$ cannot be chaotic if

$$\sigma_p(A^*) = \emptyset.$$

Proof. The proof is based on the following observation. Let $0 \neq \Phi \in X^*$ be such that $\|G^*(t)\Phi\|$ is bounded. Consider

$$<G^*(t)\Phi, x> = <\Phi, G(t)x> .$$

Along a dense trajectory $\{G(t)x_0\}_{t\geq 0}$ (for a fixed x_0) we can find $x = T(t_\epsilon)x_0$ for which $\|x\| < \epsilon$ and so the right hand side can be made arbitrarily small. This shows (modulo some limiting argument) that Φ is orthogonal to the span of $\{G(t)x_0\}_{t\geq 0}$. Similar argument works for (ii). $\square$

We use this result to show that some classes of semigroups are not hypercyclic.

Corollary 2. *[25] Let $(G(t))_{t\geq 0}$ be a strongly continuous semigroup generated by A in a Banach space X. Assume that $(G(t))_{t\geq 0}$ is eventually uniformly continuous and that $R(\lambda, A)$ is compact. Then $(G(t))_{t\geq 0}$ is not hypercyclic.*

Proof. Indeed, a hypercyclic semigroup must have positive growth bound $\omega_0(G)$. Since $(G(t))_{t\geq 0}$ is eventually uniformly continuous, $s(A)=\omega_0(A)>-\infty$ by Theorem 19. Since $R(\lambda, A)$ is compact, $R(\lambda, A^*)$ is also compact and, since $s(A) > -\infty$, $\sigma(A^*)$ is not empty and consists solely of eigenvalues. $\square$

For instance, the diffusion semigroup on a bounded domain is analytic with compact resolvent and thus cannot be chaotic.

Recent Criteria

Theorem 23. *[28] Let A be the generator of a strongly continuous semigroup $(G(t))_{t\geq 0}$ on a separable Banach space X. Assume that there is $\Omega :=
(\omega_1, \omega_2) \subset \mathbb{R}$ with $\mu(\Omega) > 0$ and a strongly measurable $x : \Omega \to X$ such that $Ax(\lambda) = i\lambda x(\lambda)$ for almost any $\lambda \in \Omega$ and*

$$\overline{Span\{x(\lambda); \ \lambda \in \Omega \setminus \Omega'\}} = X \tag{61}$$

for any $\Omega' \subset \mathbb{R}$ with $\mu(\Omega') = 0$. Then $(G(t))_{t\geq 0}$ is hypercyclic in X.

Remark. If x is continuous, then (61) can be replaced by

$$\overline{Span\{x(\lambda); \ \lambda \in \Omega\}} = X \tag{62}$$

and one obtains automatically that $(G(t))_{t\geq 0}$ is chaotic in X.

Proof. Note that x is a non-zero function and, by using a scalar multiplier, we can assume that x is (Bochner) integrable.

Let $\phi(r)$ be the Fourier transform of $s \to x(s)$, with $x(s)$ extended by zero outside Ω, if necessary. Denote

$$Y_x = \overline{Span\{\phi(\mathbb{R})\}} = \overline{Span\{\phi(r); \ r \in \mathbb{R}\}}. \tag{63}$$

By Riemann-Lebesgue theorem (see e.g. [30, Lemma C.8]), $\phi \in C_0(\mathbb{R}, X)$; that is, $\lim_{|r|\to\infty} \phi(r) = 0$. Let us fix $r \in \mathbb{R}$. Since

$$[G(t)\phi](r) = \int_{-\infty}^{\infty} e^{i(t+r)s} x(s)ds,$$

we see that $\lim_{t\to\infty}[G(t)\phi](r) = 0$. Thus, $Span\{\phi(\mathbb{R})\} \subset X_0$. Similarly,

$$\phi(r) = G(t) \int_{-\infty}^{\infty} e^{i(-t+r)s}x(s)ds =: [G(t)\psi](r)$$

where $\|\psi(r)\|$ can be made as small as we wish. Hence, $Span\{\phi(\mathbb{R})\} \subset X_\infty$. The last assumption is used to show that $Y_x = X$. $\square$

A closer look at the proof shows that actually we have a stronger result:

Corollary 3. *Let X be an arbitrary (not necessarily separable) Banach space. Let all assumptions of Theorem 23 except (61) be satisfied and $s \to x(s)$ be a non-zero function. Then $(G(t))_{t\geq 0}$ is hypercyclic in Y_x.*

The chaoticity space Y_x can be precisely described. Let X be an arbitrary Banach space, (Ω, μ) be a measure space, and $f : \Omega \to X$ be a strongly measurable function. For any measurable $U \subset \Omega$ we define the *essential image* of U through f as

$$f(U)_{ess} := \{x \in X; \ \mu(\{s \in U : \|f(s) - x\| < \epsilon\}) \neq 0, \forall \epsilon > 0\},$$

Theorem 24. *[19] Let $(G(t))_{t\geq 0}$ be a C_0-semigroup generated by the operator A on an arbitrary Banach space X. Assume that $\sigma_p(A) \cap i\mathbb{R} =: i\Omega \neq \emptyset$, where $\Omega \subset \mathbb{R}$ is measurable with $\mu(\Omega) > 0$, and that there is a (strongly) measurable function $x : \Omega \to X$ such that $0 \neq x(\lambda) \in Ker(i\lambda - A)$ for any $\lambda \in \Omega$. Then $(G(t))_{t\geq 0}$ is sub-hypercyclic, with the hypercyclicity space $X_x := \overline{Span\{x(\Omega)_{ess}\}}$.*

It is often suggested that sufficiently many periodic solutions leads to chaos. For linear systems, periodic solutions are the solutions corresponding to imaginary eigenvalues, thus Theorem 24 seems to be a step in right direction. However, one can construct a semigroup of isometries whose generator has the point spectrum filling the imaginary line. Clearly, such a semigroup cannot be sub-hypercyclic, [19].

4.4 Positive Semigroups

Definition 16. *Let X be a Banach lattice. We say that the semigroup $(G(t))_{t\geq 0}$ on X is positive if for any $x \in X_+$ and $t \geq 0$,*

$$G(t)x \geq 0.$$

We say that an operator $(A, D(A))$ is resolvent positive if there is ω such that $(\omega, \infty) \subset \rho(A)$ and $R(\lambda, A) \geq 0$ for all $\lambda > \omega$.

A strongly continuous semigroup is positive if and only if its generator is resolvent positive. In fact, the positivity of the resolvent for $\lambda > \omega$ follows from (46) and closedness of the positive cone; see Proposition 1. Conversely, the latter with the exponential formula (47) shows that resolvent positive generators generate positive semigroups.

A number of spectral results for semigroups can be substantially improved if the semigroup in question is positive. The following theorem holds, [43, Theorem 1.4.1].

Theorem 25. *Let $(G(t))_{t\geq 0}$ be a positive semigroup on a Banach lattice, with generator A. Then*

$$R(\lambda, A)x = \int_0^\infty e^{-\lambda t} G(t)x \, dt \tag{64}$$

for all $\lambda \in \mathbb{C}$ with $\Re\lambda > s(A)$. Furthermore,

(i) Either $s(A) = -\infty$ or $s(A) \in \sigma(A)$ and

$$s(A) = \omega_1(G);$$

(ii) For a given $\lambda \in \rho(A)$, we have $R(\lambda, A) \geq 0$ if and only if $\lambda > s(A)$;
(iii) For all $\Re\lambda > s(A)$ and $x \in X$, we have $|R(\lambda, A)x| \leq R(\Re\lambda, A)|x|$.

From Theorem 25 we see that the spectral bound of the generator of a positive semigroup controls the growth rate of all classical solutions. However, the strict inequality $s(A) < \omega_0(G)$ can still occur, as was shown by Arendt; see [43, Example 1.4.4]. In this example $X = L_p([1, \infty)) \cap L_q([1, \infty))$, $1 \leq p < q < \infty$, and the semigroup in question is $(G(t)f)(s) := f(se^t)$, $s > 1, t > 0$. Its generator is $(Af)(s) = sf'(s)$ on the maximal domain and it can be proved that $s(A) = -1/p < -1/q = \omega_0(G)$. Interestingly enough, $s(A) = \omega_0(G)$ holds for positive semigroups on L_p-spaces. This was proved a few years ago by L. Weis, see the proof in, say, [43, Section 3.5]. However, for the case $p = 1$, which is most relevant for the applications described in this book, it can be proved with much less effort.

Theorem 26. *Let $(G(t))_{t\geq 0}$ be a positive semigroup on an AL-space and let A be its generator. Then $s(A) = \omega_0(G)$.*

The theorem is a corollary of a general result known as the Datko theorem.

Theorem 27. *Let A be the generator of $(G(t))_{t\geq 0}$. If, for some $p \in [1, \infty)$,*

$$\int_0^\infty \|G(t)x\|^p dt < \infty, \tag{65}$$

for all $x \in X$, then $\omega_0(G) < 0$.

Proof of Theorem 26. Defining $<f, x> := \|x\|$ for $x \in X_+$ we obtain a positive additive functional which can be extended to a bounded positive linear functional by Theorems 9 and 10. Let $\omega > abs(G) = s(A)$ (see Theorem 25). Then for $x \geq 0$ and $\tau > 0$, we have

$$\int_0^\tau e^{-\omega t} \|G(t)x\| dt = \left\langle f, \int_0^\tau e^{-\omega t} G(t)x \, dt \right\rangle \leq <f, R(\omega, A)x>.$$

Therefore

$$\int_0^\infty e^{-\omega t} \|G(t)x\| dt < +\infty$$

for all $x \in X_+$ and hence for all $x \in X$. Theorem 27 then implies $\|G(t)\| \leq M e^{(\omega - \mu)t}$ for some $\mu > 0$, hence $\omega_0(G) < \omega$ which yields $\omega_0(G) \leq s(A)$ and consequently $s(A) = \omega_0(G)$.   □

4.5 Generation Through Perturbation

Verifying conditions of the Hille–Yosida, or even the Lumer–Phillips, theorems for a concrete problem is quite often a formidable task. On the other hand, in many cases the operator appearing in the evolution equation at hand is built as a combination of much simpler operators that are relatively easy to analyse. The question now is to what extent the properties of these simpler operators are inherited by the full equation. More precisely, we are interested in the problem:

Problem P. *Let $(A, D(A))$ be a generator of a C_0-semigroup on a Banach space X and $(B, D(B))$ be another operator in X. Under what conditions does $A + B$ generate a C_0-semigroup on X?*

Before attempting to address this problem we point out a difficulty that arises immediately from the above formulation. As A and B are unbounded operators, we have to realize that the sum $A + B$ is, at this moment, defined only as $(A + B)x = Ax + Bx$ on $D(A + B) = D(A) \cap D(B)$, where the latter can reduce in some cases to $\{0\}$. Also, the sum of two closed operators is not necessarily closed: a trivial example is offered by $B = -A$ and $A + B = 0$, defined on $D(A)$, is not a closed operator. Thus, $A + B$ with $B = -A$ does not generate a semigroup. On the other hand, the closure of $A + B$ that is the zero operator defined on the whole space is the generator of a constant uniformly bounded semigroup. This situation happens quite often and suggests that the formulation of Problem P is too restrictive and we often restrict ourselves to the following weaker formulation of it.

Problem P′. *Let $(A, D(A))$ be a generator of a C_0-semigroup on a Banach space X and $(B, D(B))$ be another operator in X. Find conditions that ensure that there is an extension K of $A + B$ that generates a C_0-semigroup on X and characterise this extension.*

The characterisation of extensions of $A + B$ that generate a semigroup (in general, there can be many extensions having this property) provides essential information on the properties of the semigroup and plays a role of the regularity theorems in the theory of differential equations. The best situation is when $K = A + B$ or $K = \overline{A + B}$, as there is then a close link between K and A and B. However, there are cases where K is an unspecified extension of $A + B$ in which case the semigroup can display features that are rather impossible to deduct from the properties of A and B alone.

A Spectral Criterion

Usually the first step in establishing whether $A+B$, or some of its extensions, generates a semigroup is to find if $\lambda I - (A+B)$ (or its extension) is invertible for all sufficiently large λ.

In all cases discussed here we have the generator $(A, D(A))$ of a semigroup and a perturbing operator $(B, D(B))$ with $D(A) \subseteq D(B)$.

We note that B is A-bounded; that is, for some $a, b \geq 0$ we have

$$\|Bx\| \leq a\|Ax\| + b\|x\|, \qquad x \in D(A) \tag{66}$$

if and only if $BR(\lambda, A) \in \mathcal{L}(X)$ for $\lambda \in \rho(A)$.

In what follows we denote by K an extension of $A+B$. We now present an elegant result relating the invertibility properties of $\lambda I - K$ to the properties of 1 as an element of the spectrum of BL_λ, first derived in [31].

Theorem 28. *Assume that $\Lambda = \rho(A) \cap \rho(K) \neq \emptyset$.*

(a) $1 \notin \sigma_p(BR(\lambda, A))$ for any $\lambda \in \Lambda$;

(b) $1 \in \rho(BR(\lambda, A))$ for some/all $\lambda \in \Lambda$ if and only if $D(K) = D(A)$ and $K = A + B$;

(c) $1 \in \sigma_c(BR(\lambda, A))$ for some/all $\lambda \in \Lambda$ if and only if $D(A) \subsetneq D(K)$ and $K = \overline{A+B}$;

(d) $1 \in \sigma_r(BR(\lambda, A))$ for some/all $\lambda \in \Lambda$ if and only if $K \supsetneq \overline{A+B}$.

Corollary 4. *Under the assumptions of Theorem 28, $K = A + B$ if one of the following criteria is satisfied: for some $\lambda \in \rho(A)$ either*

(i) $BR(\lambda, A)$ is compact (or, if $X = L_1(\Omega, d\mu)$, weakly compact), or
(ii) the spectral radius $r(BR(\lambda, A)) < 1$.

Proof. If (ii) holds, then obviously $I - BR(\lambda, A)$ is invertible by the Neumann series theorem:

$$(I - BR(\lambda, A))^{-1} = \sum_{n=0}^{\infty} (BR(\lambda, A))^n, \tag{67}$$

giving the thesis by Proposition 28 (b). Additionally, we obtain

$$R(\lambda, A + B) = R(\lambda, A)(I - BR(\lambda, A))^{-1} = R(\lambda, A) \sum_{n=0}^{\infty} (BR(\lambda, A))^n. \tag{68}$$

If (i) holds, then either $BR(\lambda, A)$ is compact or, in L_1 setting, $(BR(\lambda, A))^2$ is compact, [27, p. 510], and therefore, if $I - BR(\lambda, A)$ is not invertible, then 1 must be an eigenvalue, which is impossible by Theorem 28(c). $\square$

If we write the resolvent equation

$$(\lambda I - (A + B))x = y, \quad y \in X, \tag{69}$$

in the (formally) equivalent form

$$x - R(\lambda, A)Bx = R(\lambda, A)y, \tag{70}$$

then we see that we can hope to recover x provided the Neumann series

$$R(\lambda)y := \sum_{n=0}^{\infty} (R(\lambda, A)B)^n R(\lambda, A)y = \sum_{n=0}^{\infty} R(\lambda, A)(BR(\lambda, A))^n y. \tag{71}$$

is convergent. Clearly, if (67) converges, then we can factor out $R(\lambda, A)$ from the series above getting again (68). However, $R(\lambda, A)$ inside acts as a regularising factor and (71) converges under weaker assumptions than (67) and this fact is frequently used to construct the resolvent of an extension of $A + B$ (see, e.g., Theorem 32, Theorem 37 or Section 4.6).

The most often used perturbation theorem is the Bounded Perturbation Theorem and Related Results, see e.g. [30, Theorem III.1.3]

Theorem 29. *Let $(A, D(A)) \in \mathcal{G}(M, \omega)$ for some $\omega \in \mathbb{R}, M \geq 1$. If $B \in \mathcal{L}(X)$, then $(K, D(K)) = (A + B, D(A)) \in \mathcal{G}(M, \omega + M\|B\|)$. Moreover, the semigroup $(G_{A+B}(t))_{t \geq 0}$ generated by $A + B$ satisfies the Duhamel equation:*

$$G_{A+B}(t)x = G_A(t)x + \int_0^t G_A(t - s)BG_{A+B}(s)x ds, \quad t \geq 0, x \in X. \tag{72}$$

where the integral is defined in the strong operator topology.

Moreover, $(G_{A+B}(t))_{t \geq 0}$ is given by the Dyson–Phillips series:

$$G_{A+B}(t) = \sum_{n=0}^{\infty} G_n(t), \tag{73}$$

obtained by iterating (72) with starting point $G_0(t) = G_A(t)$. The series converges in the operator norm of $\mathcal{L}(X)$ and uniformly for t in bounded intervals.

The assumption of boundedness of B, however, is often too restrictive. Another frequently used result uses special structure of dissipative operators.

Theorem 30. *Let A and B be linear operators in X with $D(A) \subseteq D(B)$ and $A + tB$ is dissipative for all $0 \leq t \leq 1$. If*

$$\|Bx\| \leq a\|Ax\| + b\|x\|, \tag{74}$$

for all $x \in D(A)$ with $0 \leq a < 1$ and for some $t_0 \in [0, 1]$ the operator $(A + t_0 B, D(A))$ generates a semigroup (of contractions), then $A + tB$ generates a semigroup of contractions for every $t \in [0, 1]$.

Proof. The proof consists in showing, by using Neumann expansion, that if $I - (A + t_0 B)$ is invertible, then $I - (A + tB)$ is invertible provided $|t - t_0| < 1 - a/(2a+b)$. Since the length of the interval on which $I - (A + tB)$ is invertible is independent of the starting point t_0, by using finitely many successive steps, we can cover the whole interval $[0, 1]$. Thus $(A + tB, D(A))$ is a dissipative operator such that $I - (A + tB)$ is surjective for all $t \in [0, 1]$. It is also densely defined because $D(A)$ is dense and so $(A + tB, D(A))$ generates a semigroup of contractions. $\square$

The fact that $a < 1$ in the previous theorem is crucial and a lot of work has been done to change $<$ to $=$. One result, in general setting, is given below. Some others, employing positivity, are discussed further on.

Theorem 31. *Let A be the generator of a semigroup of contractions and B, with $D(A) \subset D(B)$, is such that $A + tB$ is dissipative for all $t \in [0, 1]$. If*

$$\|Bx\| \leq \|Ax\| + b\|x\|, \tag{75}$$

for $x \in D(A)$ and B^* is densely defined, then $\overline{A + B}$ is the generator of a contractive semigroup. In particular, if B is closable and X reflexive, then B^* is densely defined.

4.6 Positive Perturbations of Positive Semigroups

In most perturbation theorems of the previous chapter an essential role was played by a strict inequality in some condition comparing A and B (or $(G_A(t))_{t\geq0}$ and B). This provided some link between the generator and both operators A and B, and ensured that the semigroup was generated by $A + B$ or, at worst, by $\overline{A + B}$. In many cases of practical importance, however, this inequality becomes a weak inequality or even an equality. We show that in such a case we can still get existence of a semigroup albeit we usually lose control over its generator which can turn to be a larger extension of $A + B$ than $\overline{A + B}$. In such a case the resulting semigroup has properties that are not 'contained' in A and B alone; these are discussed in the next chapter. Here we provide the generation theorem, obtained in [16], which is a generalisation of Kato's result from 1954, [35], as well as some of its consequences.

Theorem 32. *Let X be a KB-space. Let us assume that we have two operators $(A, D(A))$ and $(B, D(B))$ satisfying:*

(A1) A generates a positive semigroup of contractions $(G_A(t))_{t\geq0}$,
(A2) $r(BR(\lambda, A)) \leq 1$ for some $\lambda > 0 (= s(A))$,
(A3) $Bx \geq 0$ for $x \in D(A)_+$,
(A4) $<x^*, (A + B)x> \leq 0$ for any $x \in D(A)_+$, where $<x^*, x> = \|x\|$, $x^* \geq 0$.

Then there is an extension $(K, D(K))$ of $(A + B, D(A))$ generating a C_0-semigroup of contractions, say, $(G_K(t))_{t\geq0}$. The generator K satisfies

$$R(\lambda, K)x = \sum_{k=0}^{\infty} R(\lambda, A)(BR(\lambda, A))^k x, \quad \lambda > 0. \tag{76}$$

Proof. We define operators K_r, $0 \leq r < 1$ by $K_r = A + rB$, $D(K_r) = D(A)$. We see that, as by (A2) the spectral radius of $rBR(\lambda, A)$ does not exceed $r < 1$, the resolvent $(\lambda I - (A + rB))^{-1}$ exists and is given by

$$R(\lambda, K_r) := (\lambda I - (A + rB))^{-1} = R(\lambda, A) \sum_{n=0}^{\infty} r^n (BR(\lambda, A))^n, \tag{77}$$

where the series converges absolutely and each term is positive. Hence,

$$\|R(\lambda, K_r)y\| \leq \lambda^{-1}\|y\| \tag{78}$$

for all $y \in X$. Therefore, by the Lumer–Phillips theorem, for each $0 \leq r < 1$, $(K_r, D(A))$ generates a contraction semigroup which we denote $(G_r(t))_{t \geq 0}$. The net $(R(\lambda, K_r)x)_{0 \leq r < 1}$ is increasing as $r \uparrow 1$ for each $x \in X_+$ and $\{\|R(\lambda, K_r)x\|\}_{0 \leq r < 1}$ is bounded, so by assumption that X is a KB-space, there is an element $y_{\lambda, x} \in X_+$ such that

$$\lim_{r \to 1^-} R(\lambda, K_r)x = y_{\lambda, x}$$

in X. By the Banach–Steinhaus theorem we obtain the existence of a bounded positive operator on X, denoted by $R(\lambda)$, such that $R(\lambda)x = y_{\lambda, x}$. We use the Trotter–Kato theorem to obtain that $R(\lambda)$ is defined for all $\lambda > 0$ and it is the resolvent of a densely defined closed operator K which generates a semigroup of contractions $(G_K(t))_{t \geq 0}$. Moreover, for any $x \in X$,

$$\lim_{r \to 1^-} G_r(t)x = G_K(t)x, \tag{79}$$

and the limit is uniform in t on bounded intervals and, provided $x \geq 0$, monotone as $r \uparrow 1$. By the monotone convergence theorem, [8, Theorem 2.91],

$$R(\lambda, K)x = \sum_{k=0}^{\infty} R(\lambda, A)(BR(\lambda, A))^k x, \quad x \in X \tag{80}$$

and we can prove that $R(\lambda, K)(\lambda I - (A + B))x = x$ which shows that $K \supseteq A + B$. □

The semigroup $(G_K(t))_{t \geq 0}$ obtained in Theorem 32 is the smallest in the following sense.

Proposition 7. *Let D be a core of A. If $(G(t))_{t \geq 0}$ is another positive semigroup generated by an extension of $(A + B, D)$, then $G(t) \geq G_K(t)$.*

The assumption (A2) of Theorem 32 is stronger than the assumption that B is A-bounded, used in Theorem 31. Thus, it is worthwhile to compare Theorem 32 with Theorems 31 and 30.

Proposition 8. *[8, Proposition 5.5] Let $(G(t))_{t\geq0}$ be the semigroup generated by $A + B$ or $\overline{A + B}$ under conditions of Theorems 30 or 31, respectively. If A is a resolvent positive operator and B is positive, then $(G(t))_{t\geq0}$ is positive.*

Thus, if X is reflexive and B is closable, then Theorem 31 is evidently stronger than Theorem 32 as the former requires positivity of neither $(G_A(t))_{t\geq0}$ nor of B. Moreover, in Theorem 31, we obtain the full characterisation of the generator as the closure of $A + B$. However, checking the closability of the operator B in particular applications is often difficult, whereas the positivity is often obvious. Also, there is a large class of nonclosable operators which can nevertheless be positive, for example, finite-rank operators (in particular, functionals) are closable if and only if they are bounded. Moreover, Theorem 32 gives a constructive formula (76) for the resolvent of the generator, which seems to be unavailable in general case, and this, in turn, allows other representation results that are discussed below. Also, what is possibly the most important fact, in nonreflexive spaces Theorem 32 refers to a substantially different class of phenomena because, as we show in the next chapter, in many cases covered by this theorem the generator does not coincide with the closure of $A + B$. Arguments used in the proof of Theorem 32 are very powerful and can be generalized in many ways, see [16]. Theorem 32 is most often used in L_1-setting, where it can be significantly simplified:

Corollary 5. *Let $X = L_1(\Omega)$ and suppose that the operators A and B satisfy*

1. *$(A, D(A))$ generates a substochastic semigroup $(G_A(t))_{t\geq0}$;*
2. *$D(B) \supset D(A)$ and $Bu \geq 0$ for $u \in D(B)_+$;*
3. *for all $u \in D(A)_+$*

$$\int_\Omega (Au + Bu)d\mu \leq 0. \tag{81}$$

Then the assumptions of Theorem 32 are satisfied.

Proof. First, assumption (81) gives us assumption (A4), that is, dissipativity on the positive cone. Next, let us take $u = R(\lambda, A)x = (\lambda I - A)^{-1}x$ for $x \in X_+$ so that $u \in D(A)_+$. Because $R(\lambda, A)$ is a surjection from X onto $D(A)$, by

$$(A + B)u = (A + B)R(\lambda, A)x = -x + BR(\lambda, A)x + \lambda R(\lambda, A)x,$$

we have

$$-\int_\Omega x\,d\mu + \int_\Omega BR(\lambda, A)x\,d\mu + \lambda\int_\Omega R(\lambda, A)x\,d\mu \leq 0. \tag{82}$$

Rewriting the above in terms of the norm, we obtain

$$\lambda\|R(\lambda, A)x\| + \|BR(\lambda, A)x\| - \|x\| \leq 0, \qquad x \in X_+, \tag{83}$$

from which $\|BR(\lambda, A)\| \leq 1$; that is, assumption (A2) is satisfied. $\qquad\square$

5 What can go Wrong?

In this section we discuss mathematics behind the pathological phenomena discussed in the introduction and illustrate them in the context of birth-and-death models. We note that the results on dishonesty presented below are based on [8, Chapter 6].

Dishonesty

Our main focus are equations describing evolution of a function u which is the density of objects in a population structured by certain states. Such equations, called sometimes *Master Equations*, are typically constructed by balancing, for any state x, the loss of $u(x, t)$ that is due to the transfer of a part of the population to other states x', and the gain due to the transfer of parts of the population from other states x' to the state x. A general form of such equations reads,

$$\partial_t u = \mathcal{T}_0 u + \mathcal{A} u + \mathcal{B} u, \tag{84}$$

where $\mathcal{A}$ is the loss operator, $\mathcal{B}$ is the gain operator, and $\mathcal{T}_0$ may describe some transport in the state space (e.g., free streaming or diffusion). Here these operators are taken in the sense of differential or integral expressions discussed in Introduction. Similar modelling processes can be found in the lecture notes by M. Lachowicz [37] and V. Capasso [22] in this issue.

The very nature of the modelling process sketched above requires that the described quantity (total population, total mass) should be accounted for; that is, u should add up (or integrate) to an expression which is a reflection of the laws of nature used to build the model. For instance, if the model is supposed to be conservative, than this expression should be the initial number of particles. If this is the case, then the semigroup describing the evolution is conservative for positive initial data and is called a *stochastic semigroup*.

In many cases, however, the semigroup does not return the laws of nature used to derive the model; that is we have a leakage of the described quantity out of the system that is not accounted for in the modelling processes. This in turn indicates a possibility of the phase transition during evolution and shows that the model does not provide an adequate description of the full process. Roughly speaking, a semigroup which has all properties used to build the model this semigroup is the solution of is called *honest*; otherwise we say it is *dishonest*. We shall give precise definition of these properties below. Here we mention that they are closely related to the characterisation of the generator of the semigroup. To explain why, let us look at a simplified situation with $\mathcal{T}_0 = 0$ and $X = L_1(\Omega, d\mu)$. First we note that the pair of expressions $\mathcal{A}$ and $\mathcal{B}$ in a natural way defines two operators: the *minimal operator* $K_{\min} = A + B$, where $A = \mathcal{A}|_{D(A)}$,

$$D(K_{\min}) = D(A) = \{u \in X; \; \mathcal{A} u \in X\}$$

and $B = \mathcal{B}|_{D(A)}$, and the *maximal operator* $K_{\max} = \mathcal{A} + \mathcal{B}$ defined on

$$D(K_{\max}) = \{u \in X; \ \mathcal{A}u + \mathcal{B}u \in X\},$$

where the evaluation of $\mathcal{A}$ and $\mathcal{B}$ is understood in the sense of (1). The difference between these two operators is that in the minimal operator both summands $\mathcal{A}u$ and $\mathcal{B}u$ must belong to X but in the maximal operator the summands are supposed to be defined just a.e. on Ω and only their sum is required to belong to X. In general, $D(K_{\min}) = D(A) \subsetneq D(K_{\max})$ as there may be cancellation of singularities in $\mathcal{A}u$ and $\mathcal{B}u$ due to their opposite signs.

In the discussed context it can be proved that the generator K of the semigroup associated with the problem (84) satisfies $K_{\min} \subset K \subset K_{\max}$. Where K is situated on this scale determines the well-posedness of the problem. To explain why honesty of $(G_K(t))_{t \geq 0}$ should be determined by this, let us suppose for simplicity that the model is formally conservative; that is, for sufficiently regular u, say $u \in D(A)$,

$$\int_{\Omega} (\mathcal{A} + \mathcal{B})u d\mu = 0 \tag{85}$$

If A generates a substochastic semigroup and B is positive, then by Corollary 5, there is an extension K of $A + B$ generating a semigroup of contractions, say $(G_K(t))_{t \geq 0}$.

Assume now that the semigroup $(G_K(t))_{t \geq 0}$ is generated by $K = K_{\min} = A + B$ on $D(K) = D(A)$. Then the solution $u(t) = G_K(t)u_0$, emanating from $u_0 \in D(K)_+$, satisfies $u(t) \in D(A)_+$ and, therefore, because

$$\frac{d}{dt}u(t) = Ku(t) = Au(t) + Bu(t),$$

we obtain that for any $t \geq 0$

$$\frac{d}{dt}\|u(t)\| = \int_{\Omega} \frac{du(t)}{dt} d\mu = \int_{\Omega} (Au(t) + Bu(t)) d\mu = 0, \tag{86}$$

so that $\|u(t)\| = \|u_0\|$ for any $t \geq 0$ and the solutions are indeed conservative. If $K = \overline{A + B}$, then the same result can be obtained by limit argument so that the solutions are conservative as well. That $K = \overline{A + B}$ is also the necessary condition is not that clear but can be proved, see Theorem 35.

To make the above considerations precise and general, let us recall that a semigroup $(G(t))_{t \geq 0}$ is said to be a *substochastic semigroup* if for any $t \geq 0$ and $f \geq 0$, $G(t)f \geq 0$ and $\|G(t)f\| \leq \|f\|$, and a *stochastic semigroup* if additionally $\|G(t)f\| = \|f\|$ for $f \in X_+$. We consider linear operators in $X = L_1(\Omega, d\mu)$: $T \subset T_0 + A$ with $D(T) \subset D(T_0) \cap D(A)$, and B, that satisfy the assumptions of Corollary 5: $(T, D(T))$ generates a substochastic semigroup $(G_T(t))_{t \geq 0}$, $D(B) \supset D(T)$ and $Bf \geq 0$ for $f \in D(B)_+$ and for all $f \in D(T)_+$,

$$\int_{\Omega} (Tf + Bf) d\mu = -c(f) \leq 0, \tag{87}$$

where c is an integral functional; that is, for some $\varsigma > 0$

$$c(f) = \int_\Omega \varsigma(x) f(x) \, d\mu'_x. \tag{88}$$

Under these assumptions, Corollary 5, Theorem 32, and other results of the previous chapter, give the existence of a smallest substochastic semigroup $(G_K(t))_{t\geq0}$ generated by an extension K of the operator $T + B$. This semigroup, for arbitrary $f \in D(K)$ and $t > 0$, satisfies

$$\frac{d}{dt} G_K(t)f = KG_K(t)f. \tag{89}$$

It is important to distinguish the class of semigroups corresponding to $c \neq 0$, as such semigroups cannot be stochastic but their substochasticity is built into the model and not caused by the dishonesty of it.

Definition 17. *A positive semigroup $(G_K(t))_{t\geq0}$ generated by an extension K of the operator $T + B$ is said to be strictly substochastic if (87) holds with $c \neq 0$. The semigroup $(G_K(t))_{t\geq0}$ is honest if c extends to $D(K)$ and for any $0 \leq f \in D(K)$ the solution $u(t) = G_K(t)f$ of (89) satisfies*

$$\frac{d}{dt} \int_\Omega u(t) \, d\mu = \frac{d}{dt} \|u(t)\| = -c\left(u(t)\right). \tag{90}$$

It can be proved that (90) is equivalent to its 'integrated' version: $(G_K(t))_{t\geq0}$ is honest if and only if for any $f \in X_+$ and $t \geq 0$,

$$\|G_K(t)f\| = \|f\| - c\left(\int_0^t G_K(s)f \, ds\right). \tag{91}$$

This result allows for the introduction of the defect function

$$\eta_f(t) = \|G_K(t)f\| - \|f\| + \int_0^t c\left(G_K(s)f\right) ds \tag{92}$$

for $f \in X_+$ and $t \geq 0$. It follows that η_f is a nonpositive and nonincreasing function for $t \geq 0$. For $\lambda > 0$ we define $L_\lambda = R(\lambda, T) = (\lambda I - T)^{-1}$. Arguing as in (83) we obtain that condition (87) is equivalent to

$$-c(L_\lambda f) = \lambda\|L_\lambda f\| + \|BL_\lambda f\| - \|f\|, \qquad f \in X_+. \tag{93}$$

The following theorem is fundamental for analysing honesty of substochastic semigroups.

Theorem 33. *For any fixed $\lambda > 0$, there is $0 \leq \beta_\lambda \in X^*$ with $\|\beta_\lambda\| \leq 1$ such that for any $f \in X_+$,*

$$\lambda\|R(\lambda, K)f\| = \|f\| - <\beta_\lambda, f> - c\left(R(\lambda, K)f\right). \tag{94}$$

In particular, c extends to a nonnegative continuous linear functional on $D(K)$, given again by (88).

Proof. Let us fix $f \in X_+$. From (76) and nonnegativity we obtain

$$\lambda\|(\lambda I - K)^{-1}f\| = \lim_{N \to \infty} \sum_{n=0}^{N} \lambda\|L_\lambda(BL_\lambda)^n f\|.$$

By (93) we get

$$\sum_{n=0}^{N} \lambda\|L_\lambda(BL_\lambda)^n f\| = \|f\| - \|(BL_\lambda)^{N+1}f\| - c\left(\sum_{n=0}^{N} L_\lambda(BL_\lambda)^n f\right).$$

By non-negativity, the monotone convergence theorem gives

$$\lim_{N \to \infty} c\left(\sum_{n=0}^{N} L_\lambda(BL_\lambda)^n f\right) = c(R(\lambda, K)f) < +\infty.$$

This shows that c extends to a finite functional on $D(K)$, which is continuous in the graph topology. Similarly, $\|(BL_\lambda)^{N+1}f\|$ converges to some $\beta_\lambda(f) \geq 0$ and, by a similar argument, β_λ extends to a continuous linear functional on X with the norm not exceeding 1. $\square$

By taking the Laplace transform of η_f, we obtain

$$<\beta_\lambda, f> = -\lambda \int_0^\infty e^{-\lambda t} \eta_f(t)dt$$

for $f \in X_+$ and hence the following result is true

Theorem 34. *$(G_K(t))_{t \geq 0}$ is honest if and only if $\beta_\lambda \equiv 0$ for $\lambda > 0$.*

A central result on the characterization of honesty is:

Theorem 35. *[7] The semigroup $(G_K(t))_{t \geq 0}$ is honest if and only if one of the following holds:*
(a) $K = \overline{T + B}$.
(b) $\int_\Omega Ku\, d\mu \geq -c(u), \quad u \in D(K)_+$.

Proof. (a) implies honesty as in (86) and considerations below it - properties of the functional c allow passage to the limit. Conversely, if $(G_K(t))_{t \geq 0}$ is honest, then $\beta_\lambda \equiv 0$ for any $\lambda > 0$, which means, by the proof of Theorem 33, that $\lim_{n\to\infty}(BL_\lambda)^n f = 0$. Hence the series in (76) converges to $R(\lambda, \overline{T+B})$.

If $(G_K(t))_{t\geq 0}$ is honest, then the part (a) gives (b) with the equality sign. Conversely, for $u = R(\lambda, K)f$, $f \in X_+$, we have

$$\int_\Omega Ku \, d\mu = -\|f\| + \lambda\|R(\lambda, K)f\| = -c(u) - \,<\beta_\lambda, f>,$$

which implies $<\beta_\lambda, f>\, \leq 0$ for all $f \in X_+$, thus $\beta_\lambda = 0$. $\square$

Unfortunately, typically we do not know K and thus condition (c) has a limited practical value. There are two important theorems providing conditions for honesty and dishonesty in terms of known operators. The first is based on Theorem 28 which, combined with Theorem 35, shows that $(G_K(t))_{t\geq 0}$ is honest if and only if $1 \notin \sigma_p((BL_\lambda)^*)$. In particular, using the definition of β_λ, we see that

$$<\beta_\lambda, BR(\lambda, T)f> = \lim_{n\to\infty} \|(BR(\lambda, T))^{n+1}f\| = <\beta_\lambda, f>,$$

$f \in X_+$, so that

$$(BR(\lambda, T))^*\beta_\lambda = \beta_\lambda. \tag{95}$$

The other set of results is based on the fact that we know at least one extension of the generator K, namely K_{max}. Let $\mathcal{K}$ be any extension of K.

Theorem 36. [7] (a) If $\int_\Omega \mathcal{K}u \, d\mu \geq -c(u)$ for all $u \in D(\mathcal{K})_+$, then $(G_K(t))_{t\geq 0}$ is honest.
(b) $(G_K(t))_{t\geq 0}$ is not honest if there is $u \in D(\mathcal{K})_+ \cap X$ and $\lambda > 0$
  (i) $\lambda u(x) - [\mathcal{K}u](x) = g(x) \geq 0,$ a.e.,
  (ii) $c(u)$ is finite and

$$\int_\Omega \mathcal{K}u \, d\mu < -c(u). \tag{96}$$

Proof. The statement (a) is obvious from Theorem 35(c) as $\mathcal{K}$ contains K. In practice, however, we are interested to use the smallest possible extension since taking a too large one could spoil the inequality, [6]. Similarly, (b) uses Theorem 35(c) but here the function $u \in D(\mathcal{K})$, satisfying (96) may fail to belong to $D(K)$; the other two conditions allow one to prove that there is an element of $D(K)$ satisfying (96), thus proving dishonesty of $(G_K(t))_{t\geq 0}$. $\square$

Extension Techniques

For further reference we briefly sketch a particularly effective extension technique introduced in [6]. We embed $X = L_1(\Omega, d\mu)$ in the set of μ-measurable functions that are defined on Ω and take values in the extended set of real numbers, denoted by E; by E_f we denote the subspace of E consisting of functions that are finite a. e. E is a lattice with respect to the usual relation: '$\leq$ a. e', $X \subset \mathsf{E}_f \subset \mathsf{E}$ with X and E_f being sublattices of E.

Let $\mathsf{F} \subset \mathsf{E}$ be defined by the condition: $f \in \mathsf{F}$ if and only if for any nonnegative and nondecreasing sequence $(f_n)_{n\in\mathbb{N}}$ satisfying $\sup_{n\in\mathbb{N}} f_n = |f|$ we have $\sup_{n\in\mathbb{N}}(I - T)^{-1}f_n \in X$. We define mapping $\mathsf{L} : \mathsf{F}_+ \to X_+$ by

$$\mathsf{L}f := \sup_{n\in\mathbb{N}} R(1,T)f_n, \qquad f \in \mathsf{F}_+,$$

where $0 \leq f_n \leq f_{n+1}$ for any $n \in \mathbb{N}$, and $\sup_{n\in\mathbb{N}} f_n = f$ and extend it to a positive linear operators on the whole $D(\mathsf{B})$ and F, respectively, Theorem 9.

In most applications $(I - T)^{-1}$ is an integral operator with positive kernel so that, by monotone convergence theorem, F coincides is the set of $L_1(\Omega)$ functions for which the integral exists. In the same way we define B on $D(\mathsf{B})$. It turns out that L is one-to-one therefore we can define the operator T with $D(\mathsf{T}) = \mathsf{L}\mathsf{F} \subset X$ by

$$\mathsf{T}u = u - \mathsf{L}^{-1}u, \tag{97}$$

so that T is an extension of T. The central theorem of this paragraph reads:

Theorem 37. *If $(T, D(T))$ and $(B, D(B))$ are operators in X such that $(T, D(T))$ generates a substochastic semigroup $(G_T(t))_{t\geq 0}$ on X, $D(B) \supset D(T)$, $Bu \geq 0$ for $u \in D(B)_+$, and*

$$\int_\Omega (Tu + Bu)\, d\mu \leq 0, \tag{98}$$

for all $u \in D(T)_+$, then the extension K of $A + B$, that generates the smallest substochastic semigroup on X described by Corollary 5, is given by

$$Ku = \mathsf{T}u + \mathsf{B}u, \tag{99}$$
$$D(K) = \{u \in D(\mathsf{T}) \cap D(\mathsf{B}) : \mathsf{T}u + \mathsf{B}u \in X, \text{ and } \lim_{n\to+\infty} ||(\mathsf{L}\mathsf{B})^n u|| = 0\}.$$

This notion allows to give a more focused version of Theorem 36(a).

Theorem 38. *If for any $g \in \mathsf{F}_+$ such that $-g + \mathsf{B}\mathsf{L}g \in X$ and $c(\mathsf{L}g)$ exists,*

$$\int_\Omega \mathsf{L}g\, d\mu + \int_\Omega (-g + \mathsf{B}\mathsf{L}g)\, d\mu \geq -c(\mathsf{L}g), \tag{100}$$

then $K = \overline{T + B}$.

Remark 5. It makes sense to consider 'pointwise in space' honesty and say that $(G_K(t))_{t\geq 0}$ is honest along the trajectory $\{G_K(t)f\}_{t\geq 0}$ if (91) holds for this particular f and for all $t \geq 0$. Accordingly, such a trajectory is called an honest trajectory. Thus $(G_K(t))_{t\geq 0}$ is honest if and only if each trajectory $\{G_K(t)f\}_{t\geq 0}$ is honest. Moreover, honesty can also be considered to be a 'pointwise in time' phenomenon. Indeed, if $u(t_0) \in D(\overline{T + B})$ for some $t_0 > 0$ then, as by (86) and limiting argument

$$\frac{d}{dt}\|u\|\bigg|_{t=t_0} = -c(u(t_0)),$$

and therefore we can say that the trajectory $\{G_K(t)f\}_{t\geq 0}$ is honest over a time interval I if and only if $G_K(t)f \in D(\overline{T + B})$ for $t \in I$.

In general, our theory cannot determine, in general, whether a given system $(G_K(t))_{t\geq 0}$ can be dishonest along some trajectories and honest along the others. Using specific properties of birth-and-death and fragmentation models, however, we can show that, in these models, if dishonesty occurs along one trajectory, it must occur along any other; see Theorem 43.

Unfortunately, much less can be said about how dishonest trajectories behave in time. One of the reasons for this is that our theory is based on the Laplace transform which gives, in some sense, time averages of solutions which provide little information about the properties which are local in time.

Multiple Solutions

Let us return to the general Cauchy problem (37), (38). If $\mathcal{A}$ is the generator of a semigroup, then the problem is always uniquely solvable. Hence, multiple solution only can occur if the original operator is not a generator.

Assume that, for a given u_0, (37), (38) has two solutions. Then their difference is again a solution of (37) but corresponding to the null initial condition – it is called a *nul-solution*. The following theorem is fundamental here.

Theorem 39. *[33, Theorem 23.7.2] Let $\mathcal{A}$ be a closed operator. The Cauchy problem (37), (38) has an exponentially bounded nul-solution of type $\leq \omega$ if and only if the eigenvalue problem*

$$\mathcal{A}y(\lambda) = \lambda y(\lambda) \tag{101}$$

has a solution $y(\lambda) \neq 0$ that is a bounded and holomorphic function of λ in each half-plane $\Re\lambda \geq \omega + \epsilon$, $\epsilon > 0$.

Proof. The proof essentially follows by taking the Laplace transform of both sides of (37) and some careful manipulation to ensure convergence. □

Now we investigate a relation between Cauchy problems (37), (38) and (42), (43). Let $(A, D(A))$ be the generator of a C_0-semigroup $(G(t))_{t\geq 0}$ on a Banach space X. To simplify notation we assume that $(G(t))_{t\geq 0}$ is a semigroup of contractions, hence $\{\lambda;\ Re\lambda > 0\} \subset \rho(A)$. Let us further assume that there exists an extension $\mathcal{A}$ of A defined on the domain $D(\mathcal{A})$. We have the following basic result.

Lemma 3. *Under the above assumptions, for any λ with $Re\lambda > 0$,*

$$D(\mathcal{A}) = D(A) \oplus Ker(\lambda I - \mathcal{A}). \tag{102}$$

The next corollary links Theorem 39 with the above lemma.

Corollary 6. *If $D(\mathcal{A}) \setminus D(A) \neq \emptyset$, then $\sigma_p(\mathcal{A}) \supseteq \{\lambda \in \mathbb{C};\ \mathrm{Re}\lambda > 0\}$. Moreover, there exists a holomorphic (in the norm of X) function $\{\lambda \in \mathbb{C};\ \mathrm{Re}\lambda > 0\} \ni \lambda \to e_\lambda$ such that for any λ with $\mathrm{Re}\,\lambda > 0$, $e_\lambda \in Ker(\lambda I - \mathcal{A})$, which is also bounded in any closed half-plane, $\{\lambda \in \mathbb{C};\ \mathrm{Re}\lambda \geq \gamma > 0\}$.*

An important observation is that analogous considerations can be carried also for mild (or integral) solutions of (37), (38), defined in the same way as for the semigroup, see [11]. This allows to check uniqueness only for continuous solutions of the integral version of the problem, which is technically simpler.

Summary

We have seen that the generator K satisfies $K_{\min} \subset K \subset K_{\max}$. The following situations are possible

1. $K_{\min} = K = K_{\max}$,
2. $K_{\min} \subsetneq K = \overline{K_{\min}} = K_{\max}$,
3. $K_{\min} = K \subsetneq K_{\max}$,
4. $K_{\min} \subsetneq K = \overline{K_{\min}} \subsetneq K_{\max}$,
5. $\overline{K_{\min}} \subsetneq K \subsetneq K_{\max}$,

and each of them has its own specific interpretation in the model.

In all cases where $K \subsetneq K_{\max}$ we don't have uniqueness; that is, there are differentiable X-valued solutions to (84) emanating from zero and therefore they are not described by the constructed dynamical system: 'there is more to life, than meets the semigroup' [11]. To achieve uniqueness here, one has to impose additional constraints on the solution.

If $\overline{K_{\min}} \subsetneq K$, then despite the fact that the model is formally conservative, (85), the solutions are not; the described quantity leaks out from the system and the mechanism of this leakage is not present in the model. In Markov processes such a case is called dishonesty of the transition function, [4].

Finally, the condition $u(t) \in D(A)$ for any t ensures that only a finite number of state changes can happen in the system in any finite time interval.

Therefore, strictly speaking, only problems with $K = K_{\min} = K_{\max}$ are physically realistic. However, in many applications, the case $K = \overline{K_{\min}} = K_{\max}$ is considered to be acceptable.

5.1 Applications to Birth-and-Death Type Problems

Here we will consider with the system

$$u_0' = -a_0 u_0 + d_1 u_1,$$

$$\vdots$$

$$u_n' = -a_n u_n + d_{n+1} u_{n+1} + b_{n-1} u_{n-1}, \quad n \geq 1,$$

$$\vdots \ . \tag{103}$$

We assume that the rates of change are given and are denoted by d_n and b_n for changes $n \to n-1$ and $n \to n+1$, respectively. In general, we can also include a mechanism that changes a number of objects at the state n by, for example, removing them from the environment or, otherwise, introducing them. The rate of this mechanism is denoted by $\mathbf{c} = (c_n)_{n\in\mathbb{N}}$ and in such a case we have $c_n = b_n + d_n - a_n$. The classical application of this system comes from population theory, where it is a particular case of a Kolmogorov system; in this case u_n is the probability that the described population consists of n individuals and its state can change by either the death or birth of an individual thus moving the population to the state $n-1$ or $n+1$, respectively, hence the name birth-and-death system. The classical birth-and-death system is formally conservative; this is equivalent to $a_n = d_n + b_n$. However, recently a number of other important applications have emerged. For example, [36], we can consider an ensemble of cancer cells structured by the number of copies of a drug-resistant gene they contain. Here, the number of cells with n copies of the gene can change due to mutations, but the cells also undergo division without changing the number of genes in their offspring which is modelled by a nonzero vector $\mathbf{c}$. An exhaustive survey of cancer modelling can be found in M. Chaplain's notes, [23]. Similar model also can be used to describe evolution of microsatellite repeats, [20]. Finally, system (103) can be thought of as a simplified kinetic system consisting of particles labelled by internal energy n and interacting inelastically with the surrounding matter where in each interaction they can either gain or lose a unit (quantum) of energy. Some particles can decay without a trace or be removed from the system leading again to a nonzero $\mathbf{c}$.

Existence Results

In what follows the boldface letters denote sequences, e.g. $\mathbf{u} = (u_0, u_1, \ldots)$. We assume that the sequences $\mathbf{d}$, $\mathbf{b}$, and $\mathbf{a}$ are nonnegative with $b_{-1} = d_0 = 0$.

By $\mathcal{K}$ we denote the matrix of coefficients of the right-hand side of (103) and, without causing any misunderstanding, the formal operator in the space l of all sequences, acting as

$$(\mathcal{K}\mathbf{u})_n = b_{n-1}u_{n-1} - a_n u_n + d_{n+1}u_{n+1}.$$

In the same way, we define $\mathcal{A}$ and $\mathcal{B}$ as $(\mathcal{A}\mathbf{u})_n = -a_n u_n$ and $(\mathcal{B}\mathbf{u})_n = b_{n-1}u_{n-1} + d_{n+1}u_{n+1}$, respectively.

By $\mathcal{K}_p$ we denote the maximal realization of $\mathcal{K}$ in l_p, $p \in [1,\infty)$; that is,

$$\mathcal{K}_p\mathbf{u} = \mathcal{K}\mathbf{u},$$
$$D(\mathcal{K}_p) = \{\mathbf{u} \in l_p;\ \mathcal{K}\mathbf{u} \in l_p\}. \tag{104}$$

It is easy to check that the maximal operator $\mathcal{K}_p$ is closed for any $p \in [1,\infty)$. Next, define the operator A_p by restricting $\mathcal{A}$ to

$$D(A_p) = \{\mathbf{u} \in l_p;\ \mathcal{A}\mathbf{u} \in l_p\} = \{\mathbf{u} \in l_p;\ \sum_{n=0}^{\infty} a_n^p |u_n|^p < +\infty\}.$$

Again, it is standard that $(A_p, D(A_p))$ generates a semigroup of contractions in l_p. Using Theorem 32 we can prove the following result.

Theorem 40. *[18] Assume that sequences **b** and **d** are nondecreasing and there is $\alpha \in [0,1]$ such that for all n,*

$$0 \le b_n \le \alpha a_n, \qquad 0 \le d_{n+1} \le (1-\alpha)a_n. \tag{105}$$

Then there is an extension K_p of the operator $(A_p + B_p, D(A_p))$, where $B_p = \mathcal{B}|_{D(A_p)}$, generating a positive semigroup of contractions in l_p, $p \in (1, \infty)$.

Furthermore, we can prove that $\mathcal{B}$ is closed and thus B_p is closable. Then Theorem 31 implies

$$K_p = \overline{A_p + B_p}$$

provided $p \in (1, \infty)$. The situation in l_1 is completely different.

Corollary 7. *Let $p = 1$. Assume that sequences **b** and **d** are nonnegative and*

$$a_n \ge (b_n + d_n). \tag{106}$$

Then there is an extension K_1 of the operator $(A_1 + B_1, D(A_1))$, where $B_1 = \mathcal{B}|_{D(A_1)}$, which generates a positive semigroup of contractions in l_1.

Proof. Using the definition of $D(A_1)$ we see, from (106), that $0 \le b_n \le a_n$ and $0 \le d_n \le a_n$ for $n \in \mathbb{N}$. Hence, A_1 is well defined and (87) takes the form

$$\sum_{n=0}^{\infty}((A_1 + B_1)\mathbf{u})_n = -\sum_{n=0}^{\infty}a_n u_n + \sum_{n=0}^{\infty}b_n u_n + \sum_{n=0}^{\infty}d_n u_n \le 0,$$

where we used the convention $b_{-1} = d_0 = 0$. $\square$

We have also the following result.

Theorem 41. *For any $p \in [1, \infty)$ we have $K_p \subset \mathcal{K}_p$.*

For $p = 1$, this is immediate consequence of Theorem 37.

In the following two paragraphs we assume that the system (103) is formally conservative: $a_n = b_n + c_n$, $n \ge 0$.

Birth-and-Death Problem: Honesty Results

We now find whether the constructed semigroup is honest (conservative) or dishonest by means of the extension techniques of Subsection 5. In this case $E_f = m$ (the set of all bounded sequences) and

$$\mathsf{L}\mathbf{u} = \left(\frac{u_n}{1 + b_n + d_n}\right)_{n \in \mathbb{N}}$$

on $\mathsf{F} = \{\mathbf{u} \in m; \ \mathsf{L}\mathbf{u} \in l_1\}$, $\mathsf{A}\mathbf{u} = ((b_n + d_n)u_n)_{n \in \mathbb{N}}$ on $D(\mathsf{A}) = \mathsf{LF}$, and similarly for the other operators and spaces introduced in Subsection 5.

Recall that by $\mathcal{K}$ we denoted the matrix of coefficients and, at the same time, the formal operator acting on m given by multiplication by $\mathcal{K}$. It is easy to see that the maximal operator $\mathcal{K}_1$ (see (104)) is precisely

$$\mathcal{K}_1 = \mathsf{K} = \mathsf{A} + \mathsf{B}. \tag{107}$$

Note too that for $\mathbf{u} \in D(\mathsf{K})$, the integral $\int_\Omega \mathsf{K} u d\mu$, which plays an essential role in a number of theorems (e.g., Theorems 35 and 36), here is given by

$$\sum_{n=0}^{\infty} (-(b_n + d_n)u_n + b_{n-1}u_{n-1} + d_{n+1}u_{n+1}) = \lim_{n \to +\infty} (-b_n u_n + d_{n+1}u_{n+1}), \tag{108}$$

where the limit exists as $\mathbf{u} \in D(\mathsf{K})$ yields the convergence of the series.

In the theorems concerning honesty and maximality we assume, to avoid technicalities, that $b_n > 0$ for $n \geq 0$ and $d_n > 0$ for $n \geq 1$.

Theorem 42. *[13] $\mathsf{K} = \overline{\mathsf{A} + \mathsf{B}}$ if and only if*

$$\sum_{n=0}^{\infty} \frac{1}{b_n} \left(\sum_{i=0}^{\infty} \prod_{j=1}^{i} \frac{d_{n+j}}{b_{n+j}} \right) = +\infty. \tag{109}$$

Proof. To prove honesty, we use Theorem 38. Thus, by (108) it suffices to prove that for any $\mathbf{u} \in D(\mathsf{K})_+$

$$\lim_{n \to +\infty} (-b_n u_n + d_{n+1}u_{n+1}) \geq 0,$$

where we know that the sequence above converges. If we assume the contrary, that for some $0 \leq \mathbf{u} \in D(\mathsf{K})$, then limit in (108) is negative so that there exists $b > 0$ such that

$$-b_n u_n + d_{n+1}u_{n+1} \leq -b \tag{110}$$

for all $n \geq n_0$ with large enough n_0. Using (110) as a recurrence we get

$$u_n \geq \frac{b}{b_n} \left(\sum_{i=0}^{\infty} \prod_{j=1}^{i} \frac{d_{n+j}}{b_{n+j}} \right)$$

and, if the assumption (109) is satisfied, we obtain $\sum_{n=0}^{\infty} u_n = +\infty$ which contradicts the assumption of the summability of $(u_n)_{n \in \mathbb{N}}$.

The proof of necessity is an application of Theorem 36. If the series in (109) is convergent, then, by some algebra,

$$u_n = \frac{b}{b_0} \prod_{i=0}^{n-1} \frac{b_i}{d_{i+1}} \left(\sum_{l=n}^{\infty} \prod_{i=1}^{l} \frac{d_i}{b_i} \right).$$

The constructed sequence $(u_n)_{n \in \mathbb{N}}$ is summable by (109) and satisfies

$$-b = -b_n u_n + d_{n+1} u_{n+1}, \quad n \geq 0$$

so that assumption (iii) of Theorem 36 is satisfied. By construction, $\mathcal{A}u + \mathcal{B}u \in l_1$, so that $\mathbf{u} \in D(\mathsf{K})$. We must show that $\mathbf{g} = \mathbf{u} - (\mathcal{A}u + \mathcal{B}u) \geq 0$. By direct calculations, we obtain $g_0 = u_0 + b_0 u_0 - d_1 u_1 = u_0 + b$ and for $n > 0$,

$$g_n = u_n + b_n u_n + d_n u_n - b_{n-1} u_{n-1} - d_{n+1} u_{n+1} = u_n,$$

so that $0 \leq \mathbf{g} \in l_1$. Hence assumption (i) of Theorem 36 is satisfied. $\square$

Universality of Dishonesty

Theorem 43. *If $(G_K(t))_{t \geq 0}$ is dishonest, then for each $\mathbf{u}_0 \in X_+$ there is $t_0 \geq 0$ such that $\|G_K(t)\mathbf{u}_0\| < \|\mathbf{u}_0\|$ for all $t > t_0$.*

Proof. By Theorem 34, $(G_K(t))_{t \geq 0}$ is dishonest if and only if the functional β_λ, defined in Theorem 33, is not identically zero. The defect function along the trajectory originating at $\mathbf{u}_0$, which in our case is given by $\eta_{\mathbf{u}_0}(t) = \|G_K(t)\mathbf{u}_0\| - \|\mathbf{u}_0\|$, is related to β_λ by

$$\int_0^\infty e^{-\lambda t} \eta_{\mathbf{u}_0}(t) dt = -\frac{1}{\lambda} <\beta_\lambda, \mathbf{u}_0> .$$

Clearly, λ is inessential. Putting $\beta_\lambda = \beta = (\beta_n)_{n \in \mathbb{N}}$ with $\beta_n \geq 0$, we see that for universality of dishonesty we must have $\beta_n > 0$ for any $n \geq 0$. On the other hand, by (95), β_λ is an eigenvector of $(BR(\lambda, A))^*$. Any eigenvector $(\phi_n)_{n \in \mathbb{N}}$ satisfies

$$\frac{b_0}{1 + b_0} \phi_1 = \phi_0, \quad \ldots \quad \frac{d_n}{1 + b_n + d_n} \phi_{n-1} + \frac{b_n}{1 + b_n + d_n} \phi_{n+1} = \phi_n, \ldots ,$$

and, because $b_0/(1 + b_0) < 1$, we have $\phi_1 > \phi_0$. Rearranging the terms in the nth equation,

$$\phi_{n+1} = \left(1 + \frac{1}{b_n}\right) \phi_n + \frac{d_n}{b_n} (\phi_n - \phi_{n-1}).$$

Hence $\phi_{n+1} > \phi_n$ whenever $\phi_n \geq \phi_{n-1}$ we end the proof by induction. $\square$

Maximality of the Generator

Let us recall that the relation between the generator K and its extensions K and $\mathcal{K}$ is given in (107). In particular, K is the maximal operator.

Lemma 4. *If $(G_K(t))_{t \geq 0}$ is a substochastic semigroup generated by K and for some $0 \leq \mathbf{h} \in D(\mathsf{K})$,*

$$\int_\Omega \mathsf{K}\mathbf{h}d\mu > 0, \tag{111}$$

then $K \neq \mathsf{K}$; that is, the generator is not maximal.

Conversely, assume that if $0 \neq \mathbf{u} \in l$ solves the formal equation $\mathcal{K}\mathbf{u} = \lambda\mathbf{u}, \lambda > 0$, then either $\mathbf{u} \geq 0$ or $\mathbf{u} \leq 0$, and

$$\int_\Omega \mathsf{K}\mathbf{h}d\mu = 0, \tag{112}$$

for any $\mathbf{h} \in D(\mathsf{K})$. Then $\mathsf{K} = K$; that is, the generator is the maximal operator.

Proof. It follows that if $\mathbf{h} \in D(K)$, then $\int_\Omega K\mathbf{h}d\mu = 0$. Because $K \subset \mathsf{K}$, (111) shows that $\mathbf{h} \notin D(K)$.

If $\mathsf{K} \neq K$ then, by Corollary 6, we have $N(\lambda I - \mathsf{K})_+ \neq \emptyset$. Because the problem is linear, then the assumption ascertains the existence of $0 \neq \mathbf{h} \in N(\lambda I - \mathsf{K})_+$ and for such an $\mathbf{h}$

$$\int_\Omega \mathsf{K}\mathbf{h}d\mu = \lambda\int_\Omega \mathbf{h}d\mu \neq 0, \tag{113}$$

contradicting (112). $\square$

To be able to use this result, we have the following lemma.

Lemma 5. *Let $\lambda > 0$ be fixed. Any solution to*

$$\lambda u_0 = -a_0 u_0 + d_1 u_1, \tag{114}$$

$$\vdots$$

$$\lambda u_n = -a_n u_n + b_{n-1}u_{n-1} + d_{n+1}u_{n+1}, \ n \geq 1,$$

$$\vdots$$

is either nonnegative or nonpositive.

On the basis of the above lemma we obtain:

Theorem 44. *[13] $K \neq \mathsf{K}$ if and only if*

$$\sum_{n=1}^\infty \frac{1}{d_n} \prod_{j=1}^{n-1} \frac{b_j}{d_j} \left(\sum_{i=0}^{n-1} \prod_{j=1}^i \frac{d_i}{b_i}\right) < +\infty. \tag{115}$$

Proof. By Lemma 5 and Proposition 4, $K \neq \mathsf{K}$ if and only if for each $0 \leq (u_n)_{n\in\mathbb{N}} \in l_1$, such that $(-(b_n + d_n)u_n + b_{n-1}u_{n-1} + d_{n+1}u_{n+1_n})_{n\in\mathbb{N}} \in l_1$, we have

$$I = \sum_{n=0}^\infty (-(b_n + d_n)u_n + b_{n-1}u_{n-1} + d_{n+1}u_{n+1}) > 0.$$

and, similarly to the proof of Theorem 42 and (110), we need to investigate the behaviour of the sequence $(r_n)_{n\in\mathbb{N}}$ defined as

$$r_n = -b_n u_n + d_{n+1} u_{n+1}, \quad n \geq 0, \tag{116}$$

or, solving for u_n, for $n \geq 1$,

$$u_n = \frac{1}{d_n} \sum_{i=0}^{n-1} \left(r_i \prod_{j=1}^{n-1-i} \frac{b_{n-j}}{d_{n-j}} \right) + \frac{u_0 b_0}{d_n} \prod_{j=1}^{n-1} \frac{b_j}{d_j}. \tag{117}$$

If $K \neq \mathsf{K}$, then there is a nonnegative $(u_n)_{n\in\mathbb{N}} \in l_1$ for which $I = \lim_{n\to\infty} r_n > 0$ and, by some algebra, it is enough to consider a nonnegative sequence $(u_n)_{n\in\mathbb{N}} \in D(\mathsf{K})$ with the associated sequence $(r_n)_{n\in\mathbb{N}}$ satisfying $\inf_{n\in\mathbb{N}} r_n = r > 0$. Then it can be proved that the series in (115) is convergent.

To prove the converse, define u_n by (116) with arbitrary $(r_n)_{n\in\mathbb{N}}$ converging to $I > 0$ (e.g., we may take $r_n = r$ for all n for a constant positive r). By (115) $(u_n)_{n\in\mathbb{N}} \in l_1$, so that $(u_n)_{n\in\mathbb{N}} \in D(\mathsf{K})$ and because $I > 0$, the thesis follows by (111). □

Examples

We provide a few examples showing that all possible cases of relations between the generator and maximal and minimal operators can be realized.

Proposition 9. *If both sequences $(b_n^{-1})_{n\in\mathbb{N}}, (d_n^{-1})_{n\in\mathbb{N}} \notin l_1$, then $K = \overline{A+B} = \mathsf{K}$. In particular, this is true for the standard birth-and-death problem of population theory where the coefficients are affine functions of n.*

Proof. Expanding (115) we get, for a fixed n,

$$\frac{1}{d_n} \left(1 + \frac{b_{n-1}}{d_{n-1}} + \cdots + \frac{b_{n-1}\ldots b_1}{d_{n-1}\ldots d_1} \right) \geq \frac{1}{d_n}.$$

Similarly, expanding (109), we get

$$\frac{1}{b_n} \left(1 + \frac{d_{n+1}}{b_{n+1}} \cdots + \right) \geq \frac{1}{b_n}$$

which gives divergence of both series. □

The proofs of the following results are obtained in a similar way.

Proposition 10. *a) If $(d_n^{-1})_{n\in\mathbb{N}} \in l_1$ and $\lim_{n\to\infty} b_n/d_n = q < 1$, then $K = \overline{A+B} \neq \mathsf{K}$.*

b) If the sequence $(d_n)_{n\in\mathbb{N}}$ is of polynomial growth: $d_n = O(n^\beta)$ for some β as $n \to \infty$, $(b_n^{-1})_{n\in\mathbb{N}} \in l_1$ and $\lim_{n\to\infty} b_n/d_n = q > 1$, then $\overline{A+B} \subsetneq K = \mathsf{K}$.

c) There are sequences $(b_n)_{n\in\mathbb{N}}$ and $(d_n)_{n\in\mathbb{N}}$ for which $\overline{A+B} \subsetneq K \subsetneq \mathsf{K}$.

To prove c) it suffices to take $b_n = 2 \cdot 3^n$ and $d_n = 3^n$.

5.2 Chaos in Population Theory

We consider a population of cancer cells characterized by different levels of drug resistance. The cells belonging to 0–th subpopulation are sensitive to antineoplastic drugs. The cells of n–th subpopulation, $n > 0$, are resistant with the level of resistance increasing with n. Each subpopulation contains cells characterized by a number of copies of a drug resistance gene. The more copies of the gene exist, the more resistant is the cell, with the understanding that it can survive under higher concentration of the drug. Since the number of gene copies can be very large, we use a model with an infinite number of cell subpopulations. We consider a gene amplification – deamplification process characterized by two components: the conservative and the proliferative.

The conservative component of the process describes the mutations of cells modelled as in standard birth-and-death process. The proliferative component of the process is related to the assumption that the moment of death represents the moment of cell division and that the average life–span is given by the coefficient λ_n for the n–subpopulation ($n \geq 0$). This model leads to the infinite system of ordinary differential equations (103), where we denoted $a_0 = -\lambda_0 + b_0$ and $a_n = -\lambda_n + b_n + d_n$ for $n \in \mathbb{N}$. We denote by $\mathbf{u}(t) = \{u_n(t)\}_{n \geq 0}$ the distribution function and by $\mathcal{K}$ the infinite matrix of the coefficients on the right-hand side of (103). The proper Banach space for the process defined by Eq. (103) is l_1, where the norm of any element $\mathbf{u}$ in the positive cone $l_{1,+}$: $l_{1,+} = \{ \mathbf{u} \in l_1 ; \quad u_n \geq 0 \quad n = 0, 1, 2 \dots \}$ represents the total number of cells. For the sake of completeness we consider also the Banach spaces l_p, $1 \leq p < \infty$, and c_0 (the space sequences converging to 0), with natural norms.

In [12], Eq. (103) is considered under the assumption that the coefficients a_n, b_n (for $n \in \mathbb{N}_0$), d_n (for $n \in \mathbb{N}$) are nonnegative and

(A1) for some $a \geq 0$, $a_n = a + \alpha_n$, $n \in \mathbb{N}_0$, with $\lim_{n \to \infty} \alpha_n = 0$,
(A2) for some $d > 0$ $\lim_{n \to \infty} d_n = d$
(A3) $\limsup_{n \to \infty} b_n = 0$.

Let $\mathcal{K}_p$, $p \in [1, \infty[\cup \{0\}$ denote the operator realization of $\mathcal{K}$ in l_p and c_0, respectively. Since the operators $\mathcal{K}_p$ are bounded, they generate dynamical systems $(G_p(t))_{t \geq 0}$ in l_p and c_0, respectively. Consequently, $\mathcal{K}_p = K_p$.

Theorem 45. *Let the assumptions (A1), (A2) and (A3) be satisfied. There is $q > 0$ such that if $|\alpha_n| \leq dq^{n+1}$, $|b_n d_{n-1}| \leq d^2 q^{2n+4}$ and $a < d$, then the semigroup generated by $\mathcal{K}_p$ is chaotic in any l_p, $1 \leq p < \infty$, and in c_0.*

Consider the system transposed to (103) generated by the matrix $\mathcal{K}^*$. Using the fact that $c_0^* = l_1$ and $(l_p)^* = l_r$, $1/p + 1/r = 1$, by Theorem 22, if $\mathcal{K}^*$ generated chaotic dynamics in any subspace, then the codimension of the span of all eigenvectors of the operator in (103) in respective space would be finite. Since this is not true, we have

Corollary 8. *Suppose that the sequences (a_n), (b_n) and (d_n) are as in Theorem 45. Then the semigroup generated by $\mathcal{K}^*$ is chaotic in no subspace of l_p, $1 \leq p < \infty$, or of c_0.*

Theorem 45 ensures the topological chaos for large deamplification ("death") rates and small amplification ("birth") rates, i.e. for the process which is *subcritical*. On the contrary, chaos will not appear in processes with small deamplification rates and possibly large amplification rates.

The assumptions of Theorem 45 are often not realistic – in most standard applications the coefficients depend in an affine way on n. This creates numerous problems from the generation of the semigroup through to the construction of eigenvectors. To proceed, we adopt the following assumption.

Assumption AC *There exists $N_0 \geq 1$ such that*

$$-a_n = an + \alpha, \quad d_{n+1} = dn + \delta, \quad b_{n-1} = bn + \beta, \quad n \geq N_0, \tag{118}$$

with $a = -(b+d), b, d \geq 0, \alpha, \beta, \delta \in \mathbb{R}$. In this case the proliferation rate does not depend on n for large n, and equals $\gamma = \alpha + \beta + \delta + b - d$. Recall that $\mathcal{K}$ is the infinite matrix of coefficients and the maximal operator $\mathcal{K}_p$ is defined by (104).

Theorem 46. *[17] If Assumption AC is satisfied and $p \in [1; +\infty)$, then $\mathcal{K}_p$ is a unique realization of $\mathcal{K}$ that generates a C_0-semigroup $(G_p(t))_{t\geq 0}$ in l_p.*

The importance of the identification of $\mathcal{K}_p$ as the generator stems from the fact that l_p-solutions of the infinite system are the eigenvectors of the generator (see also Proposition 9).

Proposition 11. *Let Assumption AC be satisfied, $d > b$, $N_0' := \max\{n \geq 0 : d_n = 0\}$. For any $\lambda \in \mathbb{C}$ there exists a unique sequence $u(\lambda) = (u_n(\lambda))_{n\geq 0}$ satisfying (114) such that $u_n(\lambda) = 0$ for $n < N_0'$, $u_{N_0'}(\lambda) = 1$ and $u_n(\lambda)$ is a polynomial in λ of degree $n - N_0'$ for $n \geq N_0'$. Moreover, for any $\lambda_0 \in \mathbb{C}$ and $\epsilon > 0$, there exists $K > 0$ such that if $|\lambda - \lambda_0| < \epsilon$ and $n \geq N_0' + 1$, then*

$$|u_n(\lambda)| \leq K n^{-\frac{\alpha+\beta+\delta-\Re\lambda}{d-b}}. \tag{119}$$

Denote by $\Pi_p(b, d, \alpha, \beta, \delta)$ the open left half-plane defined by

$$\Pi_p(b, d, \alpha, \beta, \delta) = \{\lambda \in \mathbb{C} : \Re\lambda < \alpha + \beta + \delta - (d - b)/p\}, \tag{120}$$

Corollary 9. *Consider the operator $\mathcal{K}_p$ acting in the space l_p, $1 \leq p < \infty$. If Assumption AC holds with $d > b$, then $\Pi_p(b, d, \alpha, \beta, \delta) \subset \sigma_p(\mathcal{K}_p)$. Moreover, for any $\lambda \in \Pi_p(b, d, \alpha, \beta, \delta)$ the sequence $u(\lambda)$, given by Proposition 11, is an eigenvector of $\mathcal{K}_p$ for the eigenvalue λ, and the vector-valued function $\Pi_p(b, d, \alpha, \beta, \delta) \ni \lambda \to u(\lambda) \in l_p$ is analytic.*

This allows to formulate the main theorem of this section.

Theorem 47. *[17] Suppose that $1 \leq p < \infty$ and that Assumption AC holds with $d > b$ and $\gamma_p > 0$. Then the semigroup $(G_p(t))_{t\geq 0}$ is sub-chaotic. On the other hand, if either $b > d$, or $d_{m_0} = 0$ for some $m_0 \geq 1$, then $(G_p(t))_{t\geq 0}$ is not topologically chaotic.*

Finally, let us reflect on the relevance of chaos for natural sciences. In most biological applications only non-negative solutions make sense and it is only fair to note that the chaotic properties discussed here cannot occur for such solutions. In fact, for systems with strictly positive proliferation, the l_1 norm of a solution may only grow and hence the solution cannot wander.

On the other hand, as we are dealing with linear systems we may wish to consider the differences between two physical (i.e. non–negative) solutions and such a difference certainly need not be non–negative. In fact, we have

Proposition 12. *Let X be a Banach lattice. If $(G(t))_{t \geq 0}$ is subchaotic, then for any $\epsilon > 0$ there exist $x_1, x_2 \in X_+$ such that $\|x_1 - x_2\| < \epsilon$ and $\{G(t)x_1 - G(t)x_2\}_{t \geq 0}$ is dense in the space of chaoticity of $(G(t))_{t \geq 0}$.*

Proof. Let X_{ch} be a space of chaoticity of $(G(t))_{t \geq 0}$. There is a dense trajectory in X_{ch} so, in particular, for any $\epsilon > 0$ there is $z \in X_{ch}$ such that $\|z\| < \epsilon$ and $\{G(t)z\}_{t \geq 0}$ is dense in X_{ch}. Since the positive cone in a Banach lattice is generating, there are $x_1, x_2 \in X_+$ such that $z = x_1 - x_2$. From linearity, $G(t)z = G(t)x_1 - G(t)x_2$. □

6 Asynchronous Growth

The analysis in Subsection 4.3 gives some information about how fast a semigroup can grow but does not yield any clue as to whether there are any long term patterns of the behaviour of the semigroup. Some such patterns were discussed in Subsection 1.2. In many cases, as in the finite dimensional case, such patterns are associated with eigenvalues of largest real part. In this section we present some techniques allowing to prove existence of such eigenvalues.

The first step in this direction is to ensure that there exists a decomposition of the spectrum of the semigroup into isolated eigenvalue(s) and the rest which have real parts smaller that the eigenvalues. This of course occurs if the semigroup is compact (or even eventually compact) and, more generally, if its essential spectrum radius is strictly smaller that the spectral radius.

6.1 Essential Growth Bound

The concept of essential spectrum provides more insight into the long time behaviour of semigroups. Since $r_\Phi(G(t))$ is defined through the norm (25), we can define the Fredholm growth rate of the semigroup using the Fredholm norm $\| \cdot \|_\Phi$ and prove, in the same way as for ω_0 (see (41)), that

$$\omega_\Phi(G) = \lim_{t \to \infty} \frac{1}{t} \log \|G(t)\|_\Phi \qquad (121)$$

exists and that

$$e^{t\omega_\Phi(G)} = r_\Phi(G(t)).$$

However, using (27), we can replace r_Φ by r_e and call ω_Φ the essential growth rate and denote it by ω_e. This shows that, at the level of spectral radii and growth bounds, the distinction between Fredholm and essential spectra (and thus between approaches of [30, 42] and [5, 24]) disappears.

Clearly, $\omega_e(G) \leq \omega_0(G)$. If $\omega_e(G) < \omega_0(G)$, then there is an eigenvalue of $(G(t))_{t\geq 0}$ satisfying $|\lambda| = \exp \omega_0(G)t$ hence, by Theorem 20(2), there is $\lambda_1 \in \sigma_p(A)$ such that $\Re\lambda_1 = \omega_0$. Since $s(A) \leq \omega_0(G)$ we get

$$\omega_0(G) = \max\{\omega_e(G), s(A)\} \tag{122}$$

Let us look into implications of $\omega_e(G) < \omega_0(G)$ so that $s(A) = \omega_0(G)$.

Theorem 48. *Suppose $\omega_e(G) < \omega_0(G)$. Then $\sigma_{per,s(A)} \neq \emptyset$ and is finite. Moreover, X can be decomposed in a unique way into a sum $N \bigoplus S$ of two closed $G(t)$-invariant subspaces with one of them (say N) of finite dimension. Furthermore, $\sigma(A|_N) = \sigma_{per,s(A)}$ and $\omega_0(G|_S) < \omega_0(G)$*

Proof. First we note that, by (122) and the definition of $s(A)$, for any $\gamma \in (\omega_e(G), \omega_0(G)]$ there is $\lambda \in \sigma(A) \setminus \sigma_e(A)$ (and also $\lambda \in \sigma(A) \setminus \sigma_\Phi(A)$ with $\gamma \leq \Re\lambda \leq \omega_0(G)$. We will show that for any $\gamma > \omega_e(G)$ there are only finitely many λ satisfying $\Re\lambda \geq \gamma$. To the contrary, assume that there is an infinite sequence $(\lambda_n)_{n\in\mathbb{N}}$ satisfying $\omega_e(G) < \gamma \leq \Re\lambda_n \leq \omega_0(G)$. Since each $\lambda_n \in \sigma_p(A)$ thus, by the Spectral Mapping Theorem for the point spectrum, $\mu_n := e^{t\lambda_n} \in \sigma_p(G(t))$ for any $t \geq 0$. Assume that for some $t_0 > 0$ the sequence $(\mu_n)_{n\in\mathbb{N}}$ has an accumulation point. By the definition of the essential spectrum, this implies $r_e(G(t_0)) \geq e^{\gamma t_0}$ but then $\omega_e(G) \geq \gamma$, which is a contradiction. So, none of the sequences $(e^{t\lambda_n})_{n\in\mathbb{N}}$ has an accumulation point hence, being bounded, must be finite. Fix again $t > 0$. There may be an infinite sequence of λ_n (denoted again by $(\lambda_n)_{n\in\mathbb{N}}$) satisfying $\mu = e^{t\lambda_n}$ for each n, see (58). The eigenspaces of A corresponding to distinct λ_n are linearly independent. But then their direct sum is infinite dimensional and corresponds to the eigenspace of $G(t)$ corresponding to μ contradicting again the definition of the essential spectrum. The first two statements of the lemma follow now by specifying $\gamma = s(A)$. The other two can be obtained by defining N as the sum of $Ker_\infty(\lambda I - A)$ over $\lambda \in \sigma_{per,s(A)}$ and S as the intersection of $Im\,((\lambda I - A)^k)$ over $k \in \mathbb{N}$ and $\lambda \in \sigma_{per,s(A)}$ (or, since $\sigma_{per,s(A)}$ is isolated in $\sigma(A)$ and compact, by taking the spectral projection corresponding to $\sigma_{per,s(A)}$ and its complement). $\square$

Remark 6. Using the terminology of Subsection 1.2, we see that EAEG holds if $\omega_e(G) < \omega_0(G)$. Then, MAEG holds if, moreover, $\sigma_{per,s(A)}$ consists of a single eigenvalue. Finally, AEG holds if, in addition, this eigenvalue has multiplicity one. In fact, these conditions are necessary and sufficient, [5].

6.2 Peripheral Spectrum of Positive Semigroups

The main result of this subsection is

Theorem 49. *[24, Theorem 8.14] If $(G(t))_{t\geq 0}$ is a positive semigroup on a Banach lattice X generated by A such that $s(A) > -\infty$ is a pole of the resolvent $R(\lambda, A)$. Then $\sigma_{per,s(A)}$ is additively cyclic.*

The proof of this result is quite technical and draws on numerous results from the theory of positive operators on Banach lattices and we shall refrain from giving it in detail. It is, however, worthwhile to discuss a few salient point of the proof which use the relations between Banach lattices and the space of continuous function $C(K)$, given by Theorem 8.

We recall the signum operator S_u (see Example 3.3)) and define

$$u^k = S_u^k |u|, \qquad k \in \mathbb{Z}$$

where $u^{-1} := \bar{u}$. The crucial result is the following lemma.

Lemma 6. *If T and R are two bounded operators satisfying $|Rx| \leq T|x|$, $x \in X$, and $Ru = u$ and $T|u| = |u|$ for $u \in X$ such that $|u|$ is a quasi-interior point (see Lemma 2), then $T = S_u^{-1} R S_u$.*

Proof. The proof uses the fact that, as u is quasi-interior, X can be identified with a space of continuous functions. Details are given in [24, Lemmas 8.8-9].

To illustrate this result we shall discuss two other results which are related to Theorem 49 but which are relatively simpler.

Proposition 13. *If L is a positive operator on a Banach lattice X (and thus bounded) and suppose that $Lu = \alpha u$ and $L|u| = |u|$ for some $u \in X$ and $\alpha \in \mathbb{C}$ with $|\alpha| = 1$. Then, for every $k \in \mathbb{Z}$ we have $Lu^k = \alpha^k u^k$.*

Proof. Since L is bounded, it leaves $\overline{X_u}$ invariant. Define $T = L|_{\overline{X_u}}$ and $R = \alpha^{-1} L$. Then $Tu = Ru$, $T|u| = L|u| = |u|$, and $|Tx| = |Lx| = |Rx|$ for $x \in \overline{X_{|u|}}$. Hence, $T = S_u^{-1} R S_u = \alpha^{-1} S_u^{-1} T S_u$. Iteration yields $T = \alpha^{-k} S_u^{-k} T S_u^k$. Hence

$$Tu^{-k} = T S_u^{-k} |u| = \alpha^{-k} S_u^{-k} T |u| = \alpha^{-k} S_u^{-k} |u| = \alpha^{-k} u^{-k}, \ k \in \mathbb{Z},$$

which gives the thesis. $\square$

Corollary 10. *Let $(G(t))_{t\geq 0}$ be a positive semigroup on a Banach lattice, generated by A, and suppose that for some $u \in X$ and $\alpha, \beta \in \mathbb{R}$ we have*

$$Au = (\alpha + i\beta)u, \quad A|u| = \alpha|u|. \tag{123}$$

Then $Au^n = (\alpha + in\beta)u^n$ for $n \in \mathbb{Z}$. Moreover, if $|u|$ is a quasi-interior point of X, then $S_u D(A) = D(A)$ and $A + i\beta I = S_u^{-1} A S_u$.

Proof. We may assume $\alpha = 0$. Eq. (123) implies $G(t)|u| = |u|$ and $G(t)u = e^{i\beta t}u$, $t \geq 0$, by Theorem 20 (2). Thus, by Proposition 13, we have $G(t)u^n = e^{in\beta}u^n$ which, again by Theorem 20, is equivalent to $Au^n = i\beta nu^n$.

If $|u|$ is a quasi-interior point of X_+ then in the proof of Proposition 13 we have $X = \overline{X_{|u|}}$ so that $T = L = G(t)$ and $R = e^{-i\beta t}G(t)$. Thus yields $e^{i\beta t}G(t) = S_u^{-1}G(t)S_u$ for all $t \geq 0$ which implies $S_uD(A) = D(A)$ (by $S_ue^{i\beta t}G(t) = G(t)S_u$) and $A + i\beta I = S_u^{-1}AS_u$. $\square$

The above result allows to give a simple proof a theorem on cyclicity of point spectrum in Banach lattices with strictly monotonic norm: $0 \leq f < g$ implies $\|f\| < \|g\|$. In particular, the spaces L_p have this property.

Theorem 50. *Suppose X is a Banach lattice with strictly monotone norm. If $(G(t))_{t\geq0}$ is a positive contractive semigroup with $s(A) = 0$, then $\sigma_{per,s(A)} \cap \sigma_p(A)$ is imaginary additively cyclic.*

Proof. Suppose that $Au = i\beta u$ for some $\beta \in \mathbb{R}, u \in X$. Then $G(t)u = e^{i\beta t}u$ and $|u| \leq G(t)|u|$. Hence $\|u\| \leq \|G(t)|u|\| \leq \|u\|$ since $(G(t))_{t\geq0}$ is contractive. Hence $\|G(t)|u|\| = \|u\|$ and, by strict monotonicity of the norm, $G(t)|u| = |0|$, which implies $A|u| = 0$. Using Corollary 10 we obtain the thesis. $\square$

Corollary 11. *If assumptions of Theorem 50 are satisfied and $\omega_e(G) < \omega_0(G)$, then $\sigma_{per,s(A)} = \{s(A)\}$. Thus, $(G(t))_{t\geq0}$ has MAEG.*

Proof. Since $\omega_e(G) < \omega_0(G)$, the peripheral spectrum $\sigma_{per,s(A)}$ is the point spectrum and, by Theorem 48, must be finite and non-empty. The only way for it to be additively cyclic is to consist of one point. $\square$

In general case this result will follow from Theorem 49 whose proof is much more involved and thus it is only sketched.

Proof of Theorem 49. Under the assumption of the theorem, $s(A)$ (taken here to be 0 for simplicity) is an approximate eigenvalue. Using considerations of Paragraph 2.2, we embed the problem into the ultrapower $\hat{X}$ so that the approximate eigenvalues become eigenvalues of the extended operator. The snag is that the extended resolvent $\hat{R}(\lambda) := \widehat{R(\lambda, A)}$ is no longer the resolvent of a densely defined operator if A is unbounded. It is however, a pseudoresolvent with the same domain of definition, which satisfies the same estimate as the resolvent of the generator of a positive semigroup (Theorem 25 (iii)):

$$|\hat{R}(\lambda)\hat{x}| \leq \hat{R}(\Re\lambda)|\hat{x}|, \qquad x \in X, \Re\lambda > 0. \tag{124}$$

By Theorem 4, $(\lambda - i\nu)^{-1}$ is an eigenvalue and a pole of $\hat{R}(\lambda)$ of the same order. Fixing $\lambda \in \mathbb{C}$ with $\Re\lambda > 0$, there is $\hat{u}$ satisfying

$$\hat{R}(\lambda)\hat{u} = (\lambda - i\nu)^{-1}\hat{u}$$

and, by properties of pseudo-resolvents, the above equation is satisfied for all λ with $\Re\lambda > 0$. Summarizing, we have

$$\hat{R}(\lambda)\hat{u} = (\lambda - i\nu)^{-1}\hat{u}, \quad \Re\lambda > 0, \qquad \lambda\hat{R}(\lambda)|\hat{u}| \geq |\hat{u}|, \quad \lambda > 0, \qquad (125)$$

where the second relation follows from the first and (124). Now, if $\hat{u}$ were a quasi-interior point and if we had equality in the second relation, then we could use Lemma 6 with $R = \hat{R}(\lambda)$ and $T = R(\Re\lambda)$ to get

$$\lambda\hat{R}(\lambda) = S_{\hat{u}}^{-1}\lambda\hat{R}(\lambda + i\nu)S_{\hat{u}}$$

for all λ, $\Re\lambda > 0$ (by analytic continuation). Replacing λ by $\lambda + i\nu$ on the right hand side and iterating, we obtain $\lambda\hat{R}(\lambda) = S_{\hat{u}}^{-k}\lambda\hat{R}(\lambda + ik\nu)S_{\hat{u}}^{k}$ and, applying this to $|\hat{u}|$, we obtain

$$S_{\hat{u}}^{-k}\lambda\hat{R}(\lambda+ik\nu)S_{\hat{u}}^{k}|\hat{u}| = S_{\hat{u}}^{-k}\lambda\hat{R}(\lambda+ik\nu)\hat{u}^{k} = \lambda\hat{R}(\lambda)|\hat{u}| = |\hat{u}|, \quad \Re\lambda > 0, k \in \mathbb{Z},$$

or

$$\lambda\hat{R}(\lambda + ik\nu)\hat{u}^{k} = \hat{u}^{k}$$

so that the peripheral spectrum of $\hat{R}(\lambda)$, and thus of the generator, is cyclic.

As we noted, there are two snags. One is that $\hat{u}$ is not necessarily a quasi-interior point. This, however, can be remedied by restricting considerations to the closed ideal $\overline{X_u}$. The second is that we have inequality and not equality in the second relation of (125). This is a more serious problem and though a solution is again to pass to an appropriate closed ideal, the construction is technically much more involved. $\square$

As we said earlier, Theorem 49 yields Corollary 11 in full generality.

Corollary 12. *Let $(G(t))_{t\geq 0}$ be a positive semigroup satisfying $\omega_e(G) < \omega_0(G)$. Then $\sigma_{per,s(A)} = \{s(A)\}$. Thus, $(G(t))_{t\geq 0}$ has MAEG.*

Note that assumption $\omega_e(G) < \omega_0(G)$ ensures that $s(A)$ is a pole of $R(\lambda, A)$.
The next step towards AEG requires irreducibility of the semigroup.

Peripheral Spectrum of Irreducible Semigroups

Let $(G(t))_{t\geq 0}$ be a positive semigroup on a Banach lattice X generated by A. Recall that a closed ideal $E \subset X$ is said to be invariant under $(G(t))_{t\geq 0}$ (or $\{G(t)\}$-invariant) if it is $G(t)$-invariant for any $t \geq 0$. The semigroup $(G(t))_{t\geq 0}$ is called irreducible if $\{0\}$ and X are the only $\{G(t)\}$-invariant ideals of X. Furthermore, $(G(t))_{t\geq 0}$ is called strongly irreducible if $G(t)$ is a strongly irreducible operator for any $t \geq 0$ (that is, if $G(t)u$ is a quasi-interior point for any $0 < u \in X$). Clearly, strongly irreducible semigroup is irreducible (see Paragraph 3.3). Irreducible semigroups can be characterised as follows.

Proposition 14. *[24, Proposition 7.6] A positive semigroups $(G(t))_{t\geq 0}$ on a Banach lattice X is irreducible if and only if*

(i) For every $0 < x \in X$ and $0 < \phi \in X^*$, there exists $t \geq 0$ such that $<\phi, G(t)x>> 0$;
(ii) $R(\lambda, A)$ is strongly irreducible for all (some) $\lambda > s(A)$;
(iii) $R(\lambda, A)$ is irreducible for all (some) $\lambda > s(A)$.

If $(G(t))_{t \geq 0}$ is an irreducible semigroup (see Paragraph 3.3), then our information about its spectrum is much more complete. We have the following result.

Theorem 51. *[24, Theorem 8.17] Let $(G(t))_{t \geq 0}$ be a positive irreducible semigroup generated by A and let $s(A) > -\infty$ be a pole of the resolvent $R(\lambda, A)$. Then $s(A)$ is a first-order pole with geometric multiplicity 1; moreover there exists a quasi-interior point $x_0 \in X_+$ satisfying*

$$Ax_0 = s(A)x_0,$$

and a strictly positive $x_0^* \in X_+^*$ such that

$$A^* x_0 = s(A)x_0^*.$$

Proof. As usual we assume $s(A) = 0$ and let p be the order of the pole 0 of $R(\lambda, A)$. Define

$$Q = \lim_{\lambda \to 0} \lambda^p R(\lambda, A).$$

If $p > 1$ then, by (23), $Q^2 = B_{-p}^2 = B_{-2p+1} = 0$ as $-2p + 1 < -p$. On the other hand, consider again the $(G(t))_{t \geq 0}$ invariant ideal

$$I = \{x \in X; \; Q|x| = 0\}.$$

By irreducibility, $I = \{0\}$ (as it cannot be X due to $Q \neq 0$). Thus, $Q^2 \neq 0$ and this contradiction proves $p = 1$.

The operator Q is thus a positive projection on $Ker\, A$. Let $x \in X_+$ be such that $Qx = x_0 \neq 0$. Since $AQ = QA$, we have $Ax_0 = 0$ and, by the Spectral Mapping Theorem for point spectrum, $G(t)x_0 = x_0$ for any $t \geq 0$ and $\bar{X}_{x_0}$ is a $(G(t))_{t \geq 0}$ invariant ideal yielding, by irreducibility, $X = \bar{X}_{x_0}$. Hence, x_0 is a quasi-interior point.

Since Q^* is a positive projection on $Ker\, A^*$, let us consider $x_0^* = Q^*x^*$; then $A^*x_0^* = 0$ and $G(t)^*x_0^* = x_0^*$. Consequently,

$$J = \{x \in X; \; <|x|, x_0> = 0\}$$

is a $(G(t))_{t \geq 0}$ invariant closed ideal and thus $J = \{0\}$. This means that x_0^* is strictly positive. We can normalize it so that $< x_0, x_0^* > = 1$.

To prove that 0 is simple, first let us consider $x > 0$ satisfying $Ax = 0$ and normalized to $< x, x_0^* > = 1$. Since we have

$$G(t)|x - x_0| \geq |G(t)(x - x_0)| = |x - x_0|$$

hence $A|x - x_0| = 0$. If $G(t)|x - x_0| > |x - x_0|$, then, by strict positivity of x_0^*, $< G(t)|x - x_0|, x_0^* >> < |x - x_0|, x_0^* >$. This is, however, impossible, as $< G(t)|x - x_0|, x_0^* > = < G(t)|x - x_0|, G(t)^*x_0^* > = < |x - x_0|, x_0^* >$ Thus

$G(t)|x - x_0| = |x - x_0|$ and consequently $A|x - x_0| = |x - x_0|$. Define $u = |x - x_0| + (x - x_0) = 2\sup\{(x - x_0), 0\} = 2(x - x_0)^+$ and $v == |x - x_0| - (x - x_0) = 2\inf\{(x - x_0), 0\} = 2(x - x_0)^-$. Thus, $u, v \in X_+$ and $Au = Av = 0$. By the above, u, v are quasi-interior points, or 0. If both were non-zero, then both would be weak units. On the other hand, they are disjoint, hence either $u = 0$ or $v = 0$, so one of them must be 0. Assume $v = 0$. Then $|x - x_0| = (x - x_0)$ and thus $< |x - x_0|, x_0^* > = < x, x_0^* > - < x_0, x_0^* > = 1 - 1 = 0$, which yields $x = x_0$. The case $u = 0$ is analogous.

The next step is taking arbitrary $y \in X$ satisfying $Ay = 0$. We write $y = y^+ - y^-$. Since $G(t)y = y$, we have $|y| = |G(t)y)| \leq G(t)|y|$ and, by the argument of the previous paragraph, we find $G(t)|y| = |y|$. Thus we have $G(t)(y^+ - y^-) = y^+ - y^-$ and $G(t)(y^+ + y^-) = y^+ + y^-$, yielding $G(t)y^{\pm} = y^{\pm}$. Using again the above argument, we find $y^{\pm} = < y^{\pm}, x_0^* > x_0$ which gives $y = < y_0, x_0^* > x_0$ and proves geometric simplicity of $s(A) = 0$. $\square$

Now we can state the final result in our quest for AEG.

Corollary 13. *If $(G(t))_{t \geq 0}$ is a positive and irreducible semigroup with $\omega_e(G) < \omega_0(G)$, then $\sigma_{per, s(A)} = \{s(A)\}$ and $s(A)$ is a simple eigenvalue admitting a positive eigenvector. Thus, $(G(t))_{t \geq 0}$ has positive AEG.*

The problem is to find working techniques which would allow to determine whether $(G(t))_{t \geq 0}$ satisfies the assumptions of Corollary 13. As for generation, the most fruitful approach seems to be through perturbations. We shall explore several such techniques in the next subsection.

6.3 Compactness, Positivity and Irreducibility of Perturbed Semigroups

In Subsection 4.5 we discussed various perturbation theorems ensuring the existence of the semigroup associated with $A + B$. In this section we shall discuss to which extent the asymptotic behaviour of the perturbed semigroup is related to that of the original one. We shall focus on bounded perturbations. Let us recall that, by Theorem 29, in this case the perturbed semigroup $(G_{A+B}(t))_{t \geq 0}$ is related to $(G_A(t))_{t \geq 0}$ by the Duhamel equation (72). It is also given by the Dyson–Phillips series (73), obtained by iterating (72).

Compact and Weakly Compact Perturbations

A model result related to the main question discussed here is

Theorem 52. *[30, p. 258] Let $(G_A(t))_{t \geq 0}$ be strongly continuous semigroup on a Banach space X generated by A and let B be a compact operator. If $(G_{A+B}(t))_{t \geq 0}$ is the semigroup generated by $A + B$, then $G_{A+B}(t) - G_A(t)$ is compact for all $t \geq 0$. In particular*

$$\omega_e(A + B) = \omega_e(A). \tag{126}$$

The proof of this results, as well as of the results below, heavily depends on the property of integrals (often referred to as the *convex compactness property*) which is the subject of the next theorem.

Theorem 53. *If* $\mathcal{B} : \Omega \rightarrow \mathcal{L}(X, Y)$ *is a bounded and strongly measurable function on a finite measure space* $(\Omega, d\mu)$ *such that* $\mathcal{B}(\omega)$ *is a compact operator for each* $\omega \in \Omega$, *then the integral* $\int_{\Omega} \mathcal{B}(\omega) d\mu_{\omega}$ *is compact as well.*

Proof. There are many proofs of this result. We sketch one of them, specified to our particular case: $\mathcal{B}(s) = G_A(t-s)BS(s)$, where S is a strongly continuous function, $s \in [0, t]$ and t is fixed. The function $\mathbb{R}_+ \times X \ni (t, x) \rightarrow G_A(t)x$ is jointly continuous. Furthermore, since a strongly continuous function is uniformly continuous on compact sets, the set

$$M = \{G_A(s)Bx \ s \in [0, t], \|x\| \le c\}$$

is relatively compact in X. Having in mind that the Riemann integral is the norm limit of Riemann sums, we find that $(ct)^{-1} \int_0^t G_A(t-s)BS(s)x ds$ is an element of the closed convex hull of M provided $c = \sup\{\|S(s)x\|; \ s \in [0, t], \|x\| \le 1\}$. Since the closed convex hull of a relatively compact set is compact, the statement follows. $\square$

The assumption of compactness of the perturbing operator is often too restrictive. We mentioned earlier that integral operators with natural kernels in important L_1 spaces are not compact but weakly compact. Fortunately, the convex (weak) compactness property holds in this case as well, though the proof in general case is much more delicate.

Theorem 54. *[45] If* $\mathcal{B} : \Omega \rightarrow \mathcal{L}(X, Y)$ *is a bounded strongly measurable function on a finite measure space* $(\Omega, d\mu)$ *such that* $\mathcal{B}(\omega)$ *is a weakly compact operator for each* $\omega \in \Omega$, *then the integral* $\int_{\Omega} \mathcal{B}(\omega) d\mu_{\omega}$ *is compact as well.*

In L_1 spaces the proof of this theorem can be significantly simplified, [41].

With Theorem 53, the proof of Theorem 52 is immediate, since compact perturbations do not change the essential spectrum. On the other hand, for weakly compact perturbations the situation is more delicate: clearly we know that the difference $G_{A+B}(t) - G_A(t)$ is weakly compact but this does not yield equality of essential spectral types. Restricting, however, our attention to L_1 spaces, we know that the square of a weakly compact operator is compact and we should be able to use Theorem 5 for power compact operators.

Unfortunately, the situation is still not obvious, as the relation between the spectra of A and $A+B$ is determined by properties of $R(\lambda, A)B$ (or $BR(\lambda, A)$) and not of B: for $\lambda \in \rho(A)$

$$\lambda \in \sigma(A + B) \Leftrightarrow 1 \in \sigma(BR(\lambda, A)) \Leftrightarrow 1 \in \sigma(R(\lambda, A)B).$$

This situation prompted J. Voigt [47] to introduce concepts of T-power compact and strictly power compact operators. C is said to be T power compact on $\Delta \in \rho(T)$ if there is n such that $(CR(\lambda, T))^n$ is compact for $\lambda \in \Delta$. C is strictly power compact if DC is power compact for any bounded D.

We note that J. Voigt introduces in [47] yet another 'essential spectrum' of an operator C. However, the unbounded component of Voigt's essential spectrum coincides with the unbounded component of the set of all Fredholm points of C and thus the essential spectral radii determined by all these definitions coincide. The main result needed here is

Theorem 55. *[47, Corollary 1.4] If C and T are bounded and C is T power compact on the unbounded component of $\rho(A)$, then the unbounded components of the Voigt's essential spectrum of T and $T + C$ coincide.*

By the remark above the theorem, unbounded components of essential spectra of T and $T + C$ coincide and thus $r_e(T) = r_e(T + C)$. The importance of this result here is due to the fact that weakly compact operators form an ideal; that is, if C is weakly compact, then AC and CA are weakly compact for any bounded A. Thus, in any L_1 space, weakly compact operators are strictly power compact with $(AC)^2, (CA)^2$ compact. Hence, arguing as for Theorem 52, but with the aid of Theorem 54, we have

Corollary 14. *If A is the generator of $(G_A(t))_{t \geq 0}$ on $X = L_1(\Omega)$ and B is a weakly compact operator, then $G_{A+B}(t) - G_A(t)$ is a strictly power compact for all $t \geq 0$. Hence, in particular $\omega_e(G_{A+B}) = \omega_e(G_A)$.*

Eventual Uniform Continuity of Perturbed Semigroups

If a semigroup is eventually uniformly continuous, then the Spectral Mapping Theorem is valid. Moreover, many compactness results can be proved if the underlying semigroup is eventually uniformly continuous. Hence, we shall discuss here a few relevant results for the perturbed semigroup.

If F and G are strongly continuous operator valued functions, then the convolution

$$(F * G)(t)(x) := \int_0^t F(t - s)G(s)x\,ds, \quad t \geq 0, x \in X, \tag{127}$$

is well defined. We have the following basic result:

Lemma 7. *[30, Lemma III.1.13] If F and G are strongly continuous, then*

(i) *If F is uniformly continuous (resp., compact) on $(0, \infty)$, then $F * G$ and $G * F$ are uniformly continuous (resp., compact) on $(0, \infty)$;*

(ii) *If F is uniformly continuous (resp., compact) on (α, ∞), and G is uniformly continuous (resp., compact) on (β, ∞), then $F * G$ and $G * F$ are uniformly continuous (resp., compact) on $(\alpha + \beta, \infty)$.*

Consider the semigroup $(G_A(t))_{t \geq 0}$ and the perturbed semigroup $(G_{A+B}(t))_{t \geq 0}$. From the Duhamel formula we have

$$G_{A+B} = G_A + G_A * BG_{A+B} = G_A + G_AB * G_{A+B}. \tag{128}$$

We define the Volterra operator $\mathcal{V}$ associated with this problem as

$$\mathcal{V}F = G_A * BF = G_AB * F$$

for any strongly continuous F. Then we have the following result.

Theorem 56. *Suppose B is bounded. Then*

(a) *If $(G_A(t))_{t \geq 0}$ is immediately uniformly continuous (resp., compact), then the same holds for $(G_{A+B}(t))_{t \geq 0}$;*

(b) *If $(G_A(t))_{t \geq 0}$ is uniformly continuous (resp., compact) on (α, ∞) and if there exists $k \in \mathbb{N}$ such that $\mathcal{V}^k G_A$ is uniformly continuous (resp., compact) on $(0, \infty)$, then $(G_{A+B}(t))_{t \geq 0}$ is uniformly continuous (resp., compact) on $(k\alpha, \infty)$.*

Proof. Item (i) follows from Lemma 7 (i) as $G_{A+B} = G_A + G_A * BG_{A+B}$. To prove (ii) we note that by Dyson-Phillips expansion

$$G_{A+B}(t) = \sum_{n=0}^{k} \mathcal{V}^n G_A(t) + \sum_{n=1}^{\infty} \mathcal{V}^n (\mathcal{V}^k G_A(t))$$

where the series converges in uniform operator topology on compact intervals by Theorem 29. The terms in the first part are uniformly continuous (resp., compact) on $(k\alpha, \infty)$ by Lemma 7 (ii). The second part can be written as

$$\sum_{n=1}^{\infty} \mathcal{V}^n (\mathcal{V}^k G_A) = G_A * B(\mathcal{V}^k G_A) + G_A * B(G_A * \mathcal{V}^k G_A) + \ldots$$

where each term is uniformly continuous on $(0, \infty)$ by Lemma 7 (i) and converges in uniform operator topology, as above. $\square$

Irreducibility of Perturbed Semigroups

Here we assume that $(G_A(t))_{t \geq 0}$ is a positive semigroup and B is a bounded positive operator. Then $BR(\lambda, A)$ are positive for sufficiently large λ ($\lambda > s(A)$) and the terms of Dyson-Phillips expansion $\mathcal{V}^n G_A(t)$ are positive operators for $t \geq 0$. The last statement follows from the fact that the iterates defining this expansion are positive. As a consequence, we have $0 \leq G_A(t) \leq G_{A+B}(t)$, hence

$$\omega_0(A) \leq \omega_0(A + B).$$

Formula (71) shows that for sufficiently large $\lambda \in \mathbb{R}$ we have $0 \leq R(\lambda, A) \leq R(\lambda, A + B)$ so that

$$s(A) \leq s(A + B).$$

This follows, e.g. from the fact that is we approach $s(A)$ then $R(\lambda, A)$ blows up and thus $R(\lambda, A + B)$ must blow up, hence $s(A) \notin \rho(A + B)$.

Theorem 57. *[24, Corollary 9.22] Let X be a Banach lattice, $(G_A(t))_{t\geq 0}$ a positive semigroup and B a positive bounded operator. The perturbed semigroup $(G_{A+B}(t))_{t\geq 0}$ is irreducible if and only if $I = \{0\}$ and $I = X$ are the only closed ideals satisfying $G_A(t)I \subseteq I$ for $t \geq 0$ and $BI \subseteq I$.*

A Model of Evolution of a Blood Cell Population

We consider a population of blood cells distinguished only by their size and describe the population by the density function $n(t, s)$ of cells having size s in time t. The following processes take place when the time passes:

1. Each cell grows linearly in time;
2. Each cell dies with a probability depending on size;
3. Each cell divides into two daughter cells of equal size with a probability depending on size.

Moreover, we assume that there exists a maximal cell size (here normalized to 1); also there exists a minimal cell size $s = \alpha > 0$ below which no division can occur. As a consequence of the last assumption, if we start with initial population with sizes greater that $\alpha/2$, the size of each cell in the population must satisfy $s > \alpha/2$ and we can assume the boundary condition

$$n(t, \alpha/2) = 0, \qquad t > 0.$$

These assumptions lead to the following evolution equation:

$$n_t(t, s) = -n_s(t, s) - \mu(s)n(t, s) - b(s)n(t, s) + 4b(2s)n(t, 2s)\chi_{[\alpha/2, 1/2]}(s),$$
$$n(0, s) = n_0(s), \tag{129}$$

for $s > \alpha/2, t > 0$ where χ_A is the characteristic function of the set A. We assume that the death rate μ is a positive continuous function on $[\alpha/2, 1]$. The division rate is continuous with $b(s) > 0$ on $(\alpha, 1)$ and $b(s) = 0$ elsewhere.

We consider this equation as an abstract evolution equation in $X = L_1([\alpha/2, 1], dx)$ and define operators

$$Af = -f' - (\mu + b)f \tag{130}$$

on $D(A) = W_1^1([\alpha/2, 1])$ and

$$(Bf)(s) = 4b(2s)f(t, 2s)\chi_{[\alpha/2, 1/2]}(s) \tag{131}$$

on $D(B) = X$ (since multiplication by 2 is bi-lipschitz, the composition is well-defined) as an operation in X. Hence, we can define $K = A + B$ on $D(K) = D(A)$. The following result is standard.

Lemma 8. *The operator $(A, D(A))$ generates a C_0-semigroup given by*

$$G_A(t)f = \begin{cases} e^{-\int_{s-t}^{s} (\mu(\tau)+b(\tau))d\tau} f(s - t) & \text{for} \quad s - t > \alpha/2 \\ 0 & \text{otherwise.} \end{cases} \tag{132}$$

The spectrum of A is empty and the resolvent $R(\lambda, A)$, given explicitly by

$$(R(\lambda, A)g)(s) = \int_{\alpha/2}^{s} e^{-\int_{\sigma}^{s}(\mu(\tau)+b(\tau))d\tau} g(\sigma)d\sigma \qquad (133)$$

is compact.

We have also

Lemma 9. *The semigroup $(G_K(t))_{t\geq0}$ is eventually uniformly continuous and eventually compact for $t > 1 - \alpha/2$.*

Proof. To prove eventual uniform continuity, we first note that $(G_A(t))_{t\geq0}$ is zero for $t > 1-\alpha/2$ and thus uniformly continuous on this interval. Hence, by Theorem 56 (ii), it suffices to prove immediate uniform continuity of some term of the Dyson-Phillips expansion. It turns out that $\mathcal{V}G_A$ is immediately uniform continuous and hence $(G_K(t))_{t\geq0}$ is uniformly continuous for $t > 1 - \alpha/2$.

To prove compactness, we note that $R(\lambda, A)$ is compact and, as $R(\lambda, K)$ is given by uniformly converging series (71) of compact operators, $R(\lambda, K)$ is compact as well. Hence, $R(\lambda, K)G_K(t)$ is compact for such t.

It is interesting that this implies compactness of $(G_K(t))_{t\geq0}$. Indeed, defining $R(t)x = \int_0^t G_K(s)xds$, we have $KR(t)x = G_K(t)x - x$ and

$$R(t) = R(\lambda, K)(I - G_K(t)) + \lambda R(\lambda, K)R(t)x$$

so that, for a fixed $t_0 > 1 - \alpha/2$,

$$R(t_0+h)-R(t_0)=-R(\lambda,K)[G_K(t_0+h)-G_K(t_0)]-\lambda R(\lambda,K)[R(t_0+h)-R(t_0)]. \qquad (134)$$

Since $(G_K(t))_{t\geq0}$ is uniformly continuous at t_0, we get

$$G_K(t_0) = \lim_{h\to0} h^{-1}(R(t_0 + h) - R(t_0))$$

in the uniform topology. Since the first term on the right-hand side in (134) is compact and the second converges to $\lambda R(\lambda, K)G_K(t_0)$, which is compact, $G_K(t_0)$ is compact. Thus, we get compactness for $t > 1 - \alpha/2$. $\square$

We note that eventual compactness implies $\omega_e(G_K) = -\infty$ and hence clearly $\omega_e(G_K) < \omega_0(G_K)$.

The final step is to establish irreducibility of $(G_K(t))_{t\geq0}$.

Lemma 10. *The semigroup $(G_K(t))_{t\geq0}$ is irreducible.*

Proof. Let us analyse how the resolvent

$$R(\lambda, K) = R(\lambda, A) + R(\lambda, A)BR(\lambda, A) + \dots$$

acts on functions with support (a.e) in $(s_0, 1]$ (precisely, $s_0 = \sup\{s\}$ such that $suppf \subset [s, 1]$. Then, by (133), $R(\lambda, A)f > 0$ on $(s_0, 1]$, $BR(\lambda, A)f > 0$ on $(s_0/2, 1/2]$, $R(\lambda, B)BR(\lambda, A)f > 0$ on $(s_0/2, 1]$ and, continuing, $R(\lambda, K)f > 0$ on $[\alpha/2, 1]$. Using the description of ideals in L_1, Example 4, we see that no closed non-trivial ideal can be invariant under $R(\lambda, K)$ and, by Proposition 14, we obtain irreducibility of the semigroup. $\square$

We collect all the information in the following summarising theorem.

Theorem 58. Let $(G_K(t))_{t \geq 0}$ be the C_0-semigroup corresponding to (129). Then there is a dominant eigenvalue equal to $s(A)$ and the corresponding 1-dimensional positive projection P such that

$$\|e^{-s(A)t}G_K(t) - P\| \leq Me^{-\epsilon t}$$

for some $M, \epsilon > 0$ and all $t \geq 0$. Thus, $(G_K(t))_{t \geq 0}$ has a positive AEG.

Emergence of Chaos in the Blood Cell Model

In [34] the author was interested in an abnormal model in which cells of any size can divide. While it is biologically unrealistic to have cells of size 0, the author used this size as a limiting case to describe accumulation of cells in a population of non-functional 'dwarf' cells which occurs in certain blood disorders. Let as begin with a simplified model without death and division terms. Then the cell density satisfies the transport equation

$$n_t = -sn_s + 0.5n, \quad n(s, 0) = n_0(s), \tag{135}$$

in the space $L_1([0, 1])$. This equation, was analysed in [48] in the space of continuous function and the occurrence of chaos was attributed to the insufficient supply of the most primitive blood cells. It was also investigated in the same space in [38] in the statistical framework.

Let $n(0, s) = n_0(s), 0 < s < 1$. The explicit solution to (135) is given by $n(t, s) = T(t)n_0(s) = e^{t/2}n_0(se^{-t})$. We use Theorem 20. The eigenfunctions of the generator are found to be $n_\lambda(s) = s^{-\lambda+1/2}$ provided $\lambda \in U = \{Re\lambda < 3/2\}$. Thus the first assumption of Theorem 20 is satisfied.

Consider now the function

$$F_\lambda[g] = \int\limits_0^1 s^{-\lambda+1/2}g(s)ds$$

where $g \in L_\infty([0, 1])$. By $\frac{d}{d\lambda}s^{-\lambda+1/2} = (\ln s)s^{-\lambda+1/2}$ we see that $F_\lambda[g]$ is analytic in U. Changing variable according to $z = -\ln s$ we obtain

$$0 = \int\limits_0^1 s^{-\lambda+1/2}g(s)ds = \int\limits_0^\infty e^{(-\frac{1}{2}+Re\lambda)z}\left(e^{-z}g(e^{-z})\right)e^{iIm\lambda z}dz.$$

Now, the function $F(z) = e^{(-\frac{1}{2}+Re\lambda)z}(e^{-z}g(e^{-z}))$ is in $L_1([0,\infty])$ for the stipulated range of λ and $Im\lambda$ is not restricted, thus the above integral represents the classical Fourier transform of a function extended by 0 for $z < 0$. Since the transform is zero, $g(s) = 0$ for all s.

Next we consider a combination of (135) and (129) which includes also an immigration term by allowing $\eta > 0$

$$n_t(t,s) = -sn_s(t,s) + \eta n(t,s) + 4\beta n(t,2s)\chi_{[0,1/2]}(s)$$
$$n(0,s) = \phi(s), \tag{136}$$

in $L_1([0,1], ds)$. Change of variables $s = e^{-y}$, $y \geq 0$ gives

$$v_t(t,y) = v_y(t,y) + \eta v(t,y) + 4\beta v(t, y - \ln 2)\chi_{[\ln 2,\infty]}(y)$$
$$v(0,y) = \psi(y) = \phi(e^{-y}) \tag{137}$$

in $X = L_1([0,\infty), e^{-y}dy)$. A nice way to find the eigenvectors ([29]) is to consider the recurrence for $v^n := v|_{[n\ln 2, (n+1)\ln 2)}$ which gives formal eigenvectors as

$$v^n(y) = e^{(\lambda-\eta)y}\sum_{k=0}^{n}\frac{(-4\beta e^{-(\lambda-\eta)\ln 2})^n}{n!}(y - n\ln 2)^n \tag{138}$$

for $n\ln 2 \leq y < (n+1)\ln 2$. Combining and rearranging terms (justified later by absolute convergence), we get

$$v_\lambda(y) = e^{(\lambda-\eta)y}\sum_{n=0}^{\infty}\frac{(-4\beta e^{-(\lambda-\eta)\ln 2})^n}{n!}(y - n\ln 2)^n\chi_{[n\ln 2,\infty)}(y). \tag{139}$$

Estimating the terms of the series in $L_1(\mathbb{R}_+, e^{-y}dy)$ we establish that it is convergent in the half-plane $\Re\lambda < \eta + 1 - 2\beta$ and uniformly convergent is any closed half-plane contained in it. It is easy to see that each term of the series is of the form

$$\lambda \to \phi(\lambda) = e^{\lambda(y-a)}f(y),$$

where a is a constant and f is such that the above function is integrable (with weight e^{-y}) for any fixed λ with $\Re\lambda < \eta + 1 - 2\beta$. Thus, taking such λ_1, λ_2 with, say, $\Re\lambda_1 > \Re\lambda_2$, we have

$$\|\phi(\lambda_1) - \phi(\lambda_2)\| \leq \int_0^{\infty}|1 - e^{(\lambda_2-\lambda_1)(y-\alpha)}|e^{\lambda_1(y-\alpha)}f(y)e^{-y}dy$$

we see that the term between the absolute bars is bounded. Hence, by the dominated convergence, ϕ is continuous and, by the uniform convergence of the series, v_λ is a continuous function of λ. Consequently, if $\eta+1-2\beta > 0$, the dynamics generated by (136) is subchaotic in $X_{ch} = \overline{Span\{v_\lambda, \Re\lambda < \eta + 1 - 2\beta\}}$ by Theorem 61 and subsequent considerations.

Recently, [29], the authors managed to prove that $X_{ch} = X$; that is, the dynamics is chaotic in the whole space, if and only if $\beta \leq 1/2\ln 2$.

7 Asymptotic Analysis of Singularly Perturbed Dynamical Systems

In this section we give a concise explanation of concepts of asymptotic analysis and, in particular, of the abstract form of the Chapman-Enskog procedure.

In order to introduce this procedure, we consider a special singularly perturbed abstract initial value problem in a Banach space $\mathcal{X}$

$$
\begin{cases}
\partial_t f_\epsilon = S f_\epsilon + \dfrac{1}{\epsilon} C f_\epsilon, \\
f_\epsilon(0) = f_0,
\end{cases}
\tag{140}
$$

where the presence of the small parameter ϵ indicates that the phenomenon modelled by the operator C is more relevant than that modelled by S or, in other words, they act on different time scales.

As elsewhere in these lectures, we are concerned with kinetic type problems. Hence the operator S describes some form of transport, whereas C is an interaction/transition operator describing interstate transfers, e.g., it may be a collision operator in the kinetic problems or a transition matrix in the structured population theory.

Often we are interested in situations when the transition processes between the structure states are dominant. If this is the case, the the population quickly becomes homogenised with respect to the structure and starts to behave as an unstructured population, governed by suitable 'macroscopic' equations (which in analogy with the kinetic theory will be called *hydrodynamic*). These equations should be the limiting (or approximating) equations for (140) as $\epsilon \to 0$ (here the parameter ϵ is related to the mean free time between state switches).

To put this in a mathematical framework, we can suppose to have on the right-hand side a family of operators $\{C_\epsilon\}_{\epsilon > 0} = \{S + \epsilon^{-1} C\}_{\epsilon > 0}$ in a suitable Banach space X, and a given initial datum. The classical asymptotic analysis consists in looking for a solution in the form of a truncated power series

$$
f_\epsilon^{(n)}(t) = f_0(t) + \epsilon f_1(t) + \epsilon^2 f_2(t) + \cdots + \epsilon^n f_n(t),
$$

and builds up an algorithm to determine the coefficients $f_0, f_1, f_2, \ldots, f_n$. Then $f_\epsilon^{(n)}(t)$ is an approximation of order n to the solution $f_\epsilon(t)$ of the original equation in the sense that we should have

$$
\| f_\epsilon(t) - f_\epsilon^{(n)}(t) \|_X = o(\epsilon^n),
\tag{141}
$$

for $0 \le t \le T$, where $T > 0$.

It is important to note that the zeroth-order approximation satisfies

$$
C f_0(t) = 0
$$

which is the mathematical expression of the fact that the hydrodynamic approximation should be transition-free and that is why the null-space of the dominant collision operator is called the *hydrodynamic space* of the problem.

Another important observation pertains to the fact that in most cases the limit equation involves less independent variables than the original one. Thus the solution of the former cannot satisfy all side conditions of the latter. Such problems are called *singularly perturbed*. If, for example, the approximation (141) does not hold close to $t = 0$, then it is necessary to introduce an *initial layer* correction by repeating the above procedure with rescaled time to improve the convergence for small t. The original approximation, which is valid only away from $t = 0$ is referred to as the *bulk approximation*.

Similarly, the approximation could fail close to the spatial boundary of the domain as well as close to the region where the spatial and temporal boundaries meet. To improve accuracy in such cases we have to introduce the *boundary* and *corner layer* corrections, but we will not discuss them here.

It is important to realize that usually the bulk approximation represents a new model of the process. A more detailed discussion of this aspect can be found in [37].

A first way to look at the problem from the point of view of the approximation theory is to find, in a systematic way, a new (simpler) family of operators, still depending on ϵ, say B_ϵ, generating new evolution problems

$$\partial_t \varphi_\epsilon = B_\epsilon \varphi_\epsilon,$$

supplemented by appropriate initial conditions, such that the solutions $\varphi_\epsilon(t)$ of the new evolution problem satisfy

$$\|f_\epsilon(t) - \varphi_\epsilon(t)\|_X = o(\epsilon^n),\tag{142}$$

for $0 \leq t \leq T$, where $T > 0$ and $n \geq 1$. In this case we say that B_ϵ is a hydrodynamic approximation of C_ϵ to order n.

This approach mathematically produces weaker results than solving system (140) for each ϵ and taking the limit of the solutions as $\epsilon \to 0$. But in real situation, ϵ is small but not zero, and it is interesting to find simpler operators B_ϵ for modelling a particular regime of this system of interacting objects.

A slightly different point of view consists in requiring that the limiting equation for the approximate solution does not contain ϵ. In other words, the task is now to find a new (simpler) operator, say B, and a new problem

$$\partial_t \varphi = B\varphi,$$

with an appropriate initial condition, such that the solutions $\varphi(t)$ of the new evolution problem satisfy

$$\|f_\epsilon(t) - \varphi(t)\|_X \to 0, \quad \text{as } \epsilon \to 0,\tag{143}$$

for $0 \leq t \leq T$, where $T > 0$.

In this case we say that B is the hydrodynamic limit of operators C_ϵ as $\epsilon \to 0$. This approach can be treated as (and in fact is) a particular version of the previous one as very often the operator B is obtained as the first step in the procedure leading eventually to the family $\{B_\epsilon\}_{\epsilon \geq 0}$. For instance, for the nonlinear Boltzmann equation with the original Hilbert scaling, B would correspond to the Euler system, whereas B_ϵ could correspond to the Navier-Stokes system with ϵ-dependent viscosity.

In any case, a mathematically rigorous asymptotic analysis should provide in a systematic way an algorithm to construct the approximating family B_ϵ (or the limit operator B) and prove the convergence of this algorithm.

In this review we focus on finding the approximating system for (140). This, to certain extent, is dictated by particular applications, where the scaling is given. The other may be seen as more mathematical as then we are looking for suitable scalings of independent variables and physical parameters which lead to the limiting equations not depending on ϵ, see [8,9].

We also will focus on the modification of the classical Chapman-Enskog procedure, also called the *compressed asymptotic expansion* which was adapted to a class of linear evolution equations by J.R. Mika at the end of the 1970s and later extended to singularly perturbed evolution equations arising in the kinetic theory. The advantage of this procedure is that the projection of the solution to the Boltzmann equation onto the null-space of the collision operator, that is, the hydrodynamic part of the solution, is not expanded in ϵ, and thus the whole information carried by this part is kept together. This is in contrast to the Hilbert type expansions, where only the zero order term of the expansion of the hydrodynamic part is recovered from the limit equation.

The first step of this asymptotic procedure is to find the null-space V (the hydrodynamic space) of the dominant operator C; then we decompose the solution into the hydrodynamic and kinetic part by applying the (spectral) projection $\mathcal{P}$ (onto V) and the complementary projection $\mathcal{Q} = I - \mathcal{P}$ (onto the kinetic space W) to equation (140). In this way we obtain a system of evolution equations in the subspaces V and W. At this point the kinetic part of the solution is expanded in series of ϵ, but the hydrodynamic part of the solution is left unexpanded. In other words, we keep all orders of approximation of the hydrodynamic part compressed into a single function.

To pick up transitional effects taking place when the system approaches equilibrium, we introduce two time scales. As we will see, the compressed algorithm permits to derive in a natural way the hydrodynamic equation, the initial condition to supplement it, and the initial layer corrections.

7.1 Compressed Expansion

For clarity, we present the compressed method on a simplified model (140) with the small parameter appearing only in one place. However, the analysis can be extended to more general cases.

The success of the method depends on the spectral properties of the operators $\mathcal{S}$ and $\mathcal{C}$. To start, we must assume that $\lambda = 0$ is the dominant simple eigenvalue of the operator $\mathcal{C}$.

It is easy to see that this requirement amounts to $\mathcal{C}$ being the generator of a semigroup of contractions which has AEG. The fact that $\lambda = 0$ needs to be dominant ensures an exponential decay of the initial layer. This assumption may be relaxed if we are not that interested in the properties of the layer.

Remark 7. In many cases we have several state variables and the operator $\mathcal{C}$ only acts on some of them. Then the above requirement as well as the properties discussed below refer to the action of $\mathcal{C}$ in the restricted space of kinetic variables.

The corresponding eigenspace (the hydrodynamic space of $\mathcal{C}$) is thus one dimensional; we denote by $\mathcal{P}$ the spectral projection of the state space onto this space. Let $\mathcal{Q} = \mathcal{I} - \mathcal{P}$ be the complementary projection. Accordingly, by $\mathcal{P}u = v$ we denote the hydrodynamic part of the solution u and by $\mathcal{Q}u = w$ the kinetic part. We use the fact that the projected operator $\mathcal{P}\mathcal{S}\mathcal{P}$ vanishes for many types of linear equations and, for simplicity, we perform analysis for such a case. Thus, applying the projections on both sides of (140), we obtain

$$\partial_t v = \mathcal{P}\mathcal{S}\mathcal{Q}w$$

$$\partial_t w = \mathcal{Q}\mathcal{S}\mathcal{P}v + \mathcal{Q}\mathcal{S}\mathcal{Q}w + \frac{1}{\varepsilon}\mathcal{Q}\mathcal{C}\mathcal{Q}w \qquad (144)$$

$$v(0) = \overset{o}{v}, \; w(0) = \overset{o}{w},$$

where $\overset{o}{v} = \mathcal{P}\overset{o}{u}$, $\overset{o}{w} = \mathcal{Q}\overset{o}{u}$. We have kept the superfluous symbols $\mathcal{P}v$ and $\mathcal{Q}w$ for the sake of notational symmetry. We represent the solution of (144) as a sum of the bulk and the initial layer parts:

$$v(t) = \bar{v}(t) + \tilde{v}(\tau), \qquad (145)$$

$$w(t) = \bar{w}(t) + \tilde{w}(\tau), \qquad (146)$$

where, in this case. the variable τ in the initial layer part is given by $\tau = t/\varepsilon$. Other scalings may require different formulae for τ.

The following algorithm describes the main features of the compressed asymptotic procedure:

1. The bulk part $\bar{v}$ is not expanded into powers of ε.
2. The bulk part $\bar{w}$ is written in terms of $\bar{v}$ and expanded in powers of ε.
3. The terms $\partial_t \bar{v}$ and $\bar{v}(0)$ are expanded into powers of ε.

Thus

$$\bar{w} = \bar{w}_0 + \varepsilon\bar{w}_1 + O(\varepsilon^2), \quad \tilde{v} = \tilde{v}_0 + \varepsilon\tilde{v}_1 + O(\varepsilon^2), \quad \tilde{w} = \tilde{w}_0 + \varepsilon\tilde{w}_1 + O(\varepsilon^2).$$

Substituting the expansion for $\bar{w}$ into (144) yields

$$\partial_t \bar{v} = \mathcal{P}\mathcal{S}\mathcal{Q}(\bar{w}_0 + \varepsilon\bar{w}_1 O(\varepsilon^2)). \qquad (147)$$

with $\bar{w}_0 \equiv 0$, and

$$\bar{w}_1 = -(\mathcal{Q}\mathcal{C}\mathcal{Q})^{-1}\mathcal{Q}\mathcal{S}\mathcal{P}\bar{v}.$$

Inserting these into (147) gives the approximate 'diffusion' equation

$$\partial_t \bar{v} = -\varepsilon \mathcal{P} \mathcal{S} \mathcal{Q} (\mathcal{Q} \mathcal{C} \mathcal{Q})^{-1} \mathcal{Q} \mathcal{S} \mathcal{P} \bar{v} + O(\varepsilon^2). \tag{148}$$

For the initial layer a similar procedure yields: $\tilde{v}_0(\tau) \equiv 0$ and

$$\partial_\tau \tilde{w}_0 = \mathcal{Q} \mathcal{C} \mathcal{Q} \tilde{w}_0, \tag{149}$$
$$\partial_\tau \tilde{v}_1 = \mathcal{P} \mathcal{S} \mathcal{Q} \tilde{w}_0, \tag{150}$$
$$\partial_\tau \tilde{w}_1 = \mathcal{Q} \mathcal{C} \mathcal{Q} \tilde{w}_1 + \mathcal{Q} \mathcal{S} \mathcal{P} \tilde{v}_0 + \mathcal{Q} \mathcal{S} \mathcal{Q} \tilde{w}_0. \tag{151}$$

We observe that, due to $\bar{w}_0 \equiv 0$, the initial condition for $\tilde{w}_0$ is $\tilde{w}_0(0) = \overset{o}{w}$. Solving (149) with this initial value allows to integrate (150) which gives

$$\tilde{v}_1(\tau) = \mathcal{P} \mathcal{S} \mathcal{Q} (\mathcal{Q} \mathcal{C} \mathcal{Q})^{-1} e^{\tau \mathcal{Q} \mathcal{C} \mathcal{Q}} \overset{o}{w}, \tag{152}$$

upon which $\tilde{v}_1(0) = \mathcal{P} \mathcal{S} \mathcal{Q} (\mathcal{Q} \mathcal{C} \mathcal{Q})^{-1} \overset{o}{w}$. This in turn allows us to determine the initial condition for the diffusion equation: we have from (146) that $\overset{o}{v} = \bar{v}(0) + \varepsilon \tilde{v}_1(0) + O(\varepsilon^2)$ so that

$$\bar{v}(0) = \overset{o}{v} - \varepsilon \mathcal{P} \mathcal{S} \mathcal{Q} (\mathcal{Q} \mathcal{C} \mathcal{Q})^{-1} \overset{o}{w} + O(\varepsilon^2). \tag{153}$$

In what follows, by ρ we shall denote the solution of the "diffusion" equation determined by discarding the $O(\varepsilon^2)$ terms in (148) with the corrected initial value obtained by discarding the $O(\varepsilon^2)$ terms in (153), that is,

$$\partial_t \rho = -\varepsilon \mathcal{P} \mathcal{S} \mathcal{Q} (\mathcal{Q} \mathcal{C} \mathcal{Q})^{-1} \mathcal{Q} \mathcal{S} \mathcal{P} \rho, \tag{154}$$
$$\rho(0) = \overset{o}{v} - \varepsilon \mathcal{P} \mathcal{S} \mathcal{Q} (\mathcal{Q} \mathcal{C} \mathcal{Q})^{-1} \overset{o}{w}. \tag{155}$$

As we noted earlier, for the procedure to start, we need $\lambda = 0$ to be a simple eigenvalue which thus admits a spectral projection onto its eigenspace. This condition is satisfied if, in particular, this eigenvalue is isolated. However, the existence of exponentially decaying initial layer requires the operator $\mathcal{Q} \mathcal{C} \mathcal{Q}$ to generate a semigroup of negative type in $\mathcal{Q} \mathcal{X}$. Since $\mathcal{Q}$ commutes with $\mathcal{C}$, the generation is obvious. However, to ensure the negative type, it is necessary to have $s(\mathcal{Q} \mathcal{C} \mathcal{Q}) < 0$. This condition is equivalent to $(G_{\mathcal{C}}(t))_{t \geq 0}$ having AEG.

Can we Prove the Convergence?

To this end we need to find an equation satisfied by the error defined by

$$y(t) = v(t) - [\bar{v}(t) + \epsilon \tilde{v}_1(t/\epsilon)],$$
$$z(t) = w(t) - [\tilde{w}_0(t/\epsilon) + \epsilon(\bar{w}_1(t) + \tilde{w}_1(t/\epsilon))]. \tag{156}$$

Inserting (formally) the error into (144) we obtain

$$\partial_t y = \mathcal{P} \mathcal{S} \mathcal{P} y + \mathcal{P} \mathcal{S} \mathcal{Q} z + \epsilon \mathcal{P} \mathcal{S} \mathcal{P} \tilde{v}_1 + \epsilon \mathcal{P} \mathcal{S} \mathcal{Q} \tilde{w}_1, \tag{157}$$
$$\partial_t z = \mathcal{Q} \mathcal{S} \mathcal{P} y + \mathcal{Q} \mathcal{S} \mathcal{Q} z + \frac{1}{\epsilon} \mathcal{Q} \mathcal{C} \mathcal{Q} z + \epsilon \mathcal{Q} \mathcal{S} \mathcal{Q} \tilde{w}_1 + \epsilon \mathcal{Q} \mathcal{S} \mathcal{P} \tilde{v}_1 + \epsilon \mathcal{Q} \mathcal{S} \mathcal{Q} \bar{w}_1 - \epsilon \partial_t \bar{w}_1.$$

Hence the total error $E(t) = y(t) + z(t)$ satisfies

$$\partial_t E = \left(\mathcal{S} + \epsilon^{-1}\mathcal{C}\right) E + \epsilon\bar{F} + \epsilon\tilde{F}.$$

If $(G_\epsilon(t))_{t\geq 0}$ is the contractive semigroup generated by $\mathcal{S} + \epsilon^{-1}\mathcal{C}$, we get

$$\|E(t)\| \leq \|E(0)\| + \epsilon \int\limits_0^t \|\bar{F}(s)\| ds + \epsilon \int\limits_0^t \|\tilde{F}(s)\| ds.$$

It can be proved that $E(0) = O(\epsilon^2)$ and so this equation yields the $O(\epsilon)$ error of approximation, which is not good as we have ϵ order terms in the approximation. A closer look at the term involving $\tilde{F}$ shows that it contains $e^{-t/\epsilon}$ which, upon integration, produces another ϵ so that the initial condition, and so the initial layer contribution to the error, are $O(\epsilon^2)$. The fact that the contribution of $\bar{F}$ is also $O(\epsilon^2)$ is rather nontrivial but can be proved for a large class of problems, see [40, Section 6.4].

It is important to note that the above considerations show that the presented asymptotic procedure **potentially** produces the convergence of the expected order. Since in most cases we work with unbounded operators, each step must be carefully justified.

How this Works in Practice: Diffusion Approximation of the Telegraph Equation

Here the compressed procedure is applied to the telegraph equation

$$\partial_t v + b\partial_x w = 0,$$
$$\partial_t w + c\partial_x v + \frac{d}{\varepsilon}w = 0, \tag{158}$$

with constant coefficients b, c, d and a small parameter $\varepsilon > 0$. Thus

$$\mathcal{S} = \begin{bmatrix} 0 & -b\partial_x \\ -c\partial_x & 0 \end{bmatrix}, \; \mathcal{C} = \begin{bmatrix} 0 & 0 \\ 0 & -d \end{bmatrix}.$$

The system (158) is supplemented by the initial conditions and the homogeneous Dirichlet conditions

$$v(x,0) = \overset{o}{v}(x), \qquad w(x,0) = \overset{o}{w}(x), \quad x \in (-1,1) \tag{159}$$
$$v(-1,t) = 0 \qquad v(1,t) = 0, \quad t > 0. \tag{160}$$

To avoid the effect of the boundary layer, we assume that $\overset{o}{v}$ and $\overset{o}{w}$ are three times differentiable and

$$\partial_x \overset{o}{v}(\pm 1) = 0, \; \overset{o}{w}(\pm 1) = 0, \; \partial_{xx} \overset{o}{w}(\pm 1) = 0. \tag{161}$$

This system may describe the voltage and the current in a telegraphic cable, where the a, b, c and d are the loss coefficient, the resistance, the capacity and the self induction respectively or it can be considered as a two-velocity linear Boltzmann equation because the relevant spectral properties are similar.

The diffusion approximation can be derived from (154), using the compressed asymptotic procedure, by taking

$$\mathcal{P}\begin{bmatrix} v \\ w \end{bmatrix} = \begin{bmatrix} v \\ 0 \end{bmatrix}, \qquad \mathcal{Q}\begin{bmatrix} v \\ w \end{bmatrix} = \begin{bmatrix} 0 \\ w \end{bmatrix}.$$

Then

$$\mathcal{QSP}\begin{bmatrix} v \\ 0 \end{bmatrix} = \begin{bmatrix} 0 \\ -c\partial_x v \end{bmatrix}, \qquad \mathcal{PSQ}\begin{bmatrix} 0 \\ w \end{bmatrix} = \begin{bmatrix} -b\partial_x w \\ 0 \end{bmatrix},$$

The inverse $(\mathcal{QCQ})^{-1}$ is given by

$$(\mathcal{QCQ})^{-1}\begin{bmatrix} 0 \\ w \end{bmatrix} = \begin{bmatrix} 0 \\ -w/d \end{bmatrix}.$$

Then

$$\mathcal{PSQ}(\mathcal{QCQ})^{-1}\mathcal{QSP}\begin{bmatrix} v \\ 0 \end{bmatrix} = \mathcal{PSQ}\begin{bmatrix} 0 \\ \frac{c}{d}\partial_x v \end{bmatrix} = \begin{bmatrix} -\frac{bc}{d}\partial_{xx} v \\ 0 \end{bmatrix}.$$

Hence the approximation diffusion equation, as given by (154), is

$$\partial_t \rho = \varepsilon \frac{bc}{d}\partial^2_{xx}\rho. \tag{162}$$

The corrected initial condition can be derived from (155) using $\mathcal{PSQ}$ and $(\mathcal{QCQ})^{-1}$ as calculated above, which gives

$$\rho(0) = \overset{o}{v} -\varepsilon\mathcal{PSQ}(\mathcal{QCQ})^{-1}\overset{o}{w} = \overset{o}{v} -\varepsilon\frac{b}{d}\partial_x \overset{o}{w}. \tag{163}$$

The initial layer, derived from (152), is given by

$$\tilde{v}_1(\tau) = \mathcal{PSQ}(\mathcal{QCQ})^{-1}e^{\tau\mathcal{QCQ}}\overset{o}{w} = \frac{b}{d}e^{-d\tau}\partial_x \overset{o}{w}, \tag{164}$$

where $\tau = t/\varepsilon$. Defining

$$D_3 = \{u \in W^3_2([-1,1]); \ u|_{x=\pm1} = 0, \ \partial^2_{xx}u|_{x=\pm1} = 0\}$$

we can prove the following theorem

Theorem 59. *[10] If $\overset{o}{v}, \overset{o}{w} \in D_3$ and the compatibility conditions (161) are satisfied, then there is a constant C such that, uniformly on $[0,\infty)$,*

$$\|v(t) - \rho(t) - \varepsilon\tilde{v}_1(t/\varepsilon)\| \leq C\varepsilon^2.$$

Age Structured Population Model

A seemingly similar system is offered by

$$\partial_t \mathbf{n} = \mathcal{H}\mathbf{n} + \mathcal{M}\mathbf{n} + \frac{1}{\varepsilon}\mathcal{K}\mathbf{n}, \qquad (165)$$

where $\mathbf{n} = (n_1, \ldots, n_N)$, $\mathcal{H} = diag\{-\partial_a, \ldots, -\partial_a\}$, $\mathcal{M} = diag\{-\mu_1, \ldots, -\mu_N\}$, $\mathcal{K} = \{k_{ij}\}_{1 \le i,j \le N}$. Here n_i is the population density of fish in patch i, a is the age, $\mu_i(a)$ is the mortality rate, and the coefficients k_{ij} represent the migration rates from patch j to patch i, $j \ne i$. The system was introduced to describe evolution of a continuous age-structured population of sole which, however, is further divided into patches (say, egg, larvae, juvenile and adult). The characteristic feature of the population is daily vertical migration provoked by light intensity, which is highly dependent on patches. The small parameter ε corresponds to the fact that the migration processes occur at a much faster time scale than the demographic ones (aging and death). This system is supplemented by the boundary condition of the McKendrick-Von Forester type

$$\mathbf{n}(0, t) = \int_0^\infty \mathcal{B}(a) n(a, t) da \qquad (166)$$

where $\mathcal{B}(a) = diag\{b_1(a), \ldots, b_N(a)\}$ gives the fertility at age a and in patches 1 to N. The initial condition is given by

$$\mathbf{n}(a, 0) = \Phi(a). \qquad (167)$$

The transition matrix $\mathcal{K}$ is a typical transition matrix (of a time-continuous process); that is, the off-diagonal entries are positive and columns sum up to 0. We further assume that it generates an irreducible (n-dimensional) semigroup. Thus, 0 is the dominant eigenvalue of $\mathcal{K}$ with a positive eigenvector $\mathbf{e}$ which will be fixed to satisfy $\mathbf{1} \cdot \mathbf{e} = 1$, where $\mathbf{1} = (1, 1 \ldots, 1)$. The vector $\mathbf{e} = (e_1, \cdots, e_N)$ represents the stable patch structure; that is, asymptotic distribution of the population. Thus, it is reasonable to approximate

$$e_i \approx \frac{n_i}{n}, \quad i = 1, \ldots, N$$

where $n = \sum_{i=1}^N n_i$. Adding together equations in (165) and using the above we obtain

$$\partial_t n = -\partial_a n - \mu^*(a)n, \qquad (168)$$

where $\mu^* = \mathbf{1} \cdot \mathcal{M}\mathbf{e} = \sum_{i=1}^N \mu_i e_i$ is the so-called 'aggregate' mortality. This model, supplemented with appropriate averaged boundary condition is called the aggregated model and is expected to provide averaged approximate description of the population.

The assumptions allow to perform the compressed asymptotic analysis. The spectral projections $\mathcal{P}, \mathcal{Q} : \mathbb{R}^N \to \mathbb{R}^N$ are given by

$$\mathcal{P}\mathbf{x} = (\mathbf{1} \cdot \mathbf{x})\mathbf{e}, \quad \mathcal{Q}\mathbf{x} = \mathbf{x} - (\mathbf{1} \cdot \mathbf{x})\mathbf{e}$$

which gives the hydrodynamical space $V := Span\{\mathbf{e}\}$ and the kinetic space $W = Im\mathcal{Q} = \{\mathbf{x}; \mathbf{1} \cdot \mathbf{x} = 0\}$, as well as the solution decomposition

$$\mathbf{n} = \mathcal{P}\mathbf{n} + \mathcal{Q}\mathbf{n} = v + w = p\mathbf{e} + w$$

where $p = p(a, t)$ is a scalar function. Applying the projections to both sides of (165) we get

$$\partial_t v = \mathcal{P}(\mathcal{H} + \mathcal{M})\mathcal{P}v + \mathcal{P}(\mathcal{H} + \mathcal{M})\mathcal{Q}w$$
$$\partial_t w = \mathcal{Q}(\mathcal{H} + \mathcal{M})\mathcal{Q}w + \mathcal{Q}(\mathcal{H} + \mathcal{M})\mathcal{P}v + \frac{1}{\varepsilon}\mathcal{Q}\mathcal{K}\mathcal{Q}w,$$

with projected initial conditions $v(0) = \overset{o}{v}$, $w(0) = \overset{o}{w}$.

Denoting again by $\bar{v}$ and $\bar{w}$ the bulk part of the solution and substituting the expansion for $\bar{w}$ into (169) we obtain, as before, $\bar{w}_0 \equiv 0$ and

$$\bar{w}_1 = -(\mathcal{Q}\mathcal{K}\mathcal{Q})^{-1}\mathcal{Q}(\mathcal{H} + \mathcal{M})\mathcal{P}\bar{v}.$$

Inserting the expressions for $\bar{w}_0$ and $\bar{w}_1$ into the expansion of the hydrodynamic part of the system (169) gives the approximate 'diffusion' equation

$$\partial_t \bar{v} = \mathcal{P}(\mathcal{H} + \mathcal{M})\mathcal{P}\bar{v} - \varepsilon\mathcal{P}(\mathcal{H} + \mathcal{M})\mathcal{Q}(\mathcal{Q}\mathcal{K}\mathcal{Q})^{-1}\mathcal{Q}(\mathcal{H} + \mathcal{M})\mathcal{P}\bar{v}.$$

The explicit expressions for the involved operators can be calculated as

$$\mathcal{P}(\mathcal{H} + \mathcal{M})\mathcal{P}\bar{v} = -(\partial_a p - p(\mathbf{1} \cdot \mathcal{M}\mathbf{e})\mathbf{e},$$
$$\mathcal{Q}(\mathcal{H} + \mathcal{M})\mathcal{P}\bar{v} = -p(\mathbf{1} \cdot \mathcal{M}\mathbf{e} - \mathcal{M})\mathbf{e},$$
$$\mathcal{P}(\mathcal{H} + \mathcal{M})\mathcal{Q}\mathbf{x} = -(\mathbf{1} \cdot \mathcal{M}\mathbf{x} - \mathbf{1} \cdot \mathcal{M}\mathbf{x})\mathbf{e},$$

and, denoting by $\mathbf{h}$ the unique solution in $W = \mathcal{Q}X$ of

$$\mathcal{K}\mathbf{h} = -(\mathbf{1} \cdot \mathcal{M}\mathbf{e} - \mathcal{M})\mathbf{e}$$

we obtain

$$\mathcal{P}(\mathcal{H} + \mathcal{M})\mathcal{Q}(\mathcal{Q}\mathcal{K}\mathcal{Q})^{-1}\mathcal{Q}(\mathcal{H} + \mathcal{M}\mathcal{P}\bar{v} = p(\mathbf{1} \cdot \mathcal{M}\mathbf{h}).$$

Therefore

$$\partial_t p = -\partial_a p + p(\mathbf{1} \cdot \mathcal{M}\mathbf{e} + \epsilon\mathbf{1} \cdot \mathcal{M}\mathbf{h})$$

or, taking into account the form of $\mathcal{M}$, we obtain

$$\partial_t p = -\partial_a p - \mu^* p + \epsilon(\mathbf{1} \cdot \mathcal{M}\mathbf{h})p$$

so that the asymptotic procedure recovers the aggregated model (168) as well as provides its first order correction.

We note, however, that here, contrary to the previous case, the abstract 'diffusion' is given just by an algebraic expression. The (mathematical) reason for this that in the telegraph equation the transport operator appears on the anti-diagonal and thus provides 'mixing' of the hydrodynamic and kinetic parts of the equation. Here the transport occurs only on the diagonal hence, at the transport level, the patches are not mixed and this feature is preserved in the approximating equation.

In this model it is impossible to neglect effects of the boundary conditions and thus one needs to analyse the boundary and corner layers as well as the initial layer. However, we will not discuss them here, see [39].

Fokker-Planck Equation of Brownian Motion

We conclude with a brief discussion of a more complicated example: the Fokker-Planck equation describing n-dimensional Brownian motion. The collision operator $\mathcal{C}$ now is given by the three-dimensional differential operator

$$(\mathcal{C}u)(x,\xi) = \partial_\xi(\xi + \partial_\xi)u(x,\xi), \tag{169}$$

$x, \xi \in \mathbb{R}^n$, $\partial_\zeta = (\partial_{\zeta_1}, \ldots, \partial_{\zeta_n})$, and the streaming operator $\mathcal{S}$ is of the form

$$(\mathcal{S}u)(x,\xi) = -\xi\partial_x u(x,\xi). \tag{170}$$

Here u is the particle distribution function in the phase space, x denotes the position and ξ the velocity of the particle.

The Fokker-Planck operator can be transformed to the harmonic oscillator operator: for the function $u(\xi)$ we define $\xi = \sqrt{2}\zeta \in \mathbb{R}^n$ and

$$y(\zeta) = (\mathsf{A}_n u)(\zeta) := (\sqrt{2})^{\frac{n}{2}} e^{\frac{|\zeta|^2}{2}} u(\sqrt{2}\zeta). \tag{171}$$

This is an isometry of the space $L_2(\mathbb{R}^n, e^{\frac{|\xi|^2}{2}} d\xi)$ onto $L_2(\mathbb{R}^n, d\zeta)$ which transforms the Fokker-Planck collision operator $\mathcal{C}$ into

$$\tilde{\mathcal{C}}y = \frac{1}{2(\sqrt{2})^{n/2}} e^{-\frac{|\zeta|^2}{2}} \left(\partial_\zeta^2 y - |\zeta|^2 y + ny\right). \tag{172}$$

Dropping the normalizing factor we arrive at the harmonic oscillator operator in $L_2(\mathbb{R}^n)$, denoted hereafter by H,

$$(Hy)(\zeta) = \partial_\zeta^2 y(\zeta) - |\zeta|^2 y(\zeta) + ny(\zeta). \tag{173}$$

To analyse this operator we introduce the sesquilinear form

$$h(\phi, \psi) = \int_{\mathbb{R}^n} \left(\partial_\zeta\phi\partial_\zeta\bar{\psi} + |\zeta|^2\phi\bar{\psi} + \phi\bar{\psi}\right) d\zeta, \tag{174}$$

defined originally on $C_0^\infty(\mathbb{R}^n)$ and the Hilbert space H_1 defined as the closure of $C_0^\infty(\mathbb{R}^n)$ with respect to the norm $\|\phi\|_{H_1} = \sqrt{h(\phi, \phi)}$. Let A_h denote the operator associated with h. It follows that the spectrum of A_h consists only of eigenvalues and the operator itself can be expressed in terms of the series of its eigenfunctions. Using the separation of variables and the one-dimensional theory of the harmonic oscillator we obtain the following expression for the eigenfunctions of A_h:

$$H_\alpha^{(n)}(\zeta) = \frac{(-1)^{|\alpha|}}{(2^{|\alpha|}\pi^{n/2}\alpha!)^{1/2}}e^{\frac{|\zeta|^2}{2}}\partial_\zeta^\alpha e^{|\zeta|^2} = \prod_{i=1}^{n}H_{\alpha_i}^{(1)}(\zeta_i), \qquad (175)$$

where $\zeta \in \mathbb{R}^n$ and $\alpha = (\alpha_1, \dots \alpha_n)$ is a multi-index. $H_m^{(1)}$ is the normalized one-dimensional Hermite function corresponding to the eigenvalue $\lambda_m = 2m + 1$:

$$H_m^{(1)}(\zeta) := \frac{(-1)^m}{\sqrt{\sqrt{\pi}2^m m!}}e^{\frac{\zeta^2}{2}}\partial_\zeta^m e^{-\zeta^2}. \qquad (176)$$

Let C denote the Fokker-Planck collision operator obtained from A_h by the inverse transformation (171), and thus corresponding to the differential expression (169). For $k = 1, \dots, n$ and the multi-index $\beta = (\beta_1, \dots, \beta_k)$ we define $\Phi_\beta^{(k)} := \mathsf{A}_k^{-1}H_\beta^{(k)}$; that is,

$$\Phi_\alpha^{(n)}(\xi) = \frac{(-1)^{|\alpha|}}{(2\pi)^{n/4}\sqrt{\alpha!}}\partial^\alpha e^{-\frac{|\xi|^2}{2}} = \prod_{i=1}^{n}\Phi_{\alpha_i}^{(1)}(\xi_i). \qquad (177)$$

Since A_k is an isometric isomorphism, the family $\left\{\Phi_\alpha^{(n)}\right\}_{\alpha\in\mathbb{N}^n}$ forms an orthonormal basis in $L_2(\mathbb{R}^n, e^{\frac{|\xi|^2}{2}}d\xi)$. We have therefore

$$u = \sum_{|\alpha|=0}^{\infty}u_\alpha\Phi_\alpha^{(n)} \qquad \mathsf{C}u = -\sum_{|\alpha|=1}^{\infty}|\alpha|u_\alpha\Phi_\alpha^{(n)}, \qquad (178)$$

so that it is clear that C is dissipative and satisfies all assumptions of the general theory developed in [40, Chapter 8].

To conclude we derive the diffusion equation. To this end we express the operator $\mathcal{S}$ in terms of eigenfunctions $\Phi_\alpha^{(n)}$. Let us adopt the following convention

$$\alpha(i, \pm 1) = (\alpha_1, \dots, \alpha_i \pm 1, \dots, \alpha_n).$$

For $i = 1, \dots, n$ we have

$$\xi_i\Phi_\alpha^{(n)} = \sqrt{\alpha_i + 1}\Phi_{\alpha(i,+1)}^{(n)}(\xi) + \sqrt{\alpha_i}\Phi_{\alpha(i,-1)}^{(n)}(\xi). \qquad (179)$$

If $\alpha_i = 0$, then the second summand vanishes. By Eq. (179) we obtain formally

$$\mathcal{S}u = -\sum_{k=1}^{n}\partial_{x_k}\left(\sum_{|\alpha|=0}^{\infty}\left(\sqrt{\alpha_k}u_{\alpha(k,-1)} + \sqrt{\alpha_k + 1}u_{\alpha(k,+1)}\right)\Phi_\alpha^{(n)}\right). \qquad (180)$$

The hydrodynamic space is clearly spanned by $\Phi_0^{(n)}$. Hence we denote $\bar{v} = \bar{\rho}\Phi_0^{(n)}$ and $\tilde{v}_1 = \tilde{\rho}\Phi_0^{(n)}$. Introducing the notation $\mathbf{0}(i;l) = (0,\ldots,l,\ldots,0)$ and $\mathbf{0}(i,j;k,l) = (0,\ldots,k,\ldots,l,\ldots,0)$, where l (resp. (k,l)) are in the i-th (resp. i-th and j-th) place, we get

$$\mathcal{S}\bar{v} = -\sum_{k=1}^{n} \partial_{x_k} \Phi_{\mathbf{0}(k;1)}^{(n)} \bar{\rho}$$

and further

$$\mathcal{S}\mathcal{Q}(\mathcal{Q}\mathcal{C}\mathcal{Q})^{-1}\mathcal{Q}\mathcal{S}\mathcal{P}\bar{v} = -\sum_{k=1}^{n}\partial_{x_k}\left(\sum_{l=1,l\neq k}^{n}\partial_{x_l}\Phi_{\mathbf{0}(k,l;1,1)}^{(n)} + \partial_{x_k}\left(\sqrt{2}\Phi_{\mathbf{0}(k,2)}^{(n)} + \Phi_0^{(n)}\right)\right)\bar{\rho}.$$

Projecting this onto $\Phi_0^{(n)}$ we get the diffusion operator in the form

$$\mathcal{P}\mathcal{S}\mathcal{Q}(\mathcal{Q}\mathcal{C}\mathcal{Q})^{-1}\mathcal{Q}\mathcal{S}\mathcal{P}\bar{v} = -\Delta_x\bar{v},$$

and the corrector of the initial value and the initial layer can be obtained in a similar way.

To formulate the final result of this section we introduce

$$\varrho(t,x) := \int_{\mathbb{R}^n} u(t,x,\xi)d\xi,$$

where u is the solution of the initial value problem for the Fokker-Planck equation of the Brownian motion. Let $\overset{\circ}{u}\in W_1^3(\mathbb{R}^n, L_2(\mathbb{R}^n, e^{\frac{|\xi|^2}{2}}d\xi))$, then

$$\left\|\varrho(t) - \bar{\rho}(t) - \epsilon\tilde{\rho}(t/\epsilon)\right\|_{L_2(\mathbb{R}^n\times\mathbb{R}^n, e^{\frac{|\xi|^2}{2}}dxd\xi)} = O(\epsilon^2) \tag{181}$$

uniformly for t in bounded intervals of $[0,\infty)$. Here $\bar{\rho}$ is the solution of the following initial value problem

$$\partial_t\bar{\rho} = \epsilon\Delta_x\bar{\rho},$$

$$\bar{\rho}(0) = \overset{\circ}{u}_0 - \epsilon\sum_{k=1}^{n}\partial_{x_k}\overset{\circ}{u}_{\mathbf{0}(k;1)},$$

and the function $\tilde{\rho}$ in the initial layer corrector $\tilde{v}_1 = \tilde{\rho}\Phi_0^{(n)}$ is given by

$$\tilde{\rho}(t/\epsilon) = e^{-t/\epsilon}\sum_{k=1}^{n}\partial_{x_k}\overset{\circ}{u}_{\mathbf{0}(k;1)}.$$

Acknowledgement. These lecture notes are based on the lectures given by the author at the Banach Centre/CIME Summer School "From a microscopic to a macroscopic description of complex systems", Będlewo, 4-9 September 2006. The author expresses sincere thanks to the Banach Centre and CIME for support received to participate in this School. The visit to Będlewo also was supported by the National Research Foundation of South Africa under the grant GUN 2053716.

References

1. Y. A. Abramovich, Ch.D. Aliprantis, *An Invitation to Operator Theory*, AMS, Providence, Rhode Island, 2002.
2. Y. A. Abramovich, C. D. Aliprantis, *Problems in Operator Theory*, American Mathematical Society, Providence, RI, 2002.
3. Ch.D. Aliprantis, O. Burkinshaw, *Positive Operators*, Academic Press, Inc., Orlando, 1985.
4. W. J. Anderson, *Continuous-Time Markov Chains. An Applications-Oriented Approach*, Springer Verlag, New York, 1991.
5. O. Arino, Some Spectral Properties for the Asymptotic Behaviour of Semigroups Connected to Population Dynamics, *SIAM Review*, **34**(3), 1992, 445–476.
6. L. Arlotti, A perturbation theorem for positive contraction semigroups on L_1-spaces with applications to transport equations and Kolmogorov's differential equations, *Acta Appl. Math.*, **23**(2), (1991), 129–144.
7. L. Arlotti, J. Banasiak, Strictly substochastic semigroups with application to conservative and shattering solutions to fragmentation equations with mass loss, *J. Math. Anal. Appl.*, **293**(2), (2004), 693–720.
8. J. Banasiak, L. Arlotti, *Perturbations of positive semigroups with applications*, Springer, London, 2006.
9. J. Banasiak, G. Frosali, G. Spiga, An interplay between elastic and inelastic scattering in models of extended kinetic theory and their hydrodynamic limits - reference manual, *Transport Theory Stat. Phys.*, **31**(3), 2002, 187–248.
10. J. Banasiak, Singularly perturbed linear and semilinear hyperbolic systems: kinetic theory approach to some folk theorems, *Acta Applicandae Mathematicae*, 49(2), (1997), 199–228.
11. J. Banasiak, On non-uniqueness in fragmentation models, *Math. Methods Appl. Sci.*, **25**(7), 2002, 541–556.
12. J. Banasiak and M. Lachowicz, Topological chaos for birth–and–death–type models with proliferation, *Math. Models Methods Appl. Sci.*, **12** (6), (2002), 755–775.
13. J. Banasiak, A complete description of dynamics generated by birth-and-death problems: A semigroup approach, in: *Mathematical Modelling of Population Dynamics*, R. Rudnicki (Ed.), Banach Center Publications, vol. 63, 2004, 165–176.
14. J. Banasiak and M. Lachowicz, Chaotic linear dynamical systems with applications, *Proceedings of 2nd International Conference on Semigroups of Operators: Theory and Applications SOTA2*, 2001, Eds. C. Kubrusly, N. Levan, and M. da Silveira, Optimization Software, Los Angeles, 2002, 32–44.
15. J. Banasiak, M. Moszyński, A generalization of Desch-Schappacher-Webb criteria for chaos, *Discr. Cont. Dyn. Sys.-A*, **12**(5), (2005), 959–972,
16. J. Banasiak, M. Lachowicz, A generalization of Kato's perturbation theorem in Banach lattices, *Studia Mathematica*, (2007), to appear.
17. J. Banasiak, M. Lachowicz, M. Moszyński, Chaotic behavior of semigroups related to the process of gene amplification–deamplification with cells' proliferation, *Math. Biosciences*, **206**, (2007), in print.
18. J. Banasiak, M. Lachowicz, M. Moszyński, Semigroups for generalized birth-and-death equations in l^p spaces, *Semigroup Forum*, **73**(2), (2007), 175–193.
19. J. Banasiak, M. Moszyński, Hypercyclicity spaces of C_0-semigroups, submitted.
20. A. Bobrowski, M. Kimmel, Dynamics of the life history of a DNA-repeat sequences, *Arch. Control Sci.*, **9**45, no. 1-2, (1999), 57–67.

21. F. E. Browder, On the spectral theory of elliptic differential operators, *Math. Ann*, **142**, 1961, 22–130.
22. V. Capasso, Stochastic processes and stochastic differential equations and their applications, in: *From a microscopic to a macroscopic description of complex systems*, V. Capasso and M. Lachowicz (Eds), CIME/Springer Verlag, 2007, in press.
23. M. Chaplain, Modelling Tumour Growth, in: *From a microscopic to a macroscopic description of complex systems*, V. Capasso & M. Lachowicz (Eds), CIME/Springer Verlag, 2007, in press.
24. Ph. Clément, H. J. A. M. Heijmans, S. Angenent, C. J. van Duijn, B. de Pagter, *One-parameter Semigroups*, CWI Monograph, North Holland, Amsterdam, New York, 1987.
25. W. Desch, W. Schappacher and G. F. Webb, Hypercyclic and chaotic semigroups of linear operators, *Ergodic Theory Dynam. Systems*, **17**, 1997, 793–819.
26. R. L. Devaney, *An Introduction to Chaotic Dynamical Systems*, 2nd edn., Addison-Wesley, New York, 1989.
27. N. Dunford, J. T. Schwartz, *Linear Operators I*, Wiley, New York, 1958.
28. S. El Mourchid, The imaginary point spectrum and hypercyclicity, *Semigroup Forum*, (2006), to appear.
29. S. El Mourchid, G. Metafune, A. Rhandi, J. Voigt, On the analysis of the chaotic behaviour of a size structured cell population, preprint.
30. K.-J. Engel, R. Nagel, *One-Parameter Semigroups for Linear Evolution Equations*, Springer Verlag, New York, 1999.
31. G. Frosali, C. van der Mee, F. Mugelli, A characterization theorem for the evolution semigroup generated by the sum of two unbounded operators, *Math. Methods Appl. Sci.*, **27**(6), (2004), 669–685.
32. G. Godefroy and J.H. Shapiro, Operators with dense, invariant, cyclic manifold, *J. Funct. Anal.* 98, 1991, 229–269.
33. E. Hille, R. S. Phillips, *Functional Analysis and Semi-groups*, Colloquium Publications, v. 31, AMS, Providence, RI, 1957.
34. K. E. Howard, A size structured model of cell dwarfizm, *Discr. Cont. Dyn. Sys.*, **1**, (2001), 471-484.
35. T. Kato, On the semi-groups generated by Kolmogoroff's differential equations, *J. Math. Soc. Jap.*, **6**(1), (1954), 1–15.
36. M. Kimmel, A. Świerniak and A. Polański, Infinite–dimensional model of evolution of drug resistance of cancer cells, *J. Math. Systems Estimation Control* **8**(1), 1998, 1–16.
37. M. Lachowicz, *Links between Microscopic and Macroscopic Descriptions*, in: *From a microscopic to a macroscopic description of complex systems*, V. Capasso & M. Lachowicz (Eds), CIME/Springer Verlag, 2007, in press.
38. A. Lasota, M. C. Mackey, *Chaos, Fractals and Noise, Stochastics Aspects of Dynamics*, Springer Verlag, New York, 1995.
39. M. Lisi, S. Totaro, The Chapman-Enskog procedure for an age-structured population models: initial, boundary and corner layer corrections, *Math. Biosciences*, **196**, (2005), 153–186.
40. J. R. Mika, J. Banasiak, *Singularly Perturbed Evolution Equations with Applications in KineticTheory*, World Sci., Singapore, 1995.
41. M. Mokhtar-Kharroubi, On the convex compactness property for the strong operator topology and related topics, *Math. Meth. Appl. Sci.*, **27**, 2004, 681–701.

42. R. Nagel (ed.), *One parameter semigroups of positive operators*, LNM 1184, Springer Verlag, Berlin, New York, 1986.

43. J. van Neerven, *The Asymptotic Behaviour of Semigroups of Linear Operators*, Operator Theory: Advances and Applications 88, Birkhäuser Verlag, Basel, 1996.

44. A. Pazy *Semigroups of Linear Operators and Applications to Partial Differential Equations*, Springer Verlag, New York, 1983.

45. G. Schlüchtermann, On weakly compact operators, *Math. Ann.*, **292**, 1992, 262–266.

46. J. Voigt, On substochastic C_0-semigroups and their generators, *Transport Theory Statist. Phys.*, **16** (4–6), (1987), 453–466.

47. J. Voigt, A Perturbation Theorem for the Essential Spectral Radius of Strongly Continuous Semigroups, *Monatsh. Math.*, **90**, 1980, 153–161.

48. G. F. Webb, Periodic and chaotic behavior in structured models of cell population dynamics, in: A. C. McBride, G. F. Roach (Eds.) *Recent developments in evolution equations*, Pitman Research Notes in Mathematics 134, Longman Scientific & Technical, Harlow, 1995, 40–49.

49. K. Yosida, *Functional Analysis*, 5th Ed., Springer Verlag, Berlin, 1978.

Rescaling Stochastic Processes: Asymptotics

Vincenzo Capasso and Daniela Morale

Department of Mathematics, University of Milano, Via C. Saldini, 50, Milano, Italy
vincenzo.capasso@unimi.it, daniela.morale@mat.unimi.it

Summary. In this chapter the authors investigate the links among different scales, from a probabilistic point of view. Particular attention is being paid to the mathematical modelling of the social behavior of interacting individuals in a biological population, on one hand because there is an intrinsic interest in dynamics of population herding, on the other hand since agent based models are being used in complex optimization problems. Among other interesting features, these systems lead to phenomena of self-organization, which exhibit interesting spatial patterns. Here we show how properties on the macroscopic level depend on interactions at the microscopic level; in particular suitable laws of large numbers are shown to imply convergence of the evolution equations for empirical spatial distributions of interacting individuals to nonlinear reaction–diffusion equations for a so called mean field, as the total number of individuals becomes sufficiently large. As a working example, an interacting particle system modelling social behavior has been proposed, based on a system of stochastic differential equations, driven by both aggregating/repelling and external "forces". In order to support a rigorous derivation of the asymptotic nonlinear integro-differential equation, compactness criteria for convergence in metric spaces of measures, and problems of existence of a weak/entropic solution have been analyzed. Further the temporal asymptotic behavior of the stochastic system of a fixed number of interacting particles has been discussed. This leads to the problem of the existence of nontrivial invariant probability measure.

1 Introduction

In group-living animals, a wide range of behaviors like resting, foraging or moving are usually performed collectively. Observational and empirical evidence show that animal groups move across the landscape quite cohesively, which strongly suggests that a collective decision has been taken. Thus, it could be assumed that individual decisions lead to a common decision, allowing the group to remain cohesive [38–40]. Such self-organized processes allow groups to carry out collective actions in various environments without any lead, external control or central coordination.

Over the past couple of decades, a large amount of literature has been devoted to the mathematical modelling of self-organizing populations, based on the concepts of short-range/long-range "social interaction" among different individuals of a biological population. The aim of the modelling is to catch the main features of the interaction at the lower scale of single individuals that are responsible, at a larger scale, for a more complex behavior that leads to the formation of aggregating patterns.

A classical widespread approach has been based on either linear or nonlinear PDE's [24, 27]. Such kind of models are often called *Eulerian models*; they describe the evolution of population densities by means of typically deterministic nonlinear partial differential equations of the advection-reaction-diffusion type

$$\rho_t + \nabla \cdot (\mathbf{v}\rho) = \nabla \cdot (D\nabla\rho) + \nu(\rho);$$

ρ is the population density, $\mathbf{v}$ is a velocity field and $\nu(\rho)$ is a possible additive reaction term which may include birth and death processes. The advection term may describe the interaction mechanisms among individuals (via the velocity $\mathbf{v}$), while the non-convective (diffusive) flux takes into account the spatial (random) dispersal of the population.

The advantages of the continuum approach are those of ease of analysis; it is useful in the case of large and dense populations [13]. The disadvantages of this approach include especially the fact that the identity of individuals is compromised. In many situations it is more appropriate to use discrete individual-based models, in which a finite number of individuals is considered and only a finite sequence of decisions is made by individuals. As pointed out by Durrett and Levin [13] and by Grünbaum and Okubo [18], an individual-based approach is also useful in deriving the correct limiting equation, as N increases to infinity, also in the case when the use of a continuum model can be justified.

As already mentioned, a fruitful approach suggested since long by various authors [4, 13, 25, 26, 29, 30] is based on the modelling of the given population as a system of interacting individuals; each individual "particle" is embedded in the total population of N similar particles (the so called individual based model - IBM). Each of them is treated as a discrete particle subject to simple rules of movement. This is known as *Lagrangian approach*: particles are followed in their individual motion. Possible randomness may be included, so that the variation in time of the random location of the k-th individual in the group at time $t \geq 0$, $X_N^k(t) \in \mathbb{R}^d, k = 1, \ldots, N$ is described by a system of stochastic differential equations (SDEs). On the other hand particles are subject to specific forces of interaction which are included in the advection term. In other words, from a Lagrangian point of view, the state of a system of N particles may be described as a stochastic process $\{X_N^k(t)\}_{t\in\mathbb{R}_+}$ defined on a suitable probability space $(\Omega, \mathcal{F}, P)$ and valued in $(\mathbb{R}^d, \mathcal{B}_{\mathbb{R}^d})$, where $\mathcal{B}_{\mathbb{R}^d}$ is the Borel σ-algebra generated by intervals. If the number of particles is kept constant, the time evolution of particle locations $X_N^k(t), k = 1, \ldots, N$ are subject to a system of stochastic differential equations (SDEs) of the following type

$$dX_N^k(t) = H_N^k(X_N^1(t), \ldots, X_N^N(t), t) \, dt + \sigma \left(X_N^1(t), \ldots, X_N^N(t), t \right) \, dW^k(t),$$

where, for $k = 1, \ldots, N$, the function H_N^k defined on $\mathbb{R}^{Nd} \times \mathbb{R}_+$, describes the deterministic force acting on the k-th particle, and randomness has been included via a family of independent Wiener processes. Let us remark that the choice of an additive noise, described by a Wiener process allows a rigorous mathematical analysis based on Itô calculus. This will be clarified later in these lecture notes.

We will show a possible procedure to model interaction among individuals, how to introduce and distinguish among different scales. Indeed, a mathematical way to consider different scales, in the above sense, is based on the choice of a "scaling" parameter, depending upon N; if we consider a system of N particles located in $\mathbb{R}^d$, in the macroscopic space-time coordinates the typical distance between neighboring particles is $O(N^{-1/d})$ and the order of the size of the whole space $O(1)$. In this respect, we may distinguish three main types of interactions:

a. *McKean-Vlasov interaction* (macroscale): any particle interacts with $O(N)$ other particles; collective long-range forces are predominant and the particles are weakly interacting; the range of interaction gets very large in comparison with the typical distance between neighboring particles, and its strength decreases fast, like $1/N$.

b. *hydrodynamic interaction* (microscale): any particle interacts with $O(1)$ other particles in a very small neighborhood with volume $O(1/N)$. The interaction gets short-ranged and rather strong for large N.

c. *moderate interaction* (mesoscale) [28]: any particle interacts with many $O\left(N^{1-\beta}\right)$ other particles in a small volume $O\left(N^{-\beta}\right)$.

By an Eulerian approach, the collective behavior of the discrete (in the number of particles) system, may be recovered in terms of the spatial distribution of particles at time t, expressed, in term of an empirical measure in the space of the probability measures on $\mathbb{R}^d$,

$$X_N(t) = \frac{1}{N} \sum_{k=1}^{N} \epsilon_{X_N^k(t)} \in \mathcal{M}_P(\mathbb{R}^d),$$

which measures the spatial relative frequency of the particles at time t. Again by methods of stochastic calculus, we may derive the time evolution for the empirical measure, and then analyze its asymptotic behavior as the number of individuals N tends to infinity. This is equivalent to study a "law of large numbers" for measures. In particular we may look for sufficient conditions for having a unique limit measure, which is absolutely continuous with respect to the Lebesgue measure, and determine the evolution equation for the density.

In conclusion, the Lagrangian and Eulerian approaches describe the system at different scales: the microscale, the finer scale description based on the stochastic behavior of individuals and the macroscale, the larger scale description based on the continuum behavior of population densities. "The central problem is to determine how information is transferred across scales, and what detail at fine scales is exactly necessary and sufficient for understanding patterns on averaged scales" [13].

Nowadays, one of the interesting mathematical field regards the analysis for providing a mathematically rigorous framework for bridging the gap between the micro and the macro scale.

In these lecture notes the authors discuss some aspects of the asymptotics of rescaled stochastic processes, in particular paying attention to the deterministic approximation of stochastic systems (see Section 4). In Section 4.1 they discuss a time change technique applied to a jump process, while in Sections 4.2 and 4.3 a general description is given of a particle systems via a system of SDEs, and, via an application of Itô's formula, obtain a weak formulation of a stochastic evolution equation for the empirical measure $X_N(t)$, and investigate the properties of the stochastic component. Then they introduce and discuss the concept of the weak convergence in the space $\mathcal{M}_P(\mathbb{R}^d)$. The compactness properties of $\mathcal{M}_P(\mathbb{R}^d)$ and, consequently, the space of the trajectories of the stochastic process X_N, i.e. $\mathcal{M}(C([0,T], \mathcal{M}(\mathbb{R}^d)))$ are presented. An alternative approach to handle multiple scales, including the analysis at a mesoscales is presented in [21].

In Section 5 a specific model for a stochastic interacting particle is introduced [25, 26]. It is shown a mathematical way to describe the interaction, via some interaction kernels which characterize both a McKean-Vlasov and a moderate interaction. The advection term also includes an external flow. The asymptotics of a system of N stochastic differential equations is analyzed, as the population size N increase to infinity. In this way a nonlinear partial integral differential equation is obtained for the asymptotic mean field. Then, equations for the path of each individual particle are given, driven by such mean field.

In Section 6 the long time behavior of the system of SDEs s analyzed, for a fixed N. The interest here is to study mechanisms that are responsible for stable aggregation. The concept of invariant measure is introduced. Both the pure interacting particle system and the system with both the interacting term and a suitable "confining" potential are considered; it is shown how, in the first case, the system cannot admit a nontrivial invariant distribution; in the latter, under suitable conditions on the "localizing" potential U, the system does admit a nontrivial invariant distribution to which the system converges. We notice that, by applying recent results by Veretennikov [35,36], the requirement on U are less restrictive about its convexity with respect to the requirements of previous literature [9,22,36]. Our interest about these topics has been addressed by a wide-through-years literature by several authors interested in modelling aggregation behavior, and studying existence and convergence to an invariant distribution [5,22,36,37].

In order to make the lecture notes self consistent, in Sections 2 and 3 we introduce the stochastic processes of our interest, i.e. Markov processes and in particular diffusion, the Brownian motion and the stochatic Itô calculus.

Now we consider some simple examples of scaling of stochastic process very well know since the first courses in probability.

1.1 First Examples of Rescaling

In the theory of stochastic processes there are some very well-known rescalings which are useful to describe and introduce the fundamental stochastic processes, i.e. the Binomial, the Poisson and the Wiener Processes. Here we introduce the basic ideas.

The Binomial Process

Take $T = \mathbb{N}$. Let $(Y_j)_{j \in \mathbb{N} \setminus \{0\}}$ be a sequence of independent and identically distributed (i.i.d.) Bernoulli random variables, having common distribution $B(1, p)$, the Binomial distribution with parameters 1 and $p \in [0, 1]$. The Bernoulli process can be defined as the discrete time process $(X(n))_{n \in \mathbb{N}}$, such that

1. $X(0) = 0$;
2. $X(n) = \displaystyle\sum_{j=1}^{n} Y_j$, for $n \geq 1$.

As a consequence we know that $X(n)$ has a $B(n, p)$ distribution, for any $n \in \mathbb{N}$, and the following properties hold

i) $X(0) = 0$;
ii) $X(n + r) - X(n)$ is independent of $\mathcal{F}_n = \sigma(X(1), \ldots, X(n))$, the σ-algebra generated by the process till time n, for any $n, r \in \mathbb{N}$; (property of independent increments);
iii) $X(n + r) - X(n)$ has a $B(r, p)$ distribution, i.e. a Binomial distribution with parameters r and p.

Rescaling the Binomial Process: The Poisson Process

Let $t \in \mathbb{R}_+$, $t > 0$, and let $n \in \mathbb{N} \setminus \{0\}$. We may rescale the Binomial process, $B(n, p)$ by choosing for the probability p the following one

$$p = \lambda \frac{t}{n} + o\left(\frac{1}{n}\right),$$

with $\lambda > 0$.

Consider the process

1. $Y^{(n)}(0) = 0$;
2. $Y^{(n)}(t) = \sum_{j=1}^{n} Y_j,$

where now $(Y_j)_{j\in\mathbb{N}\setminus\{0\}}$ is a sequence of i.i.d. Bernoulli random variables, having common distribution $B\left(1, \lambda\frac{t}{n} + o\left(\frac{1}{n}\right)\right)$.

It is well known that, for n tending to infinity, i.e. by considering an infinite sum of Bernoulli trials during the time interval $[0, t]$, the process $(Y^{(n)}(t))_{t\in\mathbb{R}+}$ "converges" to a Poisson process $(Y(t))_{t\in\mathbb{R}+}$, such that

i) $Y(0) = 0$;
ii) $Y(t + s) - Y(s)$ is independent of $\mathcal{F}_s$ for any $s, t \in \mathbb{R}+$; (property of independent increments);
iii) $Y(t+s) - Y(s)$ has a $P(\lambda t)$ distribution for any $s, t \in \mathbb{R}+$, that is a Poisson distribution with parameter λt.

Rescaling the Poisson Process: The Wiener Process

Take the standard Poisson process $(Y(t))_{t\in\mathbb{R}+}$, $(\lambda = 1)$, that is let $Y(t)$ have a $P(t)$ distribution. Let us re-scale both the time and the jump of this process by $N \in \mathbb{N} \setminus \{0\}$ so to obtain the process

$$\tilde{Y}(t) = \frac{1}{N}Y(Nt), \quad t \in \mathbb{R}_+.$$

The strong law of large numbers implies that

$$\lim_{N\to\infty} \tilde{Y}(t) - t = \lim_{N\to\infty} \frac{1}{N}[Y(Nt) - Nt] = 0, \quad \text{a.s.}$$

Indeed, thanks to the martingale properties of the Poisson process, and consequently to Doob's inequality, we have both a functional law of large numbers,

$$\lim_{N\to\infty} \sup_{u\leq v} \frac{1}{N}|Y(Nt) - Nt| = 0, \quad \text{a.s.},$$

and a functional Central Limit Theorem; i.e. the rescaled process

$$\left(\frac{1}{\sqrt{N}}[Y(Nt) - Nt]\right)_{t\in\mathbb{R}+},$$

for N tending to infinity, "converges" to the standard Wiener process $(W(t))_{t\in\mathbb{R}+}$, which is defined by the following

i) $W(0) = 0$;
ii) $W(t + s) - W(s)$ is independent of $\mathcal{F}_s$ for any $s, t \in \mathbb{R}+$; (property of independent increments);
iii) $W(t)$ has a $N(0, t)$ distribution, that is a zero mean normal distribution with variance t, for any $t \in \mathbb{R}+$.

The Wiener process can also be obtained by rescaling another discrete time process of great interest in applications.

The Simple Random Walk

Let $(Y_j)_{j \in \mathbb{N} \setminus \{0\}}$ be a sequence of i.i.d. dichotomic random variables, having common distribution $P(Y_j = 1) = P(Y_j = -1) = 1/2$. The simple random walk is defined as the discrete time process $(S_n)_{n \in \mathbb{N}}$, such that

1. $S_0 = 0$;
2. $S_n = \sum_{j=1}^{n} Y_j$, for $n \geq 1$.

Clearly, for each $n \in \mathbb{N}$, we have that $E[X_n] = 0$ and $Var[X_n] = n$.

Rescaling the Simple Random Walk

Let $t \in \mathbb{R}_+$, $t > 0$, and let $n \in \mathbb{N} - \{0\}$. Let $S_n, n \geq 1$ and $Y_j, j = 1, \ldots, n$ be defined as above. Define now the following rescaled and linearly interpolated process

$$X_n(t) := \frac{1}{\sqrt{n}} S_{[nt]} + (nt - [nt]) \frac{1}{\sqrt{n}} Y_{[nt]+1}, \quad t \in \mathbb{R}_+.$$

Donsker's theorem [33] states that the process $(X_n(t))_{t \in \mathbb{R}+}$, for n tending to infinity, "converges" to the standard Wiener process $(W(t))_{t \in \mathbb{R}+}$.

Our aim below is to provide a framework that makes the above statements mathematically rigorous. In particular the construction of continuous-time stochastic processes, valued in metric spaces, from their finite dimensional distributions; and later the convergence of stochastic processes on the relevant metric spaces.

2 Stochastic Processes

Stochastic processes generalize the notion of (finite-dimensional) vectors of random variables to the case of any family of random variables indexed in a general set T. Typically, the latter represents "time" and is an interval of $\mathbb{R}$ (in the continuous case) or $\mathbb{N}$ (in the discrete case).

Definition 1. Let $(\Omega, \mathcal{F}, P)$ be a probability space, T an index set, and (E, B) a measurable space. An $(E, \mathcal{B})$-valued *stochastic process* on $(\Omega, \mathcal{F}, P)$ is a family $(X_t)_{t \in T}$ of random variables $X_t : (\Omega, \mathcal{F}) \to (E, \mathcal{B})$ for $t \in T$.

The triple $(\Omega, \mathcal{F}, P)$ is called the underlying *probability space* of the process $(X_t)_{t \in T}$, while $(E, \mathcal{B})$ is known as the *state space* or *phase space*. For $t \in T$, the random variable X_t is the *state of the process at "time" t*. Moreover, for all $\omega \in \Omega$, the mapping $X(\cdot, \omega) : t \in T \to X_t(\omega) \in E$ is called the *trajectory* or *path of the process* corresponding to ω. Any trajectory $X(\cdot, \omega)$ of the process belongs to the space E^T of functions defined in T and valued in E.

In order to introduce a suitable probability space $(E^T, \mathcal{B}^T, P^T)$ for the trajectories, we consider the σ-algebra $\mathcal{B}^T$ on E^T defined as the σ-algebra generated by the algebra of the set of cylinders with finite-dimensional base, i.e.

$$\mathcal{B}^T = \sigma \left(\pi_{ST}^{-1}(A) : A \in \mathcal{B}^S = \bigotimes_{t \in S} \mathcal{B}_t, \mathcal{S} = \{ S \subset T | \text{ S is finite} \} \right),$$

where π_{ST} is the canonical projection of E^T on E^S [8, 23].

By the Carathéodory's extension theorem [33], it is possible to extend on $\mathcal{B}^T$, in a unique way, a probability measure defined on an algebra which generates $\mathcal{B}^T$, i.e. on the family of cylinders with the finite rectangles $B_1 \times \cdots \times B_n \in \mathcal{B}^n$ as bases, by

$$P^T(\pi_{ST}^{-1}(B_1 \times \cdots \times B_n)) = P^S(B_1 \times \cdots \times B_n) = P(X_{t_1} \in B_1, \ldots, X_{t_n} \in B_n).$$

As a consequence $\pi_{SS'}(P^{S'}) = P^S$, $S \subset S'$ and then $\left(P^S \right)_{S \in \mathcal{S}}$ is a *compatible system* of measures, as defined below.

Definition 2. If, for all $(S, S') \in \mathcal{S} \times \mathcal{S}'$, with $S \subset S'$, we have that $\pi_{SS'}(P^{S'}) = P^S$, then $(E^S, \mathcal{B}^S, P^S, \pi_{SS'})_{S,S' \in \mathcal{S}; S \subset S'})$ is called a *projective system* of measurable spaces and $(P^S)_{S \in \mathcal{S}}$ is called a *compatible system* of measures on the finite products $(E^S, \mathcal{B}^S)_{S \in \mathcal{S}}$.

We may mention this fundamental result, which characterizes the probability measures P^T [8, 23].

Theorem 1 (Kolmogorov–Bochner). *Let $(E_t, \mathcal{B}_t)_{t \in T}$ be a family of Polish spaces (i.e., metric, complete, separable) endowed with their respective Borel σ-algebras, and let $\mathcal{S}$ be the collection of finite subsets of T and, for all $S \in \mathcal{S}$ with $W^S = \prod_{t \in S} E_t$ and $\mathcal{B}^S = \bigotimes_{t \in S} \mathcal{B}_t$, let μ_S be a finite measure on $(W^S, \mathcal{B}^S)$. Under these assumptions the following two statements are equivalent:*

1. *there exists a μ_T measure on $(W^T, \mathcal{B}^T)$ such that for all $S \in \mathcal{S}$: $P^S = \pi_{ST}(P^T)$;*
2. *the system $(W^S, \mathcal{B}^S, P^S, \pi_{SS'})_{S,S' \in \mathcal{S}; S \subset S'}$ is projective.*

Moreover, in both cases, P^T, as defined in 1, is unique.

The unique measure P^T of Theorem 1 is called the *projective limit* of the projective system $(W^S, \mathcal{B}^S, P^S, \pi_{SS'})_{S,S' \in \mathcal{S}; S \subset S'}$.

Example 1. Let $(X_t)_{t \in T}$ be a family of independent random variables defined on $(\Omega, \mathcal{F}, P)$ and valued in $(E, \mathcal{B})$. (In fact, in this case, it is sufficient to assume that all finite families of $(X_t)_{t \in T}$ are independent.) We know that for all $t \in T$ the probability $P_t = X_t(P)$ is defined on $(E, \mathcal{B})$. Then

$$\forall S = \{t_1, \ldots, t_r\} \in \mathcal{S}: \qquad P^S = \bigotimes_{k=1}^{r} P_{t_k}, \text{ for some } r \in \mathbb{N}^*,$$

and the system $(P^S)_{S \in \mathcal{S}}$ is compatible with the finite products $(E^S, \mathcal{B}^S)_{S \in \mathcal{S}}$. In fact, if B is a rectangle in $\mathcal{B}^S$, i.e., $B = B_{t_1} \times \cdots \times B_{t_r}$, and if $S \subset S'$, where $S, S' \in \mathcal{S}$, then

$$\begin{aligned}
P^S(B) &= P^S(B_{t_1} \times \cdots \times B_{t_r}) = P_{t_1}(B_{t_1}) \cdot \cdots \cdot P_{t_r}(B_{t_r}) \\
&= P_{t_1}(B_{t_1}) \cdot \cdots \cdot P_{t_r}(B_{t_r}) P_{t_{r+1}}(E) \cdot \cdots \cdot P_{t_{r'}}(E) \\
&= P^{S'}(\pi_{SS'}^{-1}(B));
\end{aligned}$$

i.e. that $P^S = \pi_{SS'}(P^{S'})$.

Definition 3. A real-valued stochastic process $(X_t)_{t \in \mathbb{R}_+}$ is *continuous in probability* if

$$P - \lim_{s \to t} X_s = X_t, \qquad s, t \in \mathbb{R}_+.$$

Definition 4. A *filtration* $(\mathcal{F}_t)_{t \in \mathbb{R}_+}$ is an increasing family of sub-algebras of $\mathcal{F}$. The filtration $\mathcal{F}_t = \sigma(X(s), 0 \le s \le t)$, $t \in \mathbb{R}_+$ is called the *generated* or *natural* filtration of the process X_t.

We may refer to the natural filtration of the process X_t, also as the history of the process.

Definition 5. A stochastic process $(X_t)_{t \in \mathbb{R}_+}$ is *right-(left-)continuous* if its trajectories are right-(left-)continuous almost surely.

Definition 6. A stochastic process $(X_t)_{t \in \mathbb{R}_+}$ is said to be *right-continuous with left limits* (RCLL) or *continu à droite avec limite à gauche* (càdlàg) if, almost surely, it has trajectories that are RCLL. The latter is denoted $X_{t-} = \lim_{s \uparrow t} X_s$.

Definition 7. A filtered complete probability space $(\Omega, \mathcal{F}, P, (\mathcal{F}_t)_{t \in \mathbb{R}_+})$ is said to satisfy the *usual hypotheses* if

1. $\mathcal{F}_0$ contains all the P-null sets of $\mathcal{F}$,
2. $\mathcal{F}_t = \bigcap_{s > t} \mathcal{F}_s$, for all $t \in \mathbb{R}_+$; i.e., the filtration $(\mathcal{F}_t)_{t \in \mathbb{R}_+}$ is right-continuous.

Henceforth we will always assume that the usual hypotheses hold, unless specified otherwise.

Definition 8. The process $(X_t)_{t\in\mathbb{R}_+}$ is said to be *progressively measurable* with respect to the filtration $(\mathcal{F}_t)_{t\in\mathbb{R}_+}$, if, for all $t \in \mathbb{R}_+$, the mapping $(s,\omega) \in [0,t] \times \Omega \to X(s,\omega) \in E$ is $(\mathcal{B}_{[0,t]} \otimes \mathcal{F}_t)$-measurable.

Definition 9. A random variable T defined on Ω (endowed with the σ-algebra $\mathcal{F}$) and valued in $\bar{\mathbb{R}}_+$ is called a *stopping time* (or *Markov time*) with respect to the filtration $(\mathcal{F}_t)_{t\in\mathbb{R}_+}$, or simply an $\mathcal{F}_t$-*stopping time*, if

$$\forall t \in \mathbb{R}_+: \qquad \{\omega \in \Omega | T(\omega) \leq t\} \in \mathcal{F}_t.$$

The stopping time is said to be finite if $P(T = \infty) = 0$.

In the following, we consider some very important stochastic processes, which will be used in the main part of these lecture notes.

2.1 Processes with Independent Increments

The stochastic process $(\Omega, \mathcal{F}, P, (X_t)_{t\in\mathbb{R}_+})$, with state space $(E, \mathcal{B})$, is called a *process with independent increments* if, for all $n \in \mathbb{N}$ and for all $(t_1, \ldots, t_n) \in \mathbb{R}_+^n$, where $t_1 < \cdots < t_n$, the random variables $X_{t_1}, X_{t_2} - X_{t_1}, \ldots, X_{t_n} - X_{t_{n-1}}$ are independent.

Let us call $\mu_{t,s} = P_{X_t - X_s}$, the law of the increment $X_t - X_s$. It is possible to construct a compatible system of probability laws $(P^S)_{S\in\mathcal{S}}$, where $\mathcal{S} = \{t_1, \ldots, t_n\}$ is a collection of finite subsets of the index set by

$$P^S(B) = P((X_{t_1}, \ldots, X_{t_n}) \in B) = E[I_B(X_{t_1}, \ldots, X_{t_n})]$$

$$= \int I_B(y_0 + y_1, \ldots, y_0 + \cdots + y_n)$$

$$\mu_0 \otimes \mu_{0,t_1} \otimes \cdots \otimes \mu_{t_{n-1},t_n}(dy_0, \ldots, dy_n).$$

A process with independent increments is called *time-homogeneous* if

$$\mu_{s,t} = \mu_{s+h,t+h} \qquad \forall s,t,h \in \mathbb{R}_+, s < t. \tag{1}$$

If $(\Omega, \mathcal{F}, P, (X_t)_{t\in\mathbb{R}_+})$ is a homogeneous process with independent increments, then in particular we have

$$\mu_{s,t} = \mu_{0,t-s}, \qquad \forall s,t \in \mathbb{R}_+, s < t.$$

2.2 Martingales

Let $(X_t)_{t\in\mathbb{R}_+}$ be a real-valued family of random variables defined on the probability space $(\Omega, \mathcal{F}, P)$ and let $(\mathcal{F}_t)_{t\in\mathbb{R}_+}$ be a filtration. The stochastic process $(X_t)_{t\in\mathbb{R}_+}$ is said to be *adapted* to the family $(\mathcal{F}_t)_{t\in\mathbb{R}_+}$ if, for all $t \in \mathbb{R}_+$, X_t is $\mathcal{F}_t$-measurable.

The stochastic process $(X_t)_{t\in\mathbb{R}_+}$, adapted to the filtration $(\mathcal{F}_t)_{t\in\mathbb{R}_+}$, is a *martingale* with respect to this filtration, provided the following conditions hold:

1. X_t is P-integrable, for all $t \in \mathbb{R}_+$;
2. for all $(s,t) \in \mathbb{R}_+ \times \mathbb{R}_+, s < t : E[X_t|\mathcal{F}_s] = X_s$ almost surely.

$(X_t)_{t\in\mathbb{R}_+}$ is said to be a *submartingale* (*supermartingale*) with respect to $(\mathcal{F}_t)_{t\in\mathbb{R}_+}$ if, in addition to condition 1 and instead of condition 2, we have:

2′. for all $(s,t) \in \mathbb{R}_+ \times \mathbb{R}_+, s < t : E[X_t|\mathcal{F}_s] \geq X_s$ $(E[X_t|\mathcal{F}_s] \leq X_s)$ almost surely.

Remark 1. When the filtration $(\mathcal{F}_t)_{t\in\mathbb{R}_+}$ is not specified, it is understood to be the increasing σ-algebra generated by the random variables of the process $(\sigma(X_s, 0 \leq s \leq t))_{t\in\mathbb{R}_+}$ (suitable extended). In this case we can write $E[X_t|X_r, 0 \leq r \leq s]$, instead of $E[X_t|\mathcal{F}_s]$.

Example 2. The evolution of a gambler's wealth in a game of chance, the latter specified by the sequence of real-valued random variables $(X_n)_{n\in\mathbb{N}}$, can serve as a descriptive example of the above definitions. Suppose that two players flip a coin and the loser pays the winner (who guessed head or tail correctly) the amount α after every round. If $(X_n)_{n\in\mathbb{N}}$ represents the cumulative fortune of player 1, then after n throws he holds

$$X_n = \sum_{i=0}^{n} \Delta_i.$$

The random variables Δ_i (just like every flip of the coin) are independent and take values α and $-\alpha$ with probabilities p and q, respectively. Therefore, we see that

$$E[X_{n+1}|X_0, \ldots, X_n] = E[\Delta_{n+1} + X_n|X_0, \ldots, X_n]$$
$$= X_n + E[\Delta_{n+1}|X_0, \ldots, X_n].$$

Since Δ_{n+1} is independent of every $\sum_{i=0}^{k} \Delta_i, k = 0, \ldots, n$, we obtain

$$E[X_{n+1}|X_0, \ldots, X_n] = X_n + E[\Delta_{n+1}] = X_n + \alpha(p - q).$$

- If the game is fair, then $p = q$ and $(X_n)_{n\in\mathbb{N}}$ is a martingale.
- If the game is in player 1's favor, then $p > q$ and $(X_n)_{n\in\mathbb{N}}$ is a submartingale.
- If the game is to the disadvantage of player 1, then $p < q$ and $(X_n)_{n\in\mathbb{N}}$ is a supermartingale.

Example 3. Let $(X_t)_{t\in\mathbb{R}_+}$ be (for all $t \in \mathbb{R}_+$) a P-integrable stochastic process on $(\Omega, \mathcal{F}, P)$ with independent increments. Then $(X_t - E[X_t])_{t\in\mathbb{R}_+}$ is a martingale. In fact:[1]

$$E[X_t|\mathcal{F}_s] = E[X_t - X_s|\mathcal{F}_s] + E[X_s|\mathcal{F}_s], \qquad s < t,$$

and recalling that both X_s is $\mathcal{F}_s$-measurable and $(X_t - X_s)$ is independent of $\mathcal{F}_s$, we obtain that

$$E[X_t|\mathcal{F}_s] = E[X_t - X_s] + X_s = X_s, \qquad s < t.$$

Proposition 1. [8] *Let $(X_t)_{t\in\mathbb{R}_+}$ be a real-valued martingale. If the function $\phi : \mathbb{R} \to \mathbb{R}$ is both convex and measurable and such that*

$$\forall t \in \mathbb{R}_+, \qquad E[\phi(X_t)] < +\infty,$$

then $(\phi(X_t))_{t\in\mathbb{R}_+}$ is a submartingale.

Proposition 2. [8] *Let $(X_n)_{n\in\mathbb{N}\setminus\{0\}}$ be a sequence of real random variables defined on the probability space $(\Omega, \mathcal{F}, P)$, and X_n^+ the positive part of X_n.*

1. If $(X_n)_{n\in\mathbb{N}\setminus\{0\}}$ is a submartingale, then

$$P\left(\max_{1\leq k\leq n} X_k > \lambda\right) \leq \frac{1}{\lambda}E[X_n^+], \qquad \lambda > 0, n \in \mathbb{N} \setminus \{0\}.$$

2. If $(X_n)_{n\in\mathbb{N}\setminus\{0\}}$ is a martingale and if, for all $n \in \mathbb{N} \setminus \{0\}$, $X \in L^p(P)$, $p > 1$, then

$$E\left[\left(\max_{1\leq k\leq n} |X_k|\right)^p\right] \leq \left(\frac{p}{p-1}\right)^p E[|X_n|^p], \qquad n \in \mathbb{N} \setminus \{0\}.$$

(Points 1 and 2 are called Doob's inequalities.*)*

Corollary 1. [8] *If $(X_n)_{n\in\mathbb{N}\setminus\{0\}}$ is a martingale such that $X_n \in L^p(P)$ for all $n \in \mathbb{N} \setminus \{0\}$, then*

$$P\left(\max_{1\leq k\leq n} |X_k| > \lambda\right) \leq \frac{1}{\lambda^p}E[|X_n|^p], \qquad \lambda > 0.$$

---

[1] For simplicity, but without loss of generality, we will assume that $E[X_t] = 0$, for all t. In the case where $E[X_t] \neq 0$, we can always define a variable $Y_t = X_t - E[X_t]$, so that $E[Y_t] = 0$. In that case $(Y_t)_{t\in\mathbb{R}_+}$ will again be a process with independent increments, so that the analysis is analogous.

Proposition 3. [8] *Let $(X_t)_{t\in\mathbb{R}_+}$ be a stochastic process on $(\Omega, \mathcal{F}, P)$ valued in $\mathbb{R}$.*

1. If $(X_t)_{t\in\mathbb{R}_+}$ is a submartingale, then

$$P\left(\sup_{0\leq s\leq t} X_s > \lambda\right) \leq \frac{1}{\lambda}E[X_t^+], \qquad \lambda > 0, t \geq 0.$$

2. If $(X_t)_{t\in\mathbb{R}_+}$ is a martingale such that, for all $t \geq 0$, $X_t \in L^p(P)$, $p > 1$, then

$$E\left[\sup_{0\leq s\leq t} |X_s|^p\right] \leq \left(\frac{p}{p-1}\right)^p E[|X_t|^p].$$

Definition 10. *A stochastic process $X = (X_n, \mathcal{F}_n)$ is a local martingale if there is a sequence $(\tau_k)_{k\geq 1}$ of stopping times such that $\tau_k \leq \tau_{k+1}$ ($\mathbb{P} - a.s.$), and $\tau_k \to \infty$ ($\mathbb{P} - a.s.$) as $k \to \infty$, and every stopped sequence $X^{\tau_k} = (X_{\min(\tau_k, n)} I_{\{\tau_k \geq 0\}}, \mathcal{F}_n)$ is a martingale.*

2.3 Markov Processes

Let $(X_t)_{t\in\mathbb{R}_+}$ be a stochastic process on a probability space, valued in $(E, \mathcal{B})$ and adapted to the increasing family $(\mathcal{F}_t)_{t\in\mathbb{R}_+}$ of σ-algebras of subsets of $\mathcal{F}$. $(X_t)_{t\in\mathbb{R}_+}$ is a *Markov process* with respect to $(\mathcal{F}_t)_{t\in\mathbb{R}_+}$ if for any $B \in \mathcal{B}$, and for any $(s, t) \in \mathbb{R}_+ \times \mathbb{R}_+, s < t$, the following condition is satisfied:

$$P(X_t \in B|\mathcal{F}_s) = P(X_t \in B|X_s) \qquad \text{a.s.} \tag{2}$$

If, for all $t \in \mathbb{R}_+$, $\mathcal{F}_t = \sigma(X_r, 0 \leq r \leq t)$, then condition (2) becomes

$$P(X_t \in B|X_r, 0 \leq r \leq s) = P(X_t \in B|X_s) \qquad \text{a.s.}$$

for all $B \in \mathcal{B}$, for all $(s, t) \in \mathbb{R}_+ \times \mathbb{R}_+$, and $s < t$.

Condition (2) states that for a Markov process the future depends only on the present and not on the past history.

Theorem 2. *Every real valued stochastic process $(X_t)_{t\in\mathbb{R}_+}$ with independent increments is a Markov process.*

A Markov process $(X_t)_{t\in[t_0, T]}$ on $\mathbb{R}^d$ is well-defined by an *initial distribution* P_0, the distribution of $X(t_0)$ and by a *Markov transition distribution*, that is a non negative function $p(s, x, t, A)$, defined for $0 \leq s < t < \infty, x \in \mathbb{R}, A \in \mathcal{B}_\mathbb{R}$ such that it satisfies the following conditions

1. for all $0 \leq s < t < \infty$, for all $A \in \mathcal{B}_\mathbb{R}$, $p(s, \cdot, t, A)$ is $\mathcal{B}_\mathbb{R}$-measurable;
2. for all $0 \leq s < t < \infty$, for all $x \in \mathbb{R}$, $p(s, x, t, \cdot)$ is a probability measure on $\mathcal{B}_\mathbb{R}$;

3. p satisfies the Chapman–Kolmogorov equation (compatibility):

$$p(s, x, t, A) = \int_{\mathbb{R}} p(s, x, r, dy) p(r, y, t, A) \qquad \forall x \in \mathbb{R}, s < r < t. \qquad (3)$$

Indeed, Theorem 3 holds [1, 14], from which we can deduce that

$$p(s, x, t, A) = P(X_t \in A | X_s = x), \qquad 0 \leq s < t < \infty, x \in \mathbb{R}, A \in \mathcal{B}_{\mathbb{R}}.$$

Theorem 3. *[1, 14] Let E be a Polish space endowed with the σ-algebra $\mathcal{B}_E$ of its Borel sets, P_0 a probability measure on $\mathcal{B}_E$, and $p(r, x, s, A), t_0 \leq r < s \leq T, x \in E, A \in \mathcal{B}_E$ a Markov transition probability function. Then there exists a unique (in the sense of equivalence) Markov process $(X_t)_{t \in [t_0, T]}$ valued in E, with P_0 as its initial distribution and p as its transition probability.*

A Markov process $(X_t)_{t \in [t_0, T]}$ is *homogeneous* if the transition probability functions $p(s, x, t, A)$ depend on t and s only through their difference $t - s$. Therefore, for all $(s, t) \in [t_0, T]^2$, $s < t$, for all $u \in [0, T - t]$, for all $A \in \mathcal{B}_{\mathbb{R}}$, and for all $x \in \mathbb{R}$:

$$p(s, x, t, A) = p(s + u, x, t + u, A) \qquad \text{a.s.}$$

Semigroups Associated with Markov Transition Probability Functions.

To a Markov transition probability function $p(s, x, t, A)$ (or with its corresponding Markov process), one may associate a semigroup of operators $\{T_{s,t}\}_{0 \leq s \leq t \leq T}$ such that for any $0 \leq s < t \leq T$, $T_{s,t} : C_b(\mathbb{R}) \to C_b(\mathbb{R})$, is defined by assigning, for all $f \in C_b(\mathbb{R})$,

$$(T_{s,t} f)(x) = \int_{\mathbb{R}} f(y) p(s, x, t, dy) = E[f(X_t) | X_s = x], \qquad x \in \mathbb{R}, \qquad (4)$$

where $C_b(\mathbb{R})$ is the space of all continuous and bounded functions on $\mathbb{R}$, endowed with the norm $\|f\| = \sup_{x \in \mathbb{R}} |f(x)| (< \infty)$. It is clear that (4) is a semigroup; indeed

i) if $s = t$, then $T_{t,t} = I$ (identity), because

$$p(s, x, s, A) = \begin{cases} 1 \text{ if } x \in A, \\ 0 \text{ if } x \notin A; \end{cases}$$

ii) moreover, $T_{s,t} T_{t,u} = T_{s,u}$, $0 \leq s < t < u$. In fact, if $f \in C_b(\mathbb{R})$ and $x \in \mathbb{R}$,

$$(T_{s,t}(T_{t,u} f))(x)$$
$$= \int_{\mathbb{R}} (T_{t,u} f)(y) p(s, x, t, dy)$$
$$= \int_{\mathbb{R}} \int_{\mathbb{R}^2} f(z) p(t, y, u, dz) p(s, x, t, dy)$$
$$= \int_{\mathbb{R}} f(z) \int_{\mathbb{R}} p(t, y, u, dz) p(s, x, t, dy) \text{ (by Fubini's theorem)}$$
$$= \int_{\mathbb{R}} f(z) p(s, x, u, dz) \text{ (by the Chapman–Kolmogorov equation)}$$
$$= (T_{s,u} f)(x).$$

$\square$

If $(X_t)_{t \in \mathbb{R}_+}$ is a Markov process with transition probability function p and associated semigroup $\{T_{s,t}\}$, then the operator

$$A_s f = \lim_{h \downarrow 0} \frac{T_{s,s+h} f - f}{h}, \qquad s \geq 0, f \in C_b(\mathbb{R})$$

is called the *infinitesimal generator of the Markov process* $(X_t)_{t \geq 0}$. Its domain $\mathcal{D}_{A_s}$ consists of all $f \in C_b(\mathbb{R})$ for which the above limit exists uniformly (and therefore in the norm of $C_b(\mathbb{R})$) [16]. We may observe that

$$(A_s f)(x) = \lim_{h \downarrow 0} \frac{1}{h} \int_{\mathbb{R}} [f(y) - f(x)] p(s, x, s + h, dy).$$

The above results may be extended to more general, possibly uncountable, state spaces [2]. In particular, we will assume that E is a subset of $\mathbb{R}^d$ for $d \in \mathbb{N} \setminus \{0\}$. If we consider the time-homogeneous case, a Markov process $(X_t)_{t \in \mathbb{R}_+}$ on $(E, \mathcal{B}_E)$ will be defined in terms of a transition kernel $p(t, x, B)$ for $t \in \mathbb{R}_+$, $x \in E$, $B \in \mathcal{B}_E$, such that

$$p(h, X_t, B) = P(X_{t+h} \in B | \mathcal{F}_t) \qquad \forall t, h \in \mathbb{R}_+, B \in \mathcal{B}_E,$$

given that $(\mathcal{F}_t)_{t \in \mathbb{R}_+}$ is the natural filtration of the process. Equivalently we may state

$$E[g(X_{t+h}) | \mathcal{F}_t] = \int_E g(y) p(h, X_t, dy) \qquad \forall t, h \in \mathbb{R}_+, g \in C_b(E).$$

In this case the transition semigroup of the process is a one-parameter contraction semigroup $(T(t), t \in \mathbb{R}_+)$ on $C_b(E)$ defined by

$$(T(t)g)(x) := \int_E g(y) p(t, x, dy) = E[g(X_t) | X_0 = x], \qquad x \in E,$$

for any $g \in C_b(E)$. The infinitesimal generator will be time independent. It is defined as

$$\mathcal{A}g = \lim_{t \to 0+} \frac{1}{t} (T(t)g - g)$$

for $g \in \mathcal{D}(\mathcal{A})$, the subset of $C_b(E)$ for which the above limit exists, in $C_b(E)$, with respect to the sup norm. Given the above definitions, it is obvious that for all $g \in \mathcal{D}(\mathcal{A})$,

$$\mathcal{A}g(x) = \lim_{t \to 0+} \frac{1}{t} E[g(X_t) | X_0 = x], \qquad x \in E.$$

If $(T(t), t \in \mathbb{R}_+)$ is the contraction semigroup associated with a Markov process, it is not difficult to show that the mapping $t \to T(t)g$ is right-continuous in $t \in \mathbb{R}_+$ provided that $g \in C_b(E)$ is such that the mapping $t \to T(t)g$ is right continuous in $t = 0$. Then, for all $g \in \mathcal{D}(\mathcal{A})$ and $t \in \mathbb{R}_+$,

$$\int_0^t T(s)g ds \in \mathcal{D}(\mathcal{A})$$

and

$$T(t)g - g = \mathcal{A} \int_0^t T(s)g ds = \int_0^t \mathcal{A}T(s)g ds = \int_0^t T(s)\mathcal{A}g ds$$

by considering Riemann integrals. The following, so-called *Dynkin's formula*, establishes a fundamental link between Markov processes and martingales [8, 15].

Theorem 4. *Assume* $(X_t)_{t \in \mathbb{R}_+}$ *is a Markov process on* $(E, \mathcal{B}_E)$, *with transition kernel* $p(t, x, B)$, $t \in \mathbb{R}_+$, $x \in E$, $B \in \mathcal{B}_E$. *Let* $(T(t), t \in \mathbb{R}_+)$ *denote its transition semigroup and* $\mathcal{A}$ *its infinitesimal generator. Then, for any* $g \in \mathcal{D}(\mathcal{A})$, *the stochastic process*

$$M(t) := g(X_t) - g(X_0) - \int_0^t \mathcal{A}g(X_s)ds$$

is an $\mathcal{F}_t$-*martingale.*

The next proposition shows that a Markov process is indeed characterized by its infinitesimal generator via a martingale problem [8, 15, 32].

Theorem 5. (Martingale problem for Markov processes). *If an RCLL Markov process* $(X_t)_{t \in \mathbb{R}_+}$ *is such that*

$$g(X_t) - g(X_0) - \int_0^t \mathcal{A}g(X_s)ds$$

is an $\mathcal{F}_t$-*martingale for any function* $g \in \mathcal{D}(\mathcal{A})$, *where* $\mathcal{A}$ *is the infinitesimal generator of a contraction semigroup on* E, *then* X_t *is equivalent to a Markov process having* $\mathcal{A}$ *as its infinitesimal generator.*

Example 4. A Poisson process (see the following section for more details) is an integer-valued Markov process $(N_t)_{t \in \mathbb{R}_+}$, a so called "jump process". If its intensity parameter is $\lambda > 0$, the process $(X_t)_{t \in \mathbb{R}_+}$, defined by $X_t = N_t - \lambda t$, is a stationary Markov process with independent increments. The transition kernel of X_t is

$$p(h, x, B) = \sum_{k=0}^{\infty} \frac{(\lambda h)^k}{k!} e^{-\lambda h} I_{\{x+k-\lambda h \in B\}} \text{ for } x \in \mathbb{N}, h \in \mathbb{R}_+, B \subset \mathbb{N}.$$

Its transition semigroup is then

$$T(h)g(x) = \sum_{k=0}^{\infty} \frac{(\lambda h)^k}{k!} e^{-\lambda h} g(x + k - \lambda h) \text{ for } x \in \mathbb{N}, g \in C_b(\mathbb{R}).$$

The infinitesimal generator is then

$$\mathcal{A}g(x) = \lambda(g(x + 1) - g(x)) - \lambda g'(x+).$$

According to previous theorems,

$$M(t) = g(X_t) - \int_0^t ds(\lambda(g(X_s + 1) - g(X_s)) - \lambda g'(X_s+)$$

is a martingale for any $g \in C_b(\mathbb{R})$, such that $g(0) = 0$.

Examples of Markov Processes

Markov Diffusion Processes

A very important class of Markov processes is the one of the *diffusion processes*, that are Markov processes on $\mathbb{R}$ with transition probability function $p(s, x, t, A)$ which satisfies the following properties

1. for all $\epsilon > 0$, for all $t \geq 0$, and for all $x \in \mathbb{R}$

$$\lim_{h \downarrow 0} \frac{1}{h} \int_{|x-y|>\epsilon} p(t, x, t + h, dy) = 0;$$

2. there exist $a(t, x)$ and $b(t, x)$ such that, for all $\epsilon > 0$, for all $t \geq 0$, and for all $x \in \mathbb{R}$,

$$\lim_{h \downarrow 0} \frac{1}{h} \int_{|x-y|<\epsilon} (y - x)p(t, x, t + h, dy) = a(t, x),$$

$$\lim_{h \downarrow 0} \frac{1}{h} \int_{|x-y|<\epsilon} (y - x)^2 p(t, x, t + h, dy) = b(t, x).$$

$a(t, x)$ is the *drift coefficient* and $b(t, x)$ the *diffusion coefficient* of the process.

Proposition 4. *If $(X_t)_{t \in \mathbb{R}_+}$ is a diffusion process with transition probability function p and drift and diffusion coefficients $a(x, t)$ and $b(x, t)$, respectively, and if $\mathcal{A}_s$ is the infinitesimal generator associated with p, then we have that*

$$(\mathcal{A}_s f)(x) = \frac{\partial f}{\partial x} a(s, x) + \frac{1}{2} \frac{\partial^2 f}{\partial x^2} b(s, x), \tag{5}$$

provided that f is bounded and twice continuously differentiable.

Markov Jump Processes

Consider a Markov process $(X_t)_{t \in \mathbb{R}_+}$ valued in a countable set E (say, $\mathbb{N}$ or $\mathbb{Z}$). In such a case it is sufficient (with respect to Theorem 3) to provide the so-called *one-point* transition probability function

$$p_{ij}(s,t) := p(s,i,t,j) := P(X_t = j | X_s = i)$$

for $t_0 \leq s < t$, $i, j \in E$. It follows from the general structure of Markov processes that the one-point transition probabilities satisfy the following relations:

(a) $p_{ij}(s,t) \geq 0$,
(b) $\sum_{j \in E} p_{ij}(s,t) = 1$,
(c) $p_{ij}(s,t) = \sum_{k \in E} p_{ik}(s,r) p_{kj}(r,t)$,

provided $t_0 \leq s \leq r \leq t$, in $\mathbb{R}_+$, and $i, j \in E$. To these three conditions we need to add

(d)

$$\lim_{t \to s+} p_{ij}(s,t) = p_{ij}(s,s) = \delta_{ij} = \begin{cases} 1 & \text{for } i = j, \\ 0 & \text{for } i \neq j. \end{cases}$$

The time-homogeneous case gives transition probabilities $(\tilde{p}_{ij}(t))_{t \in \mathbb{R}_+}$, such that

$$p_{ij}(s,t) = \tilde{p}_{ij}(t - s), \qquad s \leq t.$$

From now on we shall limit our analysis to the time-homogeneous case, whose transition probabilities will be denoted $(p_{ij}(t))_{t \in \mathbb{R}_+}$. We may obtain the following result [8, 17, 32]

Theorem 6. *The limits*

$$\lim_{t \to 0+} \frac{p_{ij}(t)}{t} = p'_{ij}(0) =: q_{ij} < +\infty$$

always exist (finite) for any $i \neq j$. The limits

$$q_{ii} = -\lim_{h \to 0+} \frac{1 - p_{ii}(h)}{h} \leq +\infty$$

always exists (finite or not).

Consider the family of matrices $(P(t))_{t \in \mathbb{R}_+}$, with entries $(p_{ij}(t))_{t \in \mathbb{R}_+}$, for $i, j \in E$. We may rewrite conditions (c) and (d) in matrix form as follows:

(c') $P(s + t) = P(s)P(t)$ for any $s, t \geq 0$;
(d') $\lim_{h \to 0+} P(h) = P(0) = I$.

A family of stochastic matrices fulfilling conditions (c′) and (d′) is called a *matrix transition function*. The matrix $Q = (q_{ij})_{i,j \in E}$ is called the *intensity matrix*. The transition and the intensity matrices satisfy the following system, which is the matrix form of the Kolmogorov backward equations

$$P'(t) = QP(t), \qquad t > 0,$$

subject to

$$P(0) = I.$$

If Q is a finite-dimensional matrix, the function $exp\{tQ\}$ for $t > 0$ is well defined.

Theorem 7. *[8, 20] If E is finite, the matrix transition function can be represented in terms of its intensity matrix Q via*

$$P(t) = e^{tQ}, \qquad t \geq 0.$$

Consider a time-homogeneous Markov jump process on a countable state space E with intensity matrix $Q = (q_{ij})_{i,j \in E}$. The matrix Q can be seen as a functional operator on E as follows: For any $f : E \to \mathbb{R}_+$ define

$$Q : f \to Q(f) = \sum_{j \in E} q_{ij} f(j) = \sum_{j \neq i} q_{ij}(f(j) - f(i)).$$

From [32] we obtain the following theorem.

Theorem 8. *For any function $g \in C^{1,0}(\mathbb{R}_+ \times E)$ such that the mapping*

$$t \to \frac{\partial}{\partial t} g(t, x)$$

is continuous for all $x \in E$, the process

$$\left(g(t, X(t)) - g(0, X(0)) - \int_0^t \left(\frac{\partial g}{\partial t} + Q(g(s, \cdot)) \right)(s, X(s)) ds \right)_{t \in \mathbb{R}_+}$$

is a local martingale.

2.4 Brownian Motion and the Wiener Process

A real-valued process $(W_t)_{t \in \mathbb{R}_+}$ is a *Wiener process* if it satisfies the following conditions:

1. $W_0 = 0$ almost surely;
2. $(W_t)_{t \in \mathbb{R}_+}$ is a process with independent increments;
3. $W_t - W_s$ is normally distributed with $N(0, t - s)$, $(0 \leq s < t)$.

From the definition it turns easily out that the Wiener process is both a Markov process and, since it has independent increments, a martingale. Furthermore, by simple calculations, one can easily show that $E[W_t] = 0$ and $Cov[W_t, W_s] = \min\{s, t\}$, $s, t \in \mathbb{R}_+$.

Now we deal with almost sure properties of the Brownian sample path, in particular its continuity and nowhere differentiability [8].

Theorem 9. *Let $(W_t)_{t \in \mathbb{R}_+}$ be a real-valued Wiener process. Then*

1. *it has continuous trajectories almost surely;*
2. *$P(\sup_{t \in \mathbb{R}_+} W_t = +\infty) = 1$, and $P(\inf_{t \in \mathbb{R}_+} W_t = -\infty) = 1$.*
3. *for all $h > 0$, $P(\max_{0 \leq s \leq h} W_s > 0) = P(\min_{0 \leq s \leq h} W_s < 0) = 1$; moreover, for almost every $\omega \in \Omega$ the process $(W_t)_{t \in \mathbb{R}_+}$ has a zero (i.e., crosses the spatial axis) in $]0, h]$, for all $h > 0$;*
4. *almost every trajectory of $(W_t)_{t \in \mathbb{R}_+}$ is differentiable almost nowhere.*

The real-valued process $(W_1(t), \ldots, W_n(t))'_{t \geq 0}$ is said to be an n- dimensional Wiener process (or Brownian motion) if:

1. for all $i \in \{1, \ldots, n\}$, $(W_i(t))_{t \geq 0}$ is a Wiener process,
2. the processes $(W_i(t))_{t \geq 0}$, $i = 1, \ldots, n$, are independent

(thus the σ-algebras $\sigma(W_i(t), t \geq 0)$, $i = 1, \ldots, n$, are independent).

Proposition 5. *If $(W_1(t), \ldots, W_n(t))'_{t \geq 0}$ is an n-dimensional Brownian motion, then it can be shown that:*

1. *$(W_1(0), \ldots, W_n(0)) = (0, \ldots, 0)$ almost surely;*
2. *$(W_1(t), \ldots, W_n(t))'_{t \geq 0}$ has independent increments;*
3. *$(W_1(t), \ldots, W_n(t))' - (W_1(s), \ldots, W_n(s))', 0 \leq s < t$, has a multivariate normal distribution $N(\mathbf{0}, (t - s)I)$ (where $\mathbf{0}$ is the null-vector of order n and I is the $n \times n$ identity matrix).*

3 Itô Calculus

3.1 The Itô Integral

In classical calculus the primary components were the use of differentiation to describe rates of change, the use of integration and then the fundamental theorem of calculus. From them, the concept of ordinary differential equations came out.

In stochastic calculus, the history turn the other way round: in order to make meaningful the ordinary differential equations involving continuous stochastic processes and in particular the Brownian motion which is nowhere differentiable, it was necessary to introduce first the stochastic integral and later the stochastic differential.

For example, let us consider the classical exponential growth model

$$\frac{dN(t)}{dt} = r(t)N(t)$$

and perturb the growth rate $r(t)$ by a some stochastic noise $\theta(t)$

$$\frac{dN(t)}{dt} = [r_0 + r_1\theta(t)]N(t),$$

so that the variation follows the following equation

$$\frac{dN(t)}{N(t)} = r_0 dt + r_1\theta(t)dt.$$

If we consider a Brownian model, i.e. $\theta(t)dt \simeq dW_t$, such that $\Delta W_t \sim \mathcal{N}(0, \Delta t)$, we get

$$dN_t = r_0 N_t dt + r_1 N_t dW_t.$$

More in general

$$dN(t) = a(t, N(t))dt + b(t, N(t))dW_t, \tag{6}$$

which, however, in the current form does not make sense, because the trajectories of $(W_t)_{t\geq 0}$ are not differentiable. Instead, we will try to interpret it in the form

$$\forall \omega \in \Omega: \qquad u(\omega, t) - u(\omega, 0) = \int_0^t a(s, u(\omega, s))ds + \int_0^t b(s, u(\omega, s))dW_s,$$

which requires us to give meaning to an integral $\int_a^b f(t)dW_t$ that is not of Lebesgue–Stieltjes neither of Riemann–Stieltjes type [8].

Definition 11. Let $(W_t)_{t\geq 0}$ be a Wiener process defined on the probability space $(\Omega, \mathcal{F}, P)$ and $\mathcal{C}$ the set of functions $f(t, \omega) : [a, b] \times \Omega \to \mathbb{R}$ satisfying the following conditions:

1. f is $\mathcal{B}_{[a,b]} \otimes \mathcal{F}$-measurable;
2. for all $t \in [a, b]$, $f(t, \cdot) : \Omega \to \mathbb{R}$ is $\mathcal{F}_t$-measurable, where $\mathcal{F}_t = \sigma(W_s, 0 \leq s \leq t)$;
3. for all $f \in L^2([a, b] \times \Omega)$ and

$$\int_a^b E\left[|f(t)|^2\right]^2 dt < \infty. \tag{7}$$

Condition 2 of Definition 11 stresses the non-anticipatory nature of f through the fact that it only depends on the present and the past history of the Brownian motion, but not on its future.

Let $f \in \mathcal{C}$ and consider a sequence $(\pi_n)_{n \in \mathbb{N}}$ of the partitions $\pi_n : a = t_0^{(n)} < t_1^{(n)} < \cdots < t_n^{(n)} = b$ of the interval $[a, b]$ such that

$$|\pi_n| = \sup_{k \in \{0,\ldots,n\}} \left| t_{k+1}^{(n)} - t_k^{(n)} \right| \xrightarrow{n} 0.$$

If for every $\omega \in \Omega$ $f(t, \cdot)$ is continuous, we define the *(stochastic) Itô integral of the process f* as follows

$$P - \lim_{n \to \infty} \sum_{k=0}^{n-1} f\left(t_k^{(n)}\right) \left(W_{t_{k+1}^{(n)}} - W_{t_k^{(n)}}\right) = \int_a^b f(t) dW_t. \tag{8}$$

For a more general definition (the case f non continuous everywhere), please refer to [8].

Note that for the classical Lebesgue integral results in $\int_a^b t \, dt = \frac{b^2 - a^2}{2}$. In the stochastic Itô integral we obtain $\int_a^b W_t dW_t = \frac{1}{2}(W_b^2 - W_a^2) - \frac{b-a}{2}$, i.e. we have an additional term $\left(-\frac{b-a}{2}\right)$.

An important property of the stochastic integrals is the martingality. Indeed one can prove [8] that $(X_t = \int_a^t f(s) dW_s)_{t \in [a,b]}$ is a zero mean $\mathcal{L}^2$-martingale with respect to $\mathcal{F}_t = \sigma(W_s, 0 \le s \le t)$.

The *martingale representation theorem* establishes the relationship between a martingale and the existence of a process, vice versa. Let $(M_t)_{t \in [0,T]}$ be an $\mathcal{L}^2$ martingale with respect to the Wiener process $(W_t)_{t \in [0,T]}$ and $(\mathcal{F}_t)_{t \in [0,T]}$ its natural filtration. Then there exists a unique process $(f_t)_{t \in [0,T]} \in \mathcal{C}([0,T])$ so that

$$\forall t \in [0,T], \qquad M(t) = M(0) + \int_0^t f(s) dW_s \qquad \text{a.s.} \tag{9}$$

holds.

Let $\mathcal{C}_1$ be the set of functions $f : [a, b] \times \Omega \to \mathbb{R}$ such that the conditions 1 and 2 of the characterization of the class $\mathcal{C}$ are satisfied, but, instead of condition 3, we have

$$P\left(\int_a^b |f(t)|^2 dt < \infty\right) = 1. \tag{10}$$

It is obvious that $\mathcal{C} \subset \mathcal{C}_1$ and thus $\mathcal{S} \subset \mathcal{C}_1$. It is possible to define the stochastic integral for a function $f \in \mathcal{C}_1$ as in (8).

3.2 The Stochastic Differential

Let $(u(t))_{0 \le t \le T}$ be a process such that for every $(t_1, t_2) \in [0,T] \times [0,T], t_1 < t_2$:

$$u(t_2) - u(t_1) = \int_{t_1}^{t_2} a(t) dt + \int_{t_1}^{t_2} b(t) dW_t, \tag{11}$$

where $a \in \mathcal{C}_1([0,T])$ and $b \in \mathcal{C}_1([0,T])$. Then $u(t)$ is said to have the *stochastic differential*

$$du(t) = a(t)dt + b(t)dW_t \tag{12}$$

on $[0,T]$. Hence

1. the trajectories of $(u(t))_{0 \leq t \leq T}$ are continuous almost everywhere;
2. for $t \in [0,T]$, $u(t)$ is $\mathcal{F}_t = \sigma(W_s, 0 \leq s \leq t)$-measurable, thus $u(t) \in \mathcal{C}_1([0,T])$.

Itô's Formula

As one of the most important topics on Brownian motion, Itô's formula represents the stochastic equivalent of Taylor's theorem about the expansion of functions. It is the key concept that connects classical and stochastic theory.

If $du(t) = a(t)dt + b(t)dW_t$ and if $f(t,x) : [0,T] \times \mathbb{R} \to \mathbb{R}$ is continuous with the derivatives f_x, f_{xx}, and f_t, then the stochastic differential of the process $f(t,u(t))$ is given by

$$df(t,u(t)) = \left(f_t(t,u(t)) + \frac{1}{2}f_{xx}(t,u(t))b^2(t) + f_x(t,u(t))a(t) \right) dt$$
$$+ f_x(t,u(t))b(t)dW_t. \tag{13}$$

Examples:

1. $dW_t^2 = dt + 2W_t dW_t$;
2. $d(tW_t) = W_t dt + t dW_t$;
3. If $f \in C^2(\mathbb{R})$, then $df(W_t) = f'(W_t)dW_t + \frac{1}{2}f''(W_t)dt$.
4. If $u(t,x) : [0,T] \times \mathbb{R} \to \mathbb{R}$ is continuous with the derivatives u_x, u_{xx}, and u_t, then

$$du(t,W_t) = \left(u_t(t,W_t) + \frac{1}{2}u_{xx}(t,W_t) \right) dt + u_x(t,W_t)dW_t. \tag{14}$$

3.3 Stochastic Differential Equations

Let $(W_t)_{t \in \mathbb{R}_+}$ be a Wiener process on the probability space $(\Omega, \mathcal{F}, P)$, equipped with the filtration $(\mathcal{F}_t)_{t \in \mathbb{R}_+}$, $\mathcal{F}_t = \sigma(W_s, 0 \leq s \leq t)$. Furthermore, let $a(t,x)$, $b(t,x)$ be measurable functions in $[0,T] \times \mathbb{R}$ and $(u(t))_{t \in [0,T]}$ a stochastic process. Now $u(t)$ is said to be the *solution of the stochastic differential equation*

$$du(t) = a(t,u(t))dt + b(t,u(t))dW_t, \tag{15}$$

with the initial condition

$$u(0) = u^0 \text{ a.s. } (u^0 \text{ a random variable}), \tag{16}$$

if

1. $u(0)$ is $\mathcal{F}_0$-measurable;
2. $|a(t, u(t))|^{\frac{1}{2}}, b(t, u(t)) \in C_1([0, T])$;
3. $u(t)$ is differentiable and $du(t) = a(t, u(t))dt + b(t, u(t))dW_t$,
   thus $u(t) = u(0) + \int_0^t a(s, u(s))ds + \int_0^t b(s, u(s))dW_s$, $t \in]0, T]$.

Theorem 10 (Existence and Uniqueness of the Solution). *Resorting to the notation of the preceding definition, if the following conditions are satisfied:*

1. *for all $t \in [0, T]$ and all $(x, y) \in \mathbb{R} \times \mathbb{R}$: $|a(t, x) - a(t, y)| + |b(t, x) - b(t, y)| \leq K^* |x - y|$;*
2. *for all $t \in [0, T]$ and all $x \in \mathbb{R}$: $|a(t, x)| \leq K(1 + |x|), |b(t, x)| \leq K(1 + |x|)$ (K^*, K constants);*
3. *$E[|u^0|^2] < \infty$;*
4. *u^0 is independent of $\mathcal{F}_T$ (which is equivalent to requiring u^0 to be $\mathcal{F}_0$-measurable),*

then there exists a unique $(u(t))_{t \in [0,T]}$, solution of (15), (16), such that

a. *$(u(t))_{t \in [0,T]}$ is continuous almost surely (thus almost every trajectory is continuous);*
b. *$(u(t))_{t \in [0,T]} \in C([0, T])$.*

Remark 2. If $(u_1(t))_{t \in [0,T]}$ and $(u_2(t))_{t \in [0,T]}$ are two solutions of (15), (16), belonging to $C([0, T])$, then the uniqueness of a solution is understood in the sense that

$$P \left(\sup_{0 \leq t \leq T} |u_1(t) - u_2(t)| = 0 \right) = 1.$$

Examples:

1.
$$\begin{cases} u_0(t) = u^0, \\ du(t) = g(t)u(t)dW_t \end{cases}$$

has the solution

$$u(t) = u^0 \exp \left\{ \int_0^t g(s)dW_s - \frac{1}{2} \int_0^t g^2(s)ds \right\}.$$

2. geometric Brownian motion

$$\begin{cases} u_0(t) = u^0, \\ du(t) = au(t)dt + bu(t)dW_t; \end{cases}$$

has the solution

$$u(t) = u^0 \exp \left\{ \left(a - \frac{1}{2}b^2 \right)(t - t_0) + b(W_t - W_{t_0}) \right\}.$$

3. (mean-reverting) Ornstein–Uhlenbeck process

$$\begin{cases} u_0(t) = u^0, \\ du(t) = (a - bu(t))dt + cdW_t. \end{cases}$$

has the solution

$$u(t) = \frac{a}{b} \exp\{bt_0\} + u^0 \exp\{-b(t - t_0)\} + c \int_{t_0}^t \exp\{-b(t - s)\}dW_s.$$

Property of Solutions

Now, let $t_0 \geq 0$ and c be a random variable with $u(t_0) = c$ almost surely and, moreover, c be independent of $\mathcal{F}_{t_0,T} = \sigma(W_t - W_{t_0}, t_0 \leq t \leq T)$ as well as $E[c^2] < +\infty$. Under Conditions 1 and 2 of Theorem 10 the following properties are satisfied by the unique solution $(u(t))_{t\in[t_0,T]}$ of (15)-(16).

1. *Markov Property:* The solution $(u(t))_{t\in[t_0,T]}$ is a Markov Process. Moreover its transition probability p is such that, for all $B \in \mathcal{B}_{\mathbb{R}}$ and all $t_0 \leq s < t \leq T$ and all $x \in \mathbb{R}$

$$p(s, x, t, B) = P(u(t) \in B | u(s) = x) = P(u(t, s, x) \in B),$$

   where $\{u(t, s, x), t \geq s\}$ is the unique solution of (15) subject to the initial condition $u(s) = x$.
   So what we know is that every stochastic differential equation generates Markov processes in the sense that every solution is a Markov process.

2. *Diffusion Property:* If $a(t, x)$ and $b(t, x)$ are continuous functions in $(t, x) \in [0, \infty] \times \mathbb{R}$, then the solution $u(t)$ is a diffusion process with drift coefficient $a(t, x)$ and diffusion coefficient $b^2(t, x)$.

3.4 Kolmogorov and Fokker-Planck Equations

Let $u(s, t, x), t < s$ be the solution of the following stochastic differential equation

$$du(t) = a(t, u(t))dt + b(t, u(t))dW_t \tag{17}$$

$$u(s, s, x) = x \text{ a.s. } (x \in \mathbb{R}) \tag{18}$$

and suppose that the coefficients a and b satisfy the assumptions of existence and uniqueness of Theorem 10. If $f : \mathbb{R} \to \mathbb{R}$ is a twice continuously differentiable function and if there exist $C > 0$ and $m > 0$ such that

$$|f(x)| + |f'(x)| + |f''(x)| \leq C(1 + |x|^m), \qquad x \in \mathbb{R},$$

then the function

$$q(t, x) \equiv E[f(u(s, t, x))], \qquad 0 < t < s, \qquad x \in \mathbb{R}, s \in (0, T), \tag{19}$$

satisfies the equation

$$\frac{\partial}{\partial t}q(t,x) + a(t,x)\frac{\partial}{\partial x}q(t,x) + \frac{1}{2}b^2(t,x)\frac{\partial^2}{\partial x^2}q(t,x) = 0, \tag{20}$$

with the boundary condition

$$\lim_{t\uparrow s}q(t,x) = f(x). \tag{21}$$

Equation (20) is called *backward Kolmogorov's differential equation*.

From the above we can give the solution of a Cauchy problem via an average of realizations of a Markov process.

Proposition 6. *Consider the Cauchy problem:*

$$\begin{cases} L_0[q] + \frac{\partial q}{\partial t} = 0 & in\ [0,T) \times \mathbb{R}, \\ \lim\limits_{t\uparrow T}q(t,x) = \phi(x)\ in\ \mathbb{R}, \end{cases} \tag{22}$$

where $L_0[\cdot] = \frac{1}{2}b^2(t,x)\frac{\partial^2}{\partial x^2} + a(t,x)\frac{\partial}{\partial x}$, and suppose that
(A_1) *$a(t,x)$ is strictly positive for all $(t,x) \in [0,T] \times \mathbb{R}$;*
(B_1) *$\phi(x)$ is continuous in $\mathbb{R}$, and $A > 0, a > 0$ exist such that $|\phi(x)| \le A(1+|x|^a)$;*
(B_2) *a and b are bounded in $[0,T] \times \mathbb{R}$ and uniformly Lipschitz in (t,x) on compact subsets of $[0,T] \times \mathbb{R}$;*
(B_3) *b is Hölder continuous in x and uniform with respect to (t,x) on $[0,T]\times\mathbb{R}$;*
then the Cauchy problem (22) admits a unique solution $q(t,x)$ in $[0,T] \times \mathbb{R}$ such that

$$q(t,x) = E[\phi(u(T,t,x))], \tag{23}$$

where $u(T,t,x)$ is a solution of (17) at time T, with initial location in x at time t.

Something interesting is that under the conditions (A_1) and (B_1), the transition probability $p(s,x,t,A) = P(u(t,s,x) \in A)$ of the Markov process $u(t,s,x)$ (the solution of the differential equation (17)) is endowed with density $f(x,s;y,t)$, i.e.

$$p(s,x,t,A) = \int_A f(x,s;y,t)dy \qquad (s < t),\ \text{for all } A \in \mathcal{B}_\mathbb{R}. \tag{24}$$

The density, called *transition density* of the solution $u(t)$ is the solution of the backward Kolmogorov's equation

$$\begin{cases} L_0[f] + \frac{\partial}{\partial t}f = 0, \\ \lim\limits_{t\to T}f(x,s;y,T) = \delta(x-y). \end{cases} \tag{25}$$

If one requests further regularity on the transition density, i.e. there exist continuous derivatives

$$\frac{\partial f}{\partial t}(s, x, t, y), \qquad \frac{\partial}{\partial y}(a(t, y)f(s, x, t, y)), \qquad \frac{\partial^2}{\partial y^2}(b(t, y)f(s, x, t, y)),$$

then $f(s, x, t, y)$, as a function of t and y, satisfies the equation

$$\frac{\partial f}{\partial t}(s, x, t, y) + \frac{\partial}{\partial y}(a(t, y)f(s, x, t, y)) - \frac{\partial^2}{\partial y^2}(b(t, y)f(s, x, t, y)) = 0 \qquad (26)$$

in the region $t \in (s, T]$, $y \in \mathbb{R}$. Equation (26) is known as the *forward Kolmogorov's equation* or *Fokker–Planck equation*. It is worth pointing out that while the forward equation has a more intuitive interpretation than the backward equation, the regularity conditions on the functions a and b are more stringent than those needed in the backward case. The problem of existence and uniqueness of the solution of the Fokker–Planck equation is not of an elementary nature, especially in presence of boundary conditions. This suggests that the backward approach is more convenient than the forward approach from the viewpoint of analysis.

3.5 The Multidimensional Case

If we have n independent Wiener Processes $(W_j)_{1 \le j \le n}$, we may define the Itô integral for an $f : [a, b] \times \Omega \to \mathbb{R}^{m \times n}$, with each component $f_{ij} \in \mathcal{C}_{1, W_j}$, as follows

$$\int_a^b f(t)dW_t = \left[\sum_{j=1}^n \int_a^b f_{ij}(t)dW_j(t) \right]_{1 \le i \le m}.$$

In this case the stochastic differential for an m-dimensional process $(\mathbf{u}_t)_{0 \le t \le T}$ is

$$du_i(t) = a_i(t)dt + \sum_{j=1}^n (b_{ij}(t)dW_j(t)), \quad i = 1, \dots, m,$$

or in vector form

$$d\mathbf{u}(t) = \mathbf{a}(t)dt + b(t)d\mathbf{W}(t), \qquad (27)$$

with

$$\mathbf{a} : [0, T] \times \Omega \to \mathbb{R}^m, \mathbf{a} \in \mathcal{C}_{1\mathbf{W}}([0, T]),$$
$$b : [0, T] \times \Omega \to \mathbb{R}^{mn}, b \in \mathcal{C}_{1\mathbf{W}}([0, T]).$$

If $f(t, \mathbf{x}) : \mathbb{R}_+ \times \mathbb{R}^m \to \mathbb{R}$ is a continuous function with the derivatives $f_{x_i}, f_{x_i x_j}$, and $\mathbf{u}(t)$ is an m-dimensional process, endowed with the stochastic differential (27), then $f(t, \mathbf{u}(t))$ has the stochastic differential

$$df(t, \mathbf{u}(t)) = Lf(t, \mathbf{u}(t))dt + \nabla_{\mathbf{x}} f(t, \mathbf{u}(t)) \cdot b(t)d\mathbf{W}(t), \qquad (28)$$

where $\nabla_{\mathbf{x}} f(t, \mathbf{u}(t)) \cdot b(t) d\mathbf{W}(t)$ is the scalar product of two m-dimensional vectors, $a_{ij} = (bb')_{ij}$, $i, j = 1, \ldots, m$:

$$L = \frac{1}{2} \sum_{i,j=1}^{m} a_{ij} \frac{\partial^2}{\partial x_i \partial x_j} + \sum_{i=1}^{m} a_i \frac{\partial}{\partial x_i} + \frac{\partial}{\partial t}$$

and $\nabla_{\mathbf{x}}$ is the gradient operator. Furthermore let $\mathbf{a}(t, \mathbf{x}) = (a_1(t, \mathbf{x}), \ldots, a_m(t, \mathbf{x}))'$ and $b(t, \mathbf{x}) = (b_{ij}(t, \mathbf{x}))_{i=1,\ldots,m, j=1,\ldots,n}$ be measurable functions with respect to $(t, \mathbf{x}) \in [0, T] \times \mathbb{R}^n$. An m-*dimensional stochastic differential equation* is of the form

$$d\mathbf{u}(t) = \mathbf{a}(t, \mathbf{u}(t))dt + b(t, \mathbf{u}(t))d\mathbf{W}(t), \tag{29}$$

with the initial condition

$$\mathbf{u}(0) = \mathbf{u}^0 \text{ a.s.}, \tag{30}$$

where $\mathbf{u}^0$ is a fixed m-dimensional random vector. The entire theory of the one-dimensional case translates to the multidimensional case.

4 Deterministic Approximation of Stochastic Systems

4.1 Continuous Approximation of Jump Population Processes

In Section 2.3 we have shown that a jump Markov process can be described in terms of an intensity matrix Q, made of the rates of the relevant transitions.

For a wide class of epidemic models the total population $(N_t)_{t \in \mathbb{R}_+}$ includes three subclasses: the class of susceptibles $(S_t)_{t \in \mathbb{R}_+}$, the infectives $(I_t)_{t \in \mathbb{R}_+}$, and the class of removals $(R_t)_{t \in \mathbb{R}_+}$. For a constant total population size N it is sufficient to provide a model only for the bivariate jump Markov process $(S_t, I_t)_{t \in \mathbb{R}_+}$, which is valued in $E = \mathbb{N}^2$.

For a typical model (general stochastic epidemic) the relevant elements of the intensity matrix Q are given by

- $q_{(s,i),(s,i-1)} = \delta i = \delta N \dfrac{i}{N}$, removal rate of infectives;

- $q_{(s,i),(s-1,i+1)} = \dfrac{\kappa}{N} si = N \kappa \dfrac{s}{N} \dfrac{i}{N}$, infection rate of susceptibles,

thus taking into account a rescaling with respect to the size of the total population [7].

Both transition rates are of the form $q_{k,k+l}^{(N)} = N \beta_l \left(\dfrac{k}{N} \right)$, for $k = (s, i)$, and $k + l = (s, i - 1)$, or $(s - 1, i + 1)$.

It can be shown [8,15] that we can write the evolution equation for $\hat{X}^{(N)} = (s(t), i(t))'$ as follows

$$\hat{X}^{(N)}(t) = \hat{X}^{(N)}(0) + \sum_{l \in \mathbb{Z}^d} l Y_l \left(N \int_0^t \beta_l \left(\frac{\hat{X}^{(N)}(s)}{N} \right) ds \right),$$

where the Y_l are independent standard Poisson processes.

We may rescale the process itself $\hat{X}^{(N)}(t)$ by setting

$$X_N = \frac{1}{N} \hat{X}^{(N)},$$

for which we have

$$X_N(t) = X_N(0) + \frac{1}{N} \sum_{l \in \mathbb{Z}^d} l Y_l \left(N \int_0^t \beta_l (X_N(s)) ds \right),$$

Let now

$$\tilde{Y}_l(u) = Y_l(u) - u$$
$$F(x) = \sum_{l \in \mathbb{Z}^d} l \beta_l(x), \qquad x \in \mathbb{R}^d,$$

so that

$$X_N(t) = X_N(0) + \int_0^t F(X_N(s)) ds$$
$$+ \frac{1}{N} \sum_{l \in \mathbb{Z}^d} l \tilde{Y}_l \left(N \int_0^t \beta_l (X_N(s)) ds \right). \tag{31}$$

By Doob's inequality for martingales, we get a uniform strong law of large numbers, i.e.

$$\lim_{N \to \infty} \sup_{u \le v} \left| \frac{1}{N} \tilde{Y}_l(Nu) \right| = 0, \qquad \text{a.s.}$$

for any $v \ge 0$. This is the fundamental reason why the limit evolution for large N tends to a deterministic one. Indeed, the following theorem holds ([15] page 456).

Theorem 11. *Suppose that for each compact $K \subset E$,*

$$\sum_{l \in \mathbb{Z}^d} |l| \sup_{x \in K} \beta_l(x) < +\infty,$$

and there exists an $M_K > 0$ such that

$$|F(x) - F(y)| \le M_K |x - y|, \qquad x, y \in K;$$

suppose X_N satisfies equation (31) above, with $\lim_{N \to \infty} X^{(N)}(0) = x_0 \in \mathbb{R}^d$.

Then, for every $t \geq 0$,

$$\lim_{N \to \infty} \sup_{s \leq t} |X_N(s) - x(s)| = 0 \qquad a.s.,$$

where $x(t)$, $t \in \mathbb{R}_+$ is the unique solution of

$$x(t) = x_0 + \int_0^t F(x(s))ds, \qquad t \geq 0,$$

wherever it exists.

In differential form

$$\frac{dx}{dt}(t) = F(x(t)),$$

subject to the initial condition

$$x(0) = x_0.$$

4.2 Continuous Approximation of Stochastic Interacting Particle Systems

Suppose a population is composed of $N \in \mathbb{N} \setminus \{0\}$ individuals; the random location of the k-th individual out of N be described by a stochastic process $\{X_N^k(t)\}_{t \in \mathbb{R}_+}$ defined on a suitable probability space $(\Omega, \mathcal{F}, P)$ and valued in $(\mathbb{R}^d, \mathcal{B}_{\mathbb{R}^d})$. In another way, it may be modelled as a random Dirac-measure $\epsilon_{X_N^k(t)} \in \mathcal{M}_P(\mathbb{R}^d)$, defined as follows

$$\epsilon_{X_N^k(t)}(B) = \begin{cases} 1 & \text{if } X_N^k(t) \in B \\ 0 & \text{if } X_N^k(t) \notin B \end{cases} \qquad \forall B \in \mathcal{B}_{\mathbb{R}^d}.$$

For any sufficiently smooth $f : \mathbb{R}^d \to \mathbb{R}$ the Dirac measure is such that

$$\int_{\mathbb{R}^d} f(y) \epsilon_{X_N^k(t)}(dy) = f\left(X_N^k(t)\right).$$

Correspondingly, a global description of the spatial distribution of the system of N particles at time t may be given in terms of the random measure on $\mathbb{R}^d$

$$X_N(t) = \frac{1}{N} \sum_{k=1}^{N} \epsilon_{X_N^k(t)} \in \mathcal{M}(\mathbb{R}^d), \tag{32}$$

which is known as the *empirical measure*, while the process $X_N = \{X_N(t)\}_{t \in \mathbb{R}_+}$ is called the *empirical process*.

The social evolution of the system of N individuals may be expressed via the evolution equation of each individual $k = 1, \ldots, N$,

$$dX_N^k(t) = F_N[X_N(t)](X_N^k(t)) \, dt + \sigma_N\left(X_N(t)\right) \, dW^k(t). \tag{33}$$

This is an Itô type SDE in which both the drift F_N and the diffusion coefficient σ_N^2 may depend upon the empirical measure $X_N(t)$, i.e. upon the spatial distribution of the N particles. In this way the individual behavior depends upon the distribution of the system. We suppose that W^k, $k = 1, \ldots, N$, is a family of independent standard Wiener processes which are responsible of the stochastic fluctuations in the motion of each individual. The modeller will express the specific case by introducing a suitable mathematical model for F_N and σ_N^2 in terms of X_N. In order to avoid further technical difficulties, we will consider $\sigma_N^2(X_N(t)) = \sigma_N^2$, constant in time.

An evolution equation for the empirical process $(X_N(t))_{t \in \mathbb{R}_+}$ can be obtained thanks to a nice application of the Itô's formula. Given a regular function $f \in C_b^{2,1}(\mathbb{R}^d \times \mathbb{R}_+)$, we have

$$
\begin{aligned}
f\left(X_N^k(t), t\right) = f\left(X_N^k(0), 0\right) + \int_0^t F_N[X_N(s)]\left(X_N^k(s)\right) \nabla f\left(X_N^k(s), s\right) ds \\
+ \int_0^t \left[\frac{\partial}{\partial s} f\left(X_N^k(s), s\right) + \frac{\sigma_N^2}{2} \triangle f\left(X_N^k(s), s\right)\right] ds \\
+ \sigma_N \int_0^t \nabla f\left(X_N^k(s), s\right) dW^k(s).
\end{aligned}
\tag{34}
$$

Correspondingly, for the empirical process $(X_N(t))_{t \in \mathbb{R}_+}$, we get the following weak formulation of its evolution equation. For any $f \in C_b^{2,1}(\mathbb{R}^d \times \mathbb{R}_+)$ we have

$$
\begin{aligned}
\langle X_N(t), f(\cdot, t)\rangle = \langle X_N(0), f(\cdot, 0)\rangle + \int_0^t \langle X_N(s), F_N[X_N(s)](\cdot)\nabla f(\cdot, s)\rangle ds \\
+ \int_0^t \left\langle X_N(s), \frac{\sigma_N^2}{2}\triangle f(\cdot, s) + \frac{\partial}{\partial s} f(\cdot, s)\right\rangle ds \\
+ \frac{\sigma_N}{N} \int_0^t \sum_k \nabla f\left(X_N^k(s), s\right) dW^k(s).
\end{aligned}
\tag{35}
$$

In the previous expressions, we have used the notation $\langle \mu, f\rangle = \int f(x)\mu(dx)$, for any measure μ on $(\mathbb{R}^d, \mathcal{B}_{\mathbb{R}^d})$ and any (sufficiently smooth) function $f : \mathbb{R}^d \to \mathbb{R}$. The last term

$$
M_N(f, t) = \frac{\sigma_N}{N} \int_0^t \sum_k \nabla f(X_N^k(s), s) dW^k(s);
$$

is the only explicit source of stochasticity in the equation. It is a martingale with respect to the natural filtration of the process $\{X_N(t), t \in \mathbb{R}_+\}$.

Equation (35) shows how, when the number of particles N is large but still finite, also from the Eulerian point of view the system keeps the stochasticity which characterizes each individual. This is not true anymore when the size of the system tends to infinity. The main reason is that the martingale term $M_N(f,t)$ vanishes in probability. Indeed we may apply Doob's inequality to obtain

$$E\left[\sup_{t\leq T}|M_N(f,t)|\right]^2 \leq E\left[\sup_{t\leq T}|M_N(f,t)|^2\right] \leq 4E\left[|M_N(f,T)|^2\right]$$

$$\leq \frac{4\sigma_N^2}{N^2}\sum_{k=1}^{N} E\left[\int_0^T |\nabla f(X_N^k(s),s)|^2 ds\right]$$

$$\leq \frac{4\sigma_N^2\|\nabla f\|_\infty^2 T}{N}. \tag{36}$$

Hence for the zero-mean martingale $M_N(f,t)$, the quadratic variation (36) vanishes in the limit $N\to\infty$. This implies convergence to zero in probability. This is the substantial reason of the deterministic limiting behavior of the process, as $N\to\infty$, since in this limit the evolution equation of the process will not contain the Brownian noise anymore.

4.3 Convergence of the Empirical Measure

The problem we want to deal with now is the convergence, for N tending to infinity, of the empirical process $X_N(t)\in\mathcal{M}_P(\mathbb{R}^d), t\geq 0$. So we need to introduce a concept of convergence in suitable spaces of probability measures.

Weak Convergence on a Metric Space (S,d)

Let (S,d) be a metric space, and let $\mathcal{S}$ be the σ-algebra of Borel subsets generated by the topology induced by d. Let $P, P_1, P_2, \ldots$ be probability measures on $(S,\mathcal{S})$.

By definition, a sequence of probability measures $\{P_n\}_{n\in\mathbb{N}}$ converges weakly to the probability measure P (notation $P_n \overset{w}{\to} P$) if

$$\int_E f dP_n \to \int_E f dP$$

for every function $f\in C_b(S)$.

With respect to random variables we may state the following: a sequence (X_n) of random variables with values in the common measurable space $(S,\mathcal{S})$ converges in distribution to the random variable X,

$$X_n \overset{\mathcal{D}}{\to} X$$

if the probability laws $P_n = \mathcal{L}(X_n)$ of the X_n's weakly converge to the probability law $P = \mathcal{L}(X)$ of X

$$\mathcal{L}(X_n) \overset{w}{\to} \mathcal{L}(X).$$

From the weak convergence of probability measures one can deduce also the almost sure convergence of particular random variables, as stated in the following important theorem [3].

Theorem 12 (Skorohod representation theorem). *Consider a sequence* $(P_n)_{n\in\mathbb{N}}$ *of probability measures and a probability measure* P *on a separable metric space* $(S, \mathcal{S})$, *such that* $P_n \xrightarrow[n\to\infty]{w} P$. *Then there exists a sequence of* $S-valued$ *random variables* $(Y_n)_{n\in\mathbb{N}}$ *and an* $S-valued$ *random variable* Y *defined on a common (suitably extended) measurable space* $(\Omega, \mathcal{F})$, *such that* Y_n *has probability law* P_n, Y *has probability law* P, *and*

$$Y_n \xrightarrow[n\to\infty]{a.s.} Y.$$

Metrics on $\mathcal{M}_P(S)$ *induced by d*

Let us define the following distance on $\mathcal{M}_P(S)$

$$d_P(Q, P) = \inf\{\epsilon > 0 : Q(A) \le P(A^\epsilon) + \epsilon, P(A) \le Q(A^\epsilon) + \epsilon, \forall A \in \mathcal{S}\}, \quad (37)$$

where $A^\epsilon = \{d(x, A) < \epsilon\}$. The function d_P is a metric on $\mathcal{M}_P(S)$ (induced by d), called *Prokhorov's metric*.

Instead of Prokhorov's metric, in the next sections we consider the *bounded Lipschitz metric* proposed by Dudley [11], as it is easier to work with.

Denote by $Lip_b(S)$ the set of bounded and Lipschitz real function on S. For any $P, Q \in \mathcal{M}(S)$, define the *bounded Lipschitz metric* as follows

$$d_{BL}(Q, P) = \sup\left\{\left|\int f dP - \int f dQ\right| : f \in Lip_b(S), \|f\|_{Lip} \le 1\right\}, \quad (38)$$

where

$$\|f\|_{Lip} = \|f\|_\infty + \sup_{x\ne} \frac{f(x) - f(y)}{d(x, y)}, f \in Lip_b(S).$$

Recall that the metric space (S, d) is called *separable* if it has a countable dense subset, that is, there are $\{x_1, x_2, \ldots, \}$ in S such that its closure is equal to S, $\overline{\{x_1, x_2, \ldots, \}} = S$.

Both d_P-convergence and d_{BL}-convergence imply weak convergence of probability measures [12, 34]. In the case the metric space is separable one can get an equivalence.

Furthermore, if (S, d) is a separable (and complete) metric space, then so is $\mathcal{M}_P(S)$ with the induced either Prokhorov metric and bounded Lipschitz metric. Moreover in the case of S separable, a sequence in $\mathcal{M}_P(S)$ converges in the Prokhorov metric (BL-metric) if and only if it converges weakly and then in both senses to the same limit [12, 34].

Compactness Properties

In the study of the limit behavior of stochastic processes one often needs to know when a sequence of random variables is convergent in distribution or, at least, admits a subsequence that converges in distribution. This requires a good description of the sequences in $\mathcal{M}_P(S)$ that admit a convergent subsequence, i.e. of the relatively compact sets of $\mathcal{M}_P(S)$. Recall that a subset A of a metric space is called relatively compact if its closure $\bar{A}$ is compact.

First let us recall some important definitions.

A probability measure $P \in \mathcal{M}_P(S)$ is called *tight* if for every $\epsilon > 0$ there exists a compact set $K \subset S$ such that $P(X \setminus K) < \epsilon$. Of course, if (S, d) is a compact metric space, then every measure on S is tight. If (S, d) is a complete separable metric space (usually called a Polish space), then every probability measure on S is tight.

We may extend the property of tightness to a family of probability measures. A family Π of probability measures on the general metric space $(S, \mathcal{S})$ is said to be *(uniformly) tight* if, for all $\epsilon > 0$, there exists a compact set K such that

$$P(K) > 1 - \epsilon \qquad \forall P \in \Pi.$$

Π is said to be *relatively compact* if every sequence of elements of Π contains a weakly convergent subsequence; i.e., for every sequence $\{P_n\}$ in Π there exists a subsequence $\{P_{n_k}\}$ and a probability measure P (defined on $(S, \mathcal{S})$, but not necessarily an element of Π) such that $P_{n_k} \xrightarrow{w} P$.

The following theorem by Prokhorov offers a useful equivalence of the relatively compactness in $\mathcal{M}_P(S)$, in case S is separable and complete.

Theorem 13. (Prohorov) *[3] Let Π be a family of probability measures on the probability space $(S, \mathcal{S})$. Then*

1. *if Π is tight, then it is relatively compact;*
2. *let S be separable and complete; if Π is relatively compact, then it is tight.*

One can characterize the weak convergence of a relatively compact sequence of probability measures [3, 15].

Theorem 14. *In a metric space $(S, \mathcal{S})$, let $\{P_n\}$ be a relatively compact sequence of probability measures, and P an additional probability measure on $\mathcal{S}$. Then the following propositions are equivalent*

a) $P_n \Rightarrow P$;
b) *all weakly converging subsequences of $\{P_n\}$ weakly converge to P.*

Corollary 2. *In a Polish (complete and separable) space $(S, \mathcal{S})$, let $\{P_n\}$ be a relatively compact (tight) sequence of probability measures, and P an additional probability measure on $\mathcal{S}$. Then the following propositions are equivalent*

a) $P_n \Rightarrow P$;
b) *all finite dimensional subsequences of $\{P_n\}$ weakly converge to P.*

The Space of Trajectories of the Empirical Process X_N

Let $(S, \mathcal{S}) = (\mathbb{R}^d, \mathcal{B}_{\mathbb{R}^d})$; given $X_N(t) \in \mathcal{M}_P(\mathbb{R}^d)$ by (32), the empirical process $X_N = \{X_N(t)\}_{t \in [0,T]} \in C([0, T], \mathcal{M}_P(\mathbb{R}^d))$.

In $\mathcal{M}_P(\mathbb{R}^d)$, we consider the Prokhorov metric or the BL-metric, while in the space $C([0, T], \mathcal{M}_P(\mathbb{R}^d))$ we take the uniform metric with respect to $t \in [0, T]$, so that the distance between $f, g \in C([0, T], \mathcal{M}_P(\mathbb{R}^d))$ is given by

$$d_1(f, g) := \sup_{0 \leq t \leq T} \rho(f(t), g(t)), \qquad (39)$$

where ρ is either the Prokhorov or the BL metric.

From the separability and completeness of $\mathbb{R}^d$, we can deduce the same properties for the space of probability measures $\mathcal{M}_P(C([0, T], \mathcal{M}_P(\mathbb{R}^d))$. Indeed,

Theorem 15. *Since $\mathbb{R}^d$ is a Polish space, then $\mathcal{M}_P(\mathbb{R}^d)$ is a Polish space, so that $C([0, T], \mathcal{M}_P(\mathbb{R}^d))$ is a Polish space, and consequently $\mathcal{M}_P(C([0, T], \mathcal{M}(\mathbb{R}^d)))$ is a Polish space .*

The measure-valued process X_N lives in $C = C([0, T], \mathcal{M}_P(\mathbb{R}^d))$. Although, in general, weak convergence in C does not follow from weak convergence of the finite-dimensional distributions, it does in presence of relative compactness thanks to Corollary 2 [3]. We have the following result.

Theorem 16. *Let P_n, P be probability measures on $C([0, T], \mathcal{M}_P(\mathbb{R}^d)$. If the finite-dimensional distributions of P_n weakly converge to those of P, and if $\{P_n\}$ is tight, then $P_n \overset{W}{\to} P$.*

To use this theorem we have to characterize the relative compactness on the space $C([0, T], \mathcal{M}_P(\mathbb{R}^d))$. Let us consider first $\mathcal{M}_P(\mathbb{R}^d)$.

Lemma 1. [15] *A subset $K \subset \mathcal{M}_P(\mathbb{R}^d)$ is relatively compact if and only if*

1. $\sup\{\|\mu\|_0 : \mu \in K\} < \infty$, *with* $\|\mu\|_0 = \sup\{\langle \mu, f \rangle, f \in C_b(\mathbb{R}^d), \|f\| \leq 1\}$;
2. $\lim_{n \to \infty} \sup\{\|\mathbb{I}_{B_n^c} \mu\|_0 : \mu \in K\} = 0$, *with* $B_n^c := \{x \in \mathbb{R}^d : \|x\| \leq n\}^c$.

The condition of compactness may be stated in terms of the following modulus of continuity

$$w_f(\delta) = w(f, \delta) = \sup_{|s-t| < \delta} |f(s) - f(t)|, \quad 0 < \delta \leq 1,$$

thanks to a generalization of the Ascoli-Arzelá characterization. Indeed the Ascoli-Arzelá theorem states that a sequence of continuous functions on a metric space is tight if and only it is uniformly bounded and equicontinuous; uniform boundedness and equicontinuity are equivalent to (40) and (41) below.

Theorem 17 (Ascoli-Arzelà). [3] *A subset* $A \in C$ *is relative compact if and only if*

$$\sup_{f \in A} |f(0)| < \infty \qquad (40)$$

e

$$\lim_{\delta \to 0} \sup_{f \in A} w_f(\delta) = 0. \qquad (41)$$

From the previous theorem it is possible to characterize the tightness on $\mathcal{M}_P(C)$, where C is the space of continuous function on $[0, T]$.

Theorem 18. *[3] A sequence* $\{P_n\}$ *on* C *is tight if and only if these two conditions hold:*

i) *for each positive* η, *there exists an a such that*

$$P_n\{f : |f(0)| > a\} \leq \eta, \quad n \geq 1;$$

ii) *for each positive* ϵ *and* η, *there exists a* δ, *with* $0 < \delta < 1$, *and an integer* n_0 *such that*

$$P_n\{f : w_f(\delta) \geq \epsilon\} \leq \eta, \quad n \geq n_0.$$

A useful sufficient condition for tightness is offered by the following theorem.

Theorem 19. *[3] A sequence* $\{P_n\}$ *is tight if these two conditions are satisfied:*

i) *for each positive* η, *there exists an a such that*

$$P_n\{x : |x(0)| > a\} \leq \eta \quad n \geq 1.$$

ii) *for each positive* ϵ *and* η, *there exists a* δ, *with* $0 < \delta < 1$, *and an integer* n_0 *such that*

$$\frac{1}{\delta} P_n\{x : \sup_{t \leq s \leq t+\delta} |x(s) - x(t)| \geq \epsilon\} \leq \eta, \quad n \geq n_0,$$

for all t.

Let us now specialize the previous sufficient condition on the space of probability measures on $C([0, T], \mathcal{M}(\mathbb{R}^d))$, on which we consider the distance d_1, defined by (39).

Theorem 20. *Consider a sequence* $(X_N)_{N \in \mathbb{N}}$ *of measure-valued stochastic processes in* $C([0, T], \mathcal{M}(\mathbb{R}^d))$. *Suppose that*

(i) *the initial sequence* $(X_N(0))_{N \in \mathbb{N}}$ *is tight in* $\mathcal{M}(\mathbb{R}^d)$;

(ii) for any real positive ϵ and η there exists a $\delta \in (0,1)$, and a positive integer n_0 such that

$$\frac{1}{\delta} P \left(\sup_{t \leq s \leq t+\delta} d_1(X_N(s), X_N(t)) \geq \epsilon \right) \leq \eta$$

for $N \geq n_0$ and $0 \leq t \leq T$.

Then the sequence of laws $(\mathcal{L}(X_N))_{N \in \mathbb{N}}$ in $\mathcal{M}\left(C([0,T], \mathcal{M}(\mathbb{R}^d))\right)$ is tight.

Condition *i)* states the initial tightness of the process, while condition *ii)* states the small variation of the process during small time intervals.

A convenient, though stronger, sufficient condition for compactness is given by Ethier-Kurtz [15].

Theorem 21. *Consider a sequence $(X_N)_{N \in \mathbb{N}}$ of measure-valued stochastic processes in $C([0,T], \mathcal{M}(\mathbb{R}^d))$, and let $\mathcal{F}_t^N := \sigma\{X_N(s)|s \leq t\}$ be the natural filtration associated with $\{X_N(t), t \in [0,T]\}$. Suppose that*

(i) for any real positive ϵ and for any nonnegative rational t, a compact $\Gamma_{t,\epsilon}$ exists such that

$$\inf_N P(X_N(t) \in \Gamma_{t,\epsilon}) > 1 - \epsilon;$$

(ii) let $\alpha > 0$; for any real $\delta \in (0,1)$ a sequence $(\gamma_N^T(\delta))_{N \in \mathbb{N}}$ of nonnegative real random variables exists such that

$$\lim_{\delta \to 0} \limsup_{N \to \infty} \mathbb{E}[\gamma_N^T(\delta)] = 0$$

and, for any $t \in [0,T]$,

$$\mathbb{E}[d_1(X_N(t+\delta), X_N(t)))^\alpha | \mathcal{F}_t^N] \leq \mathbb{E}[\gamma_N^T(\delta) | \mathcal{F}_t^N].$$

Then $(\mathcal{L}(X_N))_{N \in \mathbb{N}}$ is tight.

Condition *(i)* implies a pointwise control of compactness , while condition *(ii)* is again a control of the variation of the process during small time intervals. Indeed, it states that the mean distance of the process at two close times t and $t + \Delta t$ is small, and asymptotically in N, it converges to zero as the length Δt of the time interval tends to zero.

Existence and Uniqueness of a Limit Measure

The compactness properties of the laws of the process $\{X(t)\}_{t \in \mathbb{R}_+}$ allow us to prove the existence of limit measures of subsequences. In order to prove the existence of a unique limit, by the characterization given by Theorem 14, we need to prove that all the limits are equal. Usually this can be shown by characterizing the limit measure as the unique solution of a measure-valued differential equation.

As a consequence the main steps for proving the existence and uniqueness of a limit of a sequence of the laws of X_N are the following

a) prove relative compactness of the sequences $(\mathcal{L}(X_N))_{N \in \mathbb{N}}$, so to prove the existence of a limiting measure-valued process $\{X_\infty(t),\, t \in \mathbb{R}_+\}$.
b) identify any possible limit $\{X_\infty(t),\, t \in \mathbb{R}_+\}$ as the solution of a deterministic measure-valued differential equation;
c) possibly prove its absolute continuity with respect to the usual Lebesgue measure on $\mathbb{R}^d$, and identify the density $\rho(x, t)$ of $X_\infty(t)$ as a solution of a deterministic partial differential equation;
d) show uniqueness by proving the uniqueness of the solution of the deterministic partial differential equation.

In the next section we carry on the programme for a particularly interesting case.

5 A Specific Model for Interacting Particles

Here we want to present a possible model for the drift F_N in the system of stochastic differential equations (33). In particular we consider each individual subject to both an individual motion and an interaction with other individuals. Interaction is due to long range aggregation and short range repulsion forces [4, 26].

Modelling interaction at different ranges may be obtained in terms of an appropriate rescaling of a given reference kernel K. A mathematical way to distinguish among different scales is based on the choice of a "scaling" parameter in the kernel describing the interaction among individuals. As already mentioned in the introduction, if we consider a system of N particles located in $\mathbb{R}^d$, in the macroscopic space-time coordinates the *typical distance between neighboring particles* is $O(N^{-1/d})$, while the order of the size of the whole space is $O(1)$.

Let K be a sufficiently regular function, and assume that the interaction between two particles, out of N, located in x and y respectively, is modelled by

$$\frac{1}{N} K_N(x - y), \quad \text{where} \quad K_N(z) = N^\beta K(N^{\beta/d} z), \tag{42}$$

which expresses the rescaling of K with respect to the total number N of individuals, in terms of the scaling parameter $\beta \in [0, 1]$. The force exerted on the k-th single particle located at $X_N^k(t)$, due to its interaction with all the others in the population, is given by

$$I^k \equiv (X_N(t) * V_N)(X_N^k(t)) = \sum_{i=1}^N \frac{1}{N} V_N \left(X_N^k(t) - X_N^i(t) \right)$$

$$= \sum_{i=1}^N N^{\beta - 1} V_1 \left(N^{\beta/d} \left(X_N^k(t) - X_N^i(t) \right) \right). \tag{43}$$

We have a McKean-Vlasov (long range) interaction if $\beta = 0$, i.e. the range and the strength of the interaction do not depend on the scaling parameter N, and a moderate (short range) interaction if $\beta \in (0,1)$. In the latter case, as N tends to infinity, the range of the interaction kernel tends to zero, while its strength tends to infinity. For the long range aggregation we consider a symmetric kernel $G : \mathbb{R}^d \longrightarrow \mathbb{R}_+$, such that $G(x) = \tilde{G}(|x|)$, $x \in \mathbb{R}^d$, where $\tilde{G}$ is an increasing real valued function defined on $\mathbb{R}_+$; for the short range repulsion we consider a symmetric function V_N rescaled by N via a symmetric (with respect to zero) probability density V_1, i.e.

$$V_N(z) = N^\beta V_1(N^{\beta/d} z), \quad \beta \in (0,1). \tag{44}$$

So the system of SDE's becomes

$$dX_N^k(t) = - \left[\gamma_1 \nabla U(X_N^k(t)) + \gamma_2 \left(\nabla (V_N - G) * X_N \right)(X_N^k(t)) \right] dt$$
$$+ \sigma dW^k(t), \qquad k = 1, \ldots, N, \tag{45}$$

where $U : \mathbb{R}^d \to \mathbb{R}_+ \in C^2(\mathbb{R}^d)$ is a non negative smooth even potential and $\gamma_1, \gamma_2 \in \mathbb{R}_+$ are suitable weights. From the modelling point of view, a transport term including U may represent some external information from the environment, which drives any individual along the gradient of U. Figures 1, 2 and 3 show the behavior of the system for different choices of the parameters and the kernels.

Let us consider the following assumptions about the regularity of the kernels involved in system (45)

$$G, V_1 \in C_b^2(\mathbb{R}^d, \mathbb{R}_+) \cap L^1(\mathbb{R}^d, \mathbb{R}_+) \tag{46}$$
$$U \in C^2(\mathbb{R}^d, \mathbb{R}_+) \tag{47}$$
$$|\nabla U(\mathbf{x}) - \nabla U(\mathbf{y})|^2 \leq k|\mathbf{x} - \mathbf{y}|^2, \quad \forall(\mathbf{x}, \mathbf{y}) \in \mathbb{R}^d \times \mathbb{R}^d \tag{48}$$
$$|\nabla U(\mathbf{x})|^2 \leq k^*(|\mathbf{x}|^2 + 1), \quad \forall \mathbf{x} \in \mathbb{R}^d \tag{49}$$

where k and k^* are positive constants. As a consequence, System (45) admits a unique solution.

By (35), the evolution equation of the empirical measure (32), is, for any $f \in C_b^{2,1}(\mathbb{R}^d \times \mathbb{R}_+)$

$$\langle X_N(t), f(\cdot, t) \rangle = \langle X_N(0), f(\cdot, 0) \rangle$$
$$- \int_0^t \langle X_N(s), [\gamma_1 \nabla U + \gamma_2 (\nabla (V_N - G) * X_N)] (\cdot) \nabla f(\cdot, s) \rangle \, ds$$
$$+ \int_0^t \left\langle X_N(s), \frac{\sigma_N^2}{2} \triangle f(\cdot, s) + \frac{\partial}{\partial s} f(\cdot, s) \right\rangle \, ds$$
$$+ \sigma_N \int_0^t \langle X_N(s), \nabla f(\cdot, s) \rangle \, dW^k(s). \tag{50}$$

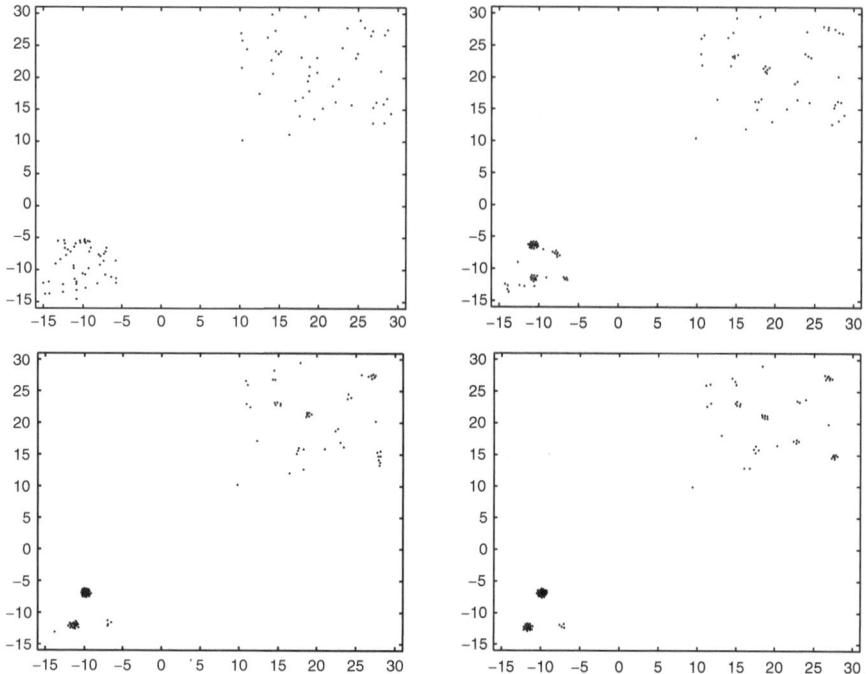

Fig. 1: Configuration of 100 particles for parameters values $\gamma_1 = 0$, $\gamma_2 = 1$, $\sigma = 0.02$, $\beta = 0.5$: (up left) $T = 0$, (up right) $T = 500$, (down left) $T = 1000$, (down right) $T = 2000$

5.1 Asymptotic Behavior of the System for Large Populations: A Heuristic Derivation

Suppose that indeed the empirical process $\{X_N(t),\ t \in \mathbb{R}_+\}$ tends, as $N \to \infty$, to a deterministic process $\{X(t),\ t \in \mathbb{R}_+\}$, and further that it admits for any $t \in \mathbb{R}_+$, a density $\rho(x,t)$ with respect to the Lebesgue measure on $\mathbb{R}^d$, so that

$$\lim_{N \to +\infty} \langle X_N(t), f(\cdot, t) \rangle = \langle X(t), f(\cdot, t) \rangle = \int_{\mathbb{R}^d} f(x,t) \rho(x,t) dx.$$

As a formal consequence, we get

$$\lim_{N \to +\infty} (X_N(t) * V_N)(x) = g_N(x) = \rho(x,t),$$

$$\lim_{N \to +\infty} (X_N(t) * \nabla V_N)(x) = \nabla g_N(x) = \nabla \rho(x,t),$$

$$\lim_{N \to +\infty} (X_N(t) * \nabla G_a(\cdot, t))(x) = (X(t) * \nabla G_a(\cdot, t))(x) = (\rho * \nabla G_a(\cdot, t))(x).$$

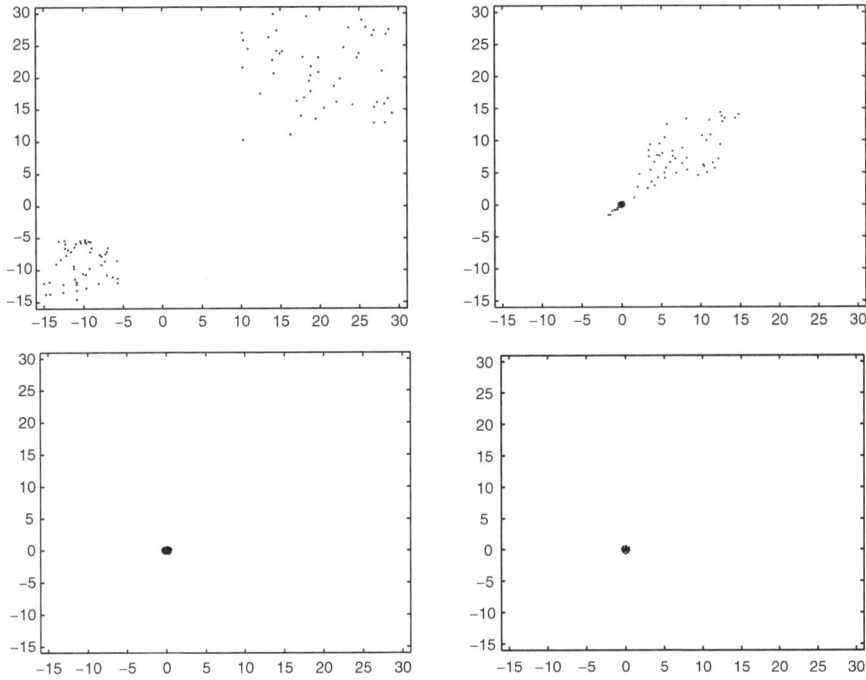

Fig. 2: Configuration of 100 particles for parameters values $\gamma_1 = 1$, $\gamma_2 = 1$, $\sigma = 0.02$, $\beta = 0.5$ with potential U such that $\nabla U(x) = x/(1 + |x|)$: (up left) $T = 0$, (up right) $T = 200$, (down left) $T = 2000$, (down right) $T = 4000$

Hence, by applying the above limits, from (36) and (50) we get

$$\int_{\mathbb{R}^d} f(x,t)\rho(x,t)dx = \int_{\mathbb{R}^d} f(x,0)\rho(x,0)dx \tag{51}$$

$$+ \int_0^t ds \int_{\mathbb{R}^d} dx \left[-\gamma_1 \nabla U(x) + \gamma_2(\nabla G_a * \rho(\cdot, s))(x)\right.$$

$$\left. -\nabla \rho(x,s)\right] \cdot \nabla f(x,s)\rho(x,s)$$

$$+ \int_0^t ds \int_{\mathbb{R}^d} dx \left[\frac{\partial}{\partial s} f(x,s)\rho(x,s) + \frac{\sigma_\infty^2}{2} \triangle f(x,s)\rho(x,s)\right],$$

where

$$\lim_{N \to \infty} \sigma_N = \sigma_\infty.$$

We recognize that (51) is the weak version of the following equation for the spatial density $\rho(x,t)$, for $x \in \mathbb{R}^d, t \geq 0$

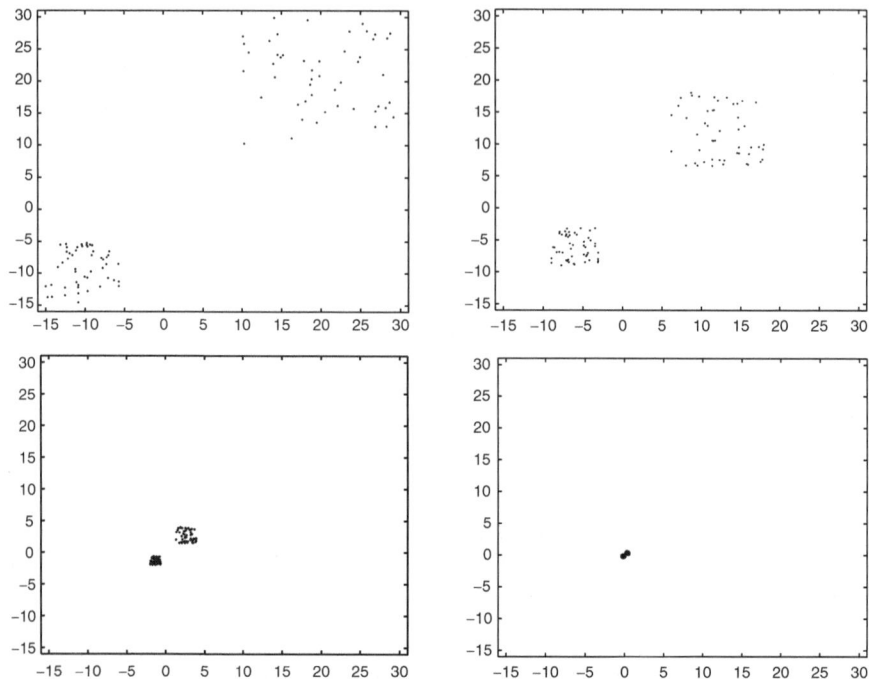

Fig. 3: Configuration of 100 particles for parameters values $\gamma_1 = 1$, $\gamma_2 = 1$, $\sigma = 0.02$, $\beta = 0.5$ with potential U such that $\nabla U(x) = |x|^2$: (up left) $T = 0$, (up right) $T = 5$, (down left) $T = 50$, (down right) $T = 100$

$$\frac{\partial}{\partial t}\rho(x,t) = \frac{\sigma_\infty^2}{2}\triangle\rho(x,t) + \gamma_1\nabla\cdot(\rho(x,t)\nabla U(x))$$

$$+\gamma_2\nabla\cdot(\rho(x,t)\left[\nabla\rho(x,t)) - (\nabla G_a * \rho(\cdot,t))(x)\right], \qquad (52)$$

$$\rho(x,0) = \rho_0(x), \quad x \in \mathbb{R}^d. \qquad (53)$$

From (52), if $\sigma_\infty > 0$ the dynamics of the density is smoothed by the diffusive term. This is due to the memory of the fluctuations existing when the number N of particles is finite. This also means that the dynamics of each individual particle is still stochastic. Indeed, it can be shown that for any k, we have $X_k(t) \sim Y^k(t)$, this last one satisfying the following SDE,

$$dY^k(t) = -\left[\nabla G_a * \rho(\cdot,t)(Y^k(t)) - \nabla\rho(Y^k(t)) - \nabla U(Y^k(t))\right]dt$$
$$+\sigma_\infty dW^k(t),$$

subject to the initial condition $Y^k(0) = X_N^k(0)$. The Brownian stochasticity of the movement of each particle is preserved.

When $\sigma_\infty = 0$ all stochasticity disappears; this leads to a degenerate equation, for $x \in \mathbb{R}^d, t \geq 0$

$$\frac{\partial}{\partial t} \rho(x,t) = \nabla \cdot (\rho(x,t) \left[\gamma_1 \nabla U(x) \right) + \gamma_2 (\nabla \rho(x,t) - (\nabla G_a * \rho(\cdot,t))(x)) \right].$$

For the individuals, for any k we have $X_k(t) \sim Y(t)$ subject to

$$dY(t) = - \left[\nabla G_a * \rho(\cdot,t)(Y(t)) - \nabla \rho(Y(t)) - \nabla U(Y(t)) \right] dt,$$

and the initial condition $Y(0) = X_N^k(0)$, so that also the dynamics of an individual particle becomes fully deterministic; there is no memory of the existing fluctuations, when the number N of particles is finite.

Existence and Uniqueness of a Solution of the Limit Equation

In the viscous case, i.e. when $\sigma_\infty > 0$, a large literature is available showing existence and uniqueness of a sufficient regular solution satisfying Equation (52). On the other hand, in the non viscous case, i.e. when $\sigma_\infty = 0$, uniqueness is not a trivial problem, the lack of uniqueness being mainly due to the nonlocal transport term [10].

A major issue is to find the right notion of solution for equations like (54). It is well-known that classical solutions do not exist in general for degenerate equations, in particular for equations like (54) one has to expect that the solution is not differentiable at the boundary of the (compact) support. A usual way to overcome such difficulties for parabolic equations is to use *weak solutions*. However, for degenerate equations of the general form

$$\frac{\partial v}{\partial t} + \text{div } f(x,t,v) - \Delta a(v) = 0, \tag{54}$$

the weak solution may not be unique, and a different concept of solutions, so-called *entropy solutions*, has to be used in order to obtain uniqueness [9]. Another example of non uniqueness has been found for transport equations with nonlocal nonlinearity [10] corresponding to (54) without the diffusion term. To our knowledge the only available uniqueness result for an equation like (54) is due to [27], but it holds only in 1D for a very special convex long-range interaction kernel.

The above issues concerning uniqueness motivate the study of weak and entropy solutions for (54). In conservation laws, entropy solutions are usually obtained as vanishing viscosity limits and well motivated from a physical point of view. In the case of biological models, entropy solutions are hardly used and not well motivated so far. In [6] we have adapted this notion to our system, closely following this approach with a simple modification enforced by the nonlocal convolution operator.

5.2 Asymptotic Behavior of the System for Large Populations: A Rigorous Derivation

In this section we present a rigorous derivation of the limit measure $X(t)$, following Section 4.3. From now on we consider the case when the individual randomness does not vanish at infinity, that is

$$\lim_{N \to \infty} \sigma_N = \sigma_\infty > 0. \tag{55}$$

Let us consider the following additional assumptions on the kernels involved in the model; for $x \in \mathbb{R}^d$

$$V_N(x) = (W_N * W_N)(x) \tag{56}$$

$$W_N(x) = \chi_N^d W_1(\chi_N x), \quad \chi_N = N^{\beta/d}, \beta \in (0, d/(d+2)) \tag{57}$$

$$W_1 \in W_2^1(\mathbb{R}^d) \cap C_0(\mathbb{R}^d), U \in C_b^1(\mathbb{R}^d) \tag{58}$$

$$\left\langle -\nabla U(x), \frac{x}{|x|} \right\rangle \le -\frac{r}{|x|}, \quad |x| \ge M_0, \tag{59}$$

where W_1 is a symmetric probability density, and define a mollified measure, i.e. a regular version of the empirical measure $X_N(t)$

$$h_N(x, t) = (W_N * X_N(t))(x), \quad x \in \mathbb{R}^d.$$

Furthermore, we consider also the following properties for the initial state

$$\sup_{N \in \mathbb{N}} \mathbb{E} \left[\int_{\mathbb{R}^d} |x| X_N(0)(dx) \right] < \infty \tag{60}$$

$$\sup_{N \in \mathbb{N}} \mathbb{E} \left[\int_{\mathbb{R}^d} |h_N(x, 0)|^2 dx \right] < \infty, \tag{61}$$

The first step for the analysis is to study the *relative compactness* of the stochastic process $\{X_N\}_{N \in \mathbb{N}}$. The main problem in the derivation of the compactness properties is due to the unboundedness of the drift term; so we have to take care of the possible explosion of the system. In order to deal with this problem, we define the following stopping time

$$\tau_N^k = \inf\{t \ge 0 \; : \; S_N(t) = ||h_N(\cdot, t)||_2^2 + A_N(t)$$

$$- \int_0^t \langle X_N(u), 2| - \nabla U(\cdot) + (\nabla G_a * X_N(u))(\cdot)|^2 \rangle du > k\},$$

where

$$A_N(t) = \int_0^t \langle X_N(s), 2(|\nabla g_N(\cdot, u)|^2 - \nabla g_N(\cdot, u)(-\nabla U(\cdot) + (\nabla G_a * X_N(u))(\cdot))$$

$$+| - \nabla U(\cdot) + (\nabla G_a * X_N(u))(\cdot)|^2) \rangle + \sigma_N^2 ||\nabla h_N(\cdot, u)||_2^2 du.$$

Following [28, 31] one can prove that the process

$$M_N(t) = ||h_N(\cdot, t)||_2^2 - \int_0^t \langle X_N(u), 2| - \nabla U(\cdot) + (\nabla G_a * X_N(u))(\cdot)|^2 \rangle du$$
$$+A_N(t) - c_1\sigma_N^2 t N^{\beta(d+2)/d-1}$$

is a martingale.

Nonexplosion in Finite Time

It is possible to prove the nonexplosion in a finite time, i.e. for any τ such that $0 < \tau < \infty$,

$$\lim_{k \to \infty} \inf_{N \in \mathbb{N}} \mathbb{P}\{\tau_N^k > \tau\} = 1. \tag{62}$$

Proof.
From the martingale property of the process $M_N(t)$, it derives that the process

$$t \mapsto S_N(t) = ||h_N(\cdot, t)||_2^2 + A_N(t) - \int_0^t \langle X_N(u), 2| - \nabla U(\cdot) + (\nabla G_a * X_N(u))(\cdot)|^2 \rangle du$$

is a submartingale. By Doob's inequality

$$\mathbb{P}\left\{\sup_{t \leq \tau} S_N(t) > k\right\} \leq \frac{1}{k}\mathbb{E}[S_N(\tau)] = \frac{1}{k}\mathbb{E}[M_N(\tau) + \tau\sigma_N^2 N^{\beta(d+2)/d-1}c_1]$$
$$= \frac{1}{k}\mathbb{E}\left[\mathbb{E}[M_N(\tau)|\mathcal{F}_0] + C\tau\sigma_N^2 N^{\beta(d+2)/d-1}\right]$$
$$= \frac{1}{k}\mathbb{E}[M_N(0) + C\tau\sigma_N^2 N^{\beta(d+2)/d-1}]$$
$$= \frac{1}{k}\left(\mathbb{E}[||h_N(\cdot, 0)||_2^2] + C\tau\sigma_N^2 N^{\beta(d+2)/d-1}\right)$$
$$\leq C(\tau)/k; \tag{63}$$

the last inequality comes from by (55),(58) and (61).

□

First we study the stopped process

$$X_{N,k}(t) = X_N(t \wedge \tau_N^k); \tag{64}$$

indeed the relative compactness of the stopped process (64) and condition (62) imply the relative compactness of the full process [28].

Relative Compactness for the Stopped Process $X_{N,k}(t)$

In order to prove the relative compactness of the laws of $X_{N,k}(t)$, we use the characterization of Theorem 21. By using the martingale property of the process (62) and Doob's inequality, one can prove that, for any $\epsilon > 0$, there exists a compact K_ϵ^k in $(\mathcal{M}_\mathcal{P}(\mathbb{R}^d), d_{BL})$ such that

$$\inf_{N \in \mathbb{N}} \mathbb{P}\{X_{N,k}(t) \in K_\epsilon^k, \ \forall t \in [0, T]\} \geq 1 - \epsilon,$$

i.e. the first sufficient condition of Theorem 21 is satisfied. Then, by the properties of the dynamics of the system, one can prove the property of small fluctuations during small time intervals, and in particular that for any $0 < \delta < 1$, there exists a sequence $\{\gamma_n^T(\delta)\}_{n \in \mathbb{N}}$ of non negative random variables such that

$$\mathbb{E}\left[d_{BL}(X_{N,k}(t + \delta), X_{N,k}(t))^4\right] \leq \mathbb{E}\left[\gamma_n^T(\delta)\right] \quad 0 \leq t \leq T,$$

and $\lim_{\delta \to 0} \lim_{n \to \infty} \sup \mathbb{E}[\gamma_n^T(\delta)] = 0$.

From the previous analysis we may then state that the sequence of laws $\{\mathcal{L}(X_N(\cdot \wedge \tau_N^k))\}_{N \in \mathbb{N}}$ is relatively compact in $\mathcal{M}_\mathcal{P}(C([0, T], \mathcal{M}_\mathcal{P}(\mathbb{R}^d)))$.

As already mentioned, the relative compactness of the sequence $\{\mathcal{L}(X_N)\}_{N \in \mathbb{N}}$ in $\mathcal{M}_\mathcal{P}(C([0, T], \mathcal{M}_\mathcal{P}(\mathbb{R}^d)))$ follows from the non explosion of the stopping time $\tau_{N,\cdot}^k$.

Finally from Skorokhod's Theorem we may assert that for the possible unique limit law we can get also an almost sure convergence, i.e.

$$\lim_{N \to \infty} \sup_{t \leq T} d_{BL}(X_N(t), X(t)) = 0 \quad \mathbb{P} - a.s. \tag{65}$$

Regularity of the Limit Measure

The next step is to study the regularity properties of the limit measure. It is possible to show that the mollified measure h_N, given by (60) is an $L^2([0, T], \mathcal{M}_P(\mathbb{R}^d))$ Cauchy sequence [28, 31]. As a consequence the sequence admits a limit $\rho \in L^2(\mathbb{R}^d, \mathbb{R}_+)$

$$\lim_{N \to \infty} \mathbb{E}\left[\int_0^T \int_{\mathbb{R}^d} |h_N(x, t) - \rho(x, t)|^2 dx dt\right] = 0. \tag{66}$$

From (66) and $\lim_{N \to \infty} W_N(\cdot) = \delta_0$ (in the sense of distributions), we have

$$\lim_{N \to \infty} \int_{\mathbb{R}^d} f(x, t) X_N(t)(dx) = \int_{\mathbb{R}^d} f(x, t) \rho(x, t) dx \quad f \in C_b(\mathbb{R}^d \times [0, T]) \quad \mathbb{P} - a.s.$$

As a consequence, because of the limit property (65), we get

$$\int_{\mathbb{R}^d} f(x, t) X(t)(dx) = \int_{\mathbb{R}^d} f(x, t) \rho(x, t) dx \quad f \in C_b(\mathbb{R}^d \times [0, T]), \mathbb{P} - a.s.$$

Therefore the measure $X(t)$ is absolutely continuous with respect to the Lebesgue measure with density $\rho(x, t)$, i.e. for any $f \in C_b(\mathbb{R}^d), t \in [0, T]$

$$\lim_{N \to \infty} \langle X_N(t), f(\cdot) \rangle = \langle X(t), f(\cdot) \rangle = \int_{\mathbb{R}^d} f(x) \rho(x, t) dx. \tag{67}$$

Uniqueness of the Limit

Uniqueness of the limit measure X is shown, by proving that the density ρ is the solution of the system (52)-(53), which admits a unique solution. We further assume that a law of large numbers applies to the initial condition, i.e.

$$\lim_{N \to \infty} \mathcal{L}(X_N(0)) = \delta_{X_0} \quad \text{in} \quad \mathcal{M}_P(\mathcal{M}_P(\mathbb{R}^d)), \tag{68}$$

where X_0 has density $\rho(x,0)$ with respect to the Lebesgue measure. As a consequence

$$\lim_{N \to \infty} \mathcal{L}(X_N(t)) = \delta_{X_N(t)} \quad \text{in} \quad \mathcal{M}_P(\mathcal{M}_P(\mathbb{R}^d)),$$

for any $f \in C_b^{2,1}(\mathbb{R}^d, \mathbb{R}_+)$, uniformly in $t \in [0,T]$, where ρ is the unique solution of (52)-(53), as proven in the Section A.

6 Long Time Behavior: Invariant Measure

Here we wish to analyze the long time behavior of the system (51) for a fixed N. Our interest is to analyze mechanisms that are responsible for stable aggregation. From the mathematical point of view this is equivalent to the study of the joint distribution law of the vector $\mathbf{X} = (X_N^1, \ldots, X_N^N) \in \mathbb{R}^n$, where $n = Nd$. Let now $P_N^{x_0}(t)$ denote the joint distribution of the N particles at time t, conditional upon a non random initial condition x_0.

Invariant Measure

Since $\mathbf{X}$ is a homogeneous Markov process, one can associate a homogeneous transition probability $p(x, t-s, dy)$ such that, with $B \in \mathcal{B}_{\mathbb{R}^n}$, for all $0 \le s < t < +\infty$,

$$P\{X(t) \in B | X(s) = x\} = \int_B p(x, t-s, dy).$$

If there exists a probability measure $\overline{P}_N$ on $\mathbb{R}^n$ independent of time t, as a solution of the integral equation

$$\forall t > 0, \quad \overline{P}_N(dy) = \int_{x \in \mathbb{R}^n} \overline{P}_N(dx) p(x; t, dy),$$

then, $\overline{P}_N(dx)$ is called an *invariant measure* associated with the Markov process X.

Let us consider some sufficient conditions for the *existence of an invariant measure*. In particular we consider the results the reader may find in [19], p. 118, that stipulate that a process $\mathbf{X}(t)$ has finite mean recurrence time for some bounded domain U, and within this domain all sample paths "mix sufficiently well". More precisely we consider the following assumption (H):

There exists a bounded domain $D \subset \mathbb{R}^n$ with regular boundary, having the following properties:

H1: in the domain D and some neighborhood thereof, the smallest eingen-value of the diffusion matrix is bounded away from zero.

H2: if $x \in D^c$, the mean time τ_{U^c} at which a path issuing from x reaches the set D is finite, i.e. $E[\tau_{D^c}|\mathbf{X}(0) = x] < \infty$, and $\sup_{x \in K} E[\tau_K|\mathbf{X}(0) = x] < \infty$, for every compact $K \subset \mathbb{R}^n$.

Theorem 22. *[19] If condition (H) holds, then the Markov process $\mathbf{X}(t)$ has a unique invariant measure $\overline{P}_N$.*

The Purely Interacting Case

First of all we consider the case of the drift due only to interaction among particles, that is $\gamma_1 = 0$.

In Figure 4 we consider some simulation results with $V_1 = \alpha_1 \mathcal{N}(0,1)$ and $G = \alpha_2 \mathcal{N}(0,1)$, for different values of α_1 and α_2. We notice that, as t increases, the mean distance between two particles and consequently the radius of clusters fluctuates around an asymptotic value which is finite, but in the case of pure repulsion.

Following [22], for the position of the center of mass of the N particles,

$$\bar{X}_N(t) = \frac{1}{N} \sum_{k=1}^{N} X_N^k(t),$$

we have

$$d\bar{X}_N(t) = -\frac{1}{N^2} \sum_{k,j=1}^{N} \nabla (V_N - G)(X_N^k(t) - X_N^j(t))dt + \sigma d\bar{W}(t), \qquad (69)$$

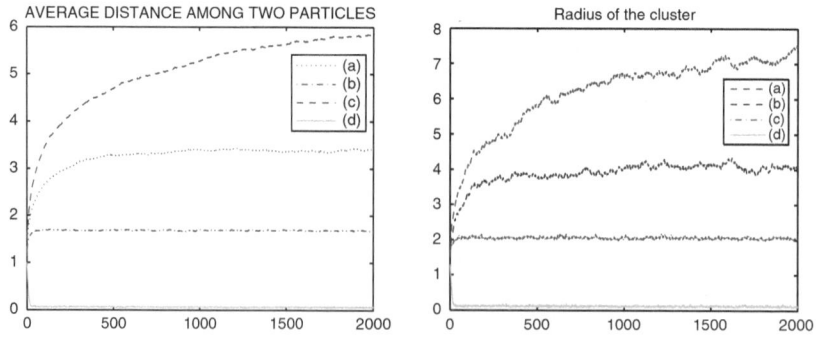

Fig. 4: Comparison among the evolution of the radius of the cluster and the average distance among particles for different values of parameters: (a) $\alpha_1 = \alpha_2 = 1$; (b) $\alpha_1 = 1; \alpha_2 = 2$; (c) $\alpha_1 = 1; \alpha_2 = 0$; (d) $\alpha_1 = 0; \alpha_2 = 1$

where $\bar{W}(t) = \frac{1}{N}\sum_{k=1}^{N} W^k(t)$; by the symmetry properties of V_1 and G, the first term on the right hand side vanishes and we get

$$d\bar{X}_N(t) = \sigma d\bar{W}(t), \tag{70}$$

i.e. $\bar{X}_N(t)$ is a Brownian motion. Hence, its distribution is

$$\mathcal{L}\left(\bar{X}_N(t)|\bar{X}_N(0)\right) = \mathcal{L}\left(\bar{X}_N(0), \sigma^2 \bar{W}(t)\right) = \mathcal{N}\left(\bar{X}_N(0), \frac{\sigma^2}{N}t\right);$$

with variance $\frac{\sigma^2}{N}t$, which, for any fixed N, increases as t tends to infinity.

Consequently we may claim that the probability law of the system does not converge to any probability measure, since otherwise the same would happen for the law of the center of mass.

In other models [22] in addition to symmetry, strictly convexity of the interaction kernel K is added. In spite of that, once again because of the symmetry of K, the system

$$d\mathbf{X}_N(t) = -\nabla \cdot (K * X_N(t))(\mathbf{X}_N(t))dt + \sigma d\mathbf{W}(t), \quad \mathbf{X}_N(t) \in \mathbb{R}^n, \forall t,$$

with

$$\exists \lambda > 0, \forall x, v \in \mathbb{R}^d, \quad \langle \mathrm{Hess}K(x)v, v\rangle \geq \lambda\langle v, v\rangle; \tag{71}$$

does not admit an invariant distribution.

The interesting feature is that, in this case, one may show stationary properties for the reduced system for the relative coordinates, the distances of the particle position with respect to the center of mass $\bar{X}_N(t)$,

$$Y_N^j(t) = X_N^j(t) - \bar{X}_N.$$

Indeed, under the assumption (71), it has been shown [22] that the system

$$Y_N^k(t) = Y_N^k(0) - \frac{1}{N}\sum_{l=1}^{N}\int_0^t \nabla K(Y_N^k(s) - Y_N^l(s))ds$$

$$+ \sigma W^k(t) - \frac{\sigma}{N}\sum_{l=1}^{N} W^l(t) \qquad k = 1, \dots, N,$$

is still a diffusion on the manifold

$$\mathcal{M} = \left\{(x_1, \dots, x_n) : \sum_i x_i = 0\right\}, \tag{72}$$

and does admit an invariant measure. Roughly speaking, adding the strictly convexity, each particle, and in particular the center of mass, attracts any other one in the whole space.

Hence, if we consider the system under study, (35), with $\gamma_1 = 0$, since the kernels G and V_N are symmetric, does not admit an invariant measure. In the next section we see a way to overcome this problem and which properties to require to the added kernel U in (35) in order to get an invariant measure for the system.

The Complete System

Consider now the potential P associated with a measure on $\mathbb{R}^d$, $\mu \in \mathcal{M}(\mathbb{R}^d)$

$$P(\mu)(x) = \gamma_1 U(x) - \gamma_2 \left((V_N - G) * \mu \right)(x), \quad x \in \mathbb{R}^d, \tag{73}$$

so that system (45) can be rewritten as

$$dX_N^k(t) = -\nabla P(X_N(t))(X_N^k(t)) + \sigma dW^k(t), \qquad k = 1, \ldots, N. \tag{74}$$

System (74) has been thoroughly analyzed in literature under the sufficient condition (71) of strict convexity on U; it has been shown [9, 22] that if this condition applies, system (74) does admit a nontrivial invariant distribution. From a biological point of view a strictly convex confining potential is difficult to explain; it would mean an infinite range of attraction, with an at least constant drift, even far from origin.

A weaker sufficient condition for the existence of a unique invariant measure has been more recently suggested by Veretennikov [36], following Has'minski [19],

V: there exist constants $M_0 \geq 0$ and $r > 0$ such that for $|x| \geq M_0$

$$\left(-\nabla P(\mu)(x), \frac{x}{|x|} \right) \leq -\frac{r}{|x|}. \tag{75}$$

Since the diffusion matrix is σI, where I is the identity matrix, then assumption (H1) for the existence of an invariant measure is satisfied. Assumption (75), is needed to show that $\mathbf{X}(t)$ has finite mean recurrence time for some bounded domain U, i.e. assumption (H2) is satisfied [36].

Without any further condition on the interaction kernels V_N and G, for (75) to hold it is sufficient to assume that there exist constants $M_0 \geq 0$ and $r > \frac{1}{N\gamma_1}(\frac{Nd}{2} + 1)$ such that for $|x| \geq M_0$, equation (59) hold. We wish to remark that condition (59) means that ∇D may decay to zero as $|x|$ tends to infinity, provided that its tails are sufficiently "fat".

Proposition 7. *If the confining potential U satisfies condition (59), then (74) admits a unique invariant measure.*

Proof.

Let $\pi_i(\mathbf{x}) = x_i, i = 1, \ldots, N$ be the i-th projection of $\mathbf{x} \in (R^d)^N$, $\tilde{U}(\mathbf{x})$ and $\tilde{K}(\mathbf{x})$ the vector function defined by

$$\tilde{U}(\mathbf{x}) = (U \circ \pi_i(\mathbf{x}))_{1 \leq i \leq N}, \quad \tilde{K}(\mathbf{x}) = \left((G - V_N) * \frac{1}{N} \sum_i \epsilon_{\pi_i(\mathbf{x})} \circ \pi_i(\mathbf{x}) \right)_{1 \leq i \leq N}$$

In order to apply Theorem 2 in [35], we have to prove that there exist constants $M \geq 0$ and $\tilde{r} > (\frac{Nd}{2} + 1)$ such that for all $\mathbf{x} \in (R^d)^N : |x| \geq M$

$$\left(-\gamma_1 \nabla \tilde{U}(\mathbf{x}) + \gamma_2 \nabla \tilde{K}(\mathbf{x}), \frac{\mathbf{x}}{|\mathbf{x}|}\right) \le -\frac{\tilde{r}}{|\mathbf{x}|}. \tag{76}$$

We have

$$\left(-\gamma_1 \nabla \tilde{U}(\mathbf{x}) + \gamma_2 \nabla \tilde{K}(\mathbf{x}), \frac{\mathbf{x}}{|\mathbf{x}|}\right) = -\gamma_1 \sum_{k=1}^{N} \nabla U(x_k) \frac{x_k}{|\mathbf{x}|}$$

$$+ \gamma_2 \frac{1}{N} \sum_{k=1}^{N} \sum_{i=1}^{N} \nabla(G - V_N)(x_i - x_k) \frac{x_k}{|\mathbf{x}|}$$

$$\le -\gamma_1 \sum_{k=1}^{N} \nabla U(x_k) \frac{x_k}{|\mathbf{x}|}$$

$$+ \gamma_2 \frac{1}{N} \sum_{k=1}^{N} \sum_{i=1}^{N} \nabla(G - V_N)(x_i - x_k)$$

$$= -\gamma_1 \sum_{k=1}^{N} \nabla U(x_k) \frac{x_k}{|\mathbf{x}|} \le -\frac{\gamma_1 r N}{|\mathbf{x}|}$$

The last two inequalities derive from the symmetry of G and V_N, together with (59). So if for $\tilde{r} = \gamma_1 r N$ and condition on r in (59), we have condition (76).

$\square$

Following [35] we may also provide an estimate of the rate of convergence. Indeed, as far as the convergence of $P_N^{x_0}(t)$, for t tending to infinity, is concerned, we may state that

Theorem 23. *Under assumption* (59), *for any* k, $0 < k < \tilde{r} - \frac{Nd}{2} - 1$ *with* $m \in (2k + 2, 2\tilde{r} - Nd)$ *and* $\tilde{r} = \gamma_1 Nr$, *a positive constant* c *exists such that*

$$\left|P_N^{x_0}(t) - P_N^S\right| \le c(1 + |x_0|^m)(1 + t)^{-(k+1)},$$

where $\left|P_N^{x_0}(t) - P_N^S\right|$ *denotes the total variation distance of the two measures, i.e.*

$$\left|P_N^{x_0}(t) - P_N^S\right| = \sup_{A \in \mathcal{B}_{\mathbb{R}^d}} \left[P_N^{x_0}(t)(A) - P_N^S(A)\right].$$

and x_0 *the initial data.*

A Proof of the Identification of the Limit ρ

Since (51) is the weak form of (52), it is sufficient to show that for any $f \in C_b^{2,1}(\mathbb{R}^d, \mathbb{R}_+)$,

$$\mathbb{E}\left[\left|\langle X(t), f(\cdot,t)\rangle - \langle \mu_0, f(\cdot,0)\rangle - \int_0^t \langle \rho(\cdot,s), \frac{1}{2}\sigma_\infty^2 \Delta f(\cdot,s) + \frac{\partial}{\partial s}f(\cdot,s)\right.\right.$$

$$\left.\left. + [(\nabla G_a * \rho(\cdot,s))(\cdot) - \nabla U(\cdot) - \nabla \rho(\cdot,s)] \cdot \nabla f(\cdot,s)\rangle ds\ \right|\right] = 0.$$

For fixed $f \in C_b^{2,1}(\mathbb{R}^d, \mathbb{R}_+)$

$$\mathbb{E}\left[\left|\langle X(t), f(\cdot,t)\rangle - \langle \mu_0, f(\cdot,0)\rangle - \int_0^t \langle \rho(\cdot,s), \frac{1}{2}\sigma_\infty^2 \Delta f(\cdot,s) + \frac{\partial}{\partial s}f(\cdot,s)\right.\right.$$

$$\left.\left. + [(\nabla G_a * \rho(\cdot,s))(\cdot) - \nabla U(\cdot) - \nabla \rho(\cdot,s)] \cdot \nabla f(\cdot,s)\rangle ds\ \right|\right]$$

$$\leq \mathbb{E}\left[|\langle X(t), f(\cdot,t)\rangle - \langle X_N(t), f(\cdot,t)\rangle|\right]$$

$$+ \mathbb{E}\left[|\langle \mu_0, f(\cdot,0)\rangle - \langle X_N(0), f(\cdot,0)\rangle|\right]$$

$$+ \frac{\sigma_\infty^2}{2}\mathbb{E}\left[\int_0^t |\langle -\rho(\cdot,s), \Delta f(\cdot,s)\rangle + \langle X_N(s), \Delta f(\cdot,s)\rangle| ds\right]$$

$$+ \mathbb{E}\left[\int_0^t |-\langle \rho(\cdot,s), \frac{\partial}{\partial s}f(\cdot,s)\rangle + \langle X_N(s), \frac{\partial}{\partial s}f(\cdot,s)\rangle| ds\right]$$

$$+ \mathbb{E}\left[\int_0^t |\langle \rho(\cdot,s), \nabla \rho(\cdot,s) \cdot \nabla f(\cdot,s)\rangle - \langle h_N(\cdot,s), \nabla h_N(\cdot,s) \cdot \nabla f(\cdot,s)\rangle| ds\right]$$

$$+ \mathbb{E}\left[\left|\int_0^t \langle h_N(\cdot,s), \nabla h_N(\cdot,s) \cdot \nabla f(\cdot,s)\rangle - \langle X_N(s), \nabla g_N \cdot \nabla f(\cdot,s)\rangle ds\right|\right]$$

$$+ \mathbb{E}\left[\int_0^t |-\langle \rho(\cdot,s), [(\nabla G_a * \rho(\cdot,s))(\cdot) - \nabla U(\cdot)] \cdot \nabla f(\cdot,s)\rangle\right.$$

$$\left. + \langle X_N(s), [(\nabla G_a * X_N(s))(\cdot) - \nabla U(\cdot)] \cdot \nabla f(\cdot,s)\rangle| ds\right]$$

$$+ \mathbb{E}\left[\left|\frac{\sigma_N}{N}\int_0^t \sum_{k=1}^N \nabla f(X_N^k(s), s)dW_k(s)\right|\right]$$

$$+ \mathbb{E}\left[\left|\langle X_N(t), f(\cdot,t)\rangle - \langle X_N(0), f(\cdot,0)\rangle\right.\right.$$

$$-\int_0^t \langle X_N(s), (\nabla G_a * X_N(s)) \cdot \nabla f(\cdot,s)\rangle ds + \int_0^t \langle X_N(s), \nabla g_N(\cdot,s) \cdot \nabla f(\cdot,s)\rangle ds$$

$$+\int_0^t \langle X_N(s), \nabla U(\cdot) \cdot \nabla f(\cdot,s)\rangle ds - \int_0^t \langle X_N(s), \frac{1}{2}\sigma_N^2 \Delta f(\cdot,s) + \frac{\partial}{\partial s}f(\cdot,s)\rangle ds$$

$$\left.\left.- \frac{\sigma_N}{N}\int_0^t \sum_{k=1}^N \nabla f(X_N^k(s), s)dW_k(s)\right|\right]$$

$$:= \sum_{i=1}^9 I_N^i(t). \tag{77}$$

Clearly
By (67) and hypothesis (68) $\sum_{i=1}^4 I_N^i(t) = 0$, by (50) $I_N^9(t) = 0$ and by (36) $I_N^8(t) = 0$. It remains to estimate the terms $I_N^5(t)$, $I_N^6(t)$ and $I_N^7(t)$.

$$I_N^5(t) = \mathbb{E}\left[\int_0^t |\langle \rho(\cdot, s), \rho(\cdot, s)\Delta f(\cdot, s)\rangle - \langle h_N(\cdot, s), h_N(\cdot, s)\Delta f(\cdot, s)\rangle|ds\right]$$

$$\leq ||\Delta f||_\infty \int_0^T \mathbb{E}\left[\int_{\mathbb{R}^d} |h_N(x,t) - \rho(x,t)||h_N(x,t) + \rho(x,t)|dx\right]dt$$

$$\leq ||\Delta f||_\infty \left(\mathbb{E}\left[\int_0^T \int_{\mathbb{R}^d} |h_N(x,t) - \rho(x,t)|^2 dx dt\right]\right)^{1/2}$$

$$\cdot \left(\mathbb{E}\left[\int_0^T \int_{\mathbb{R}^d} |h_N(x,t) + \rho(x,t)|^2 dx dt\right]\right)^{1/2};$$

by (63) and (66) we obtain

$$\lim_{N\to\infty} I_N^5(t) = 0. \tag{78}$$

By the symmetry of W_1,

$$I_N^6(t) = \mathbb{E}\left[\left|\int_0^t \langle X_N(s), W_N * (\nabla h_N(\cdot, s) \cdot \nabla f(\cdot, s))\right.\right.$$

$$\left.\left. -(W_N * \nabla h_N(\cdot, s)) \cdot \nabla f(\cdot, s)\rangle ds\right|\right]$$

$$= \mathbb{E}\left[\left|\int_0^t \left(\int_{\mathbb{R}^d} X_N(s)(dx) \int_{\mathbb{R}^d} W_N(x-y)\nabla h_N(y, s)\right.\right.\right.$$

$$\left.\left.\left. \cdot(\nabla f(y) - \nabla f(x))dy\right)ds\right|\right]. \tag{79}$$

By the definition of W_N and since W_1 has compact support, with $c = \text{diam}(\text{supp}W_1(\cdot))$ and $||D^2 f||_\infty = \sup_{i,j\leq d}||\partial_{ij}^2||_\infty$, (79) is less than or equal to

$$c\chi_N^{-1}||D^2 f||_\infty \mathbb{E}\left[\int_0^t \langle X_N(s) * W_N, |\nabla h_N(\cdot, s)|\rangle ds\right]$$

$$\leq c\chi_N^{-1}||D^2 f||_\infty \left(\mathbb{E}\left[\int_0^T ||h_N(\cdot, s)||_2^2 ds\right]\right)^{1/2} \left(\mathbb{E}\left[\int_0^T ||\nabla h_N(\cdot, s)||_2^2 ds\right]\right)^{1/2}$$

$$\leq c\chi_N^{-1}||D^2 f||_\infty.$$

It follows that

$$\lim_{N\to\infty} I_N^6(t) = 0 \tag{80}$$

.

$$
\begin{aligned}
I_N^7(t) = \mathbb{E}\Bigg[\int_0^t & \, |-\langle \rho(\cdot,s), [(\nabla G_a * \rho(\cdot,s))(\cdot) - \nabla U(\cdot)] \cdot \nabla f(\cdot,s)\rangle \\
& + \langle X_N(s), [(\nabla G_a * X_N(s))(\cdot) - \nabla U(\cdot)] \cdot \nabla f(\cdot,s)\rangle \\
& + \langle X_N(s), [(\nabla G_a * \rho(\cdot,s))(\cdot) - \nabla U(\cdot)] \cdot \nabla f(\cdot,s)\rangle \\
& - \langle X_N(s), [(\nabla G_a * \rho(\cdot,s))(\cdot) - \nabla U(\cdot)] \cdot \nabla f(\cdot,s)\rangle| \, ds \Bigg] \\
\le \mathbb{E}\Bigg[\int_0^t & \, |\langle X_N(s) - \rho(\cdot,s), [(\nabla G_a * \rho(\cdot,s))(\cdot) - \nabla U(\cdot)] \cdot \nabla f(\cdot,s)\rangle| \\
& + |\langle X_N(s), [(\nabla G_a * \rho(\cdot,s))(\cdot) - (\nabla G_a * X_N(s))(\cdot)] \cdot \nabla f(\cdot,s)\rangle| \, ds \Bigg].
\end{aligned}
$$

By (65)
$$
\lim_{N\to\infty} I_N^7(t) = 0.
$$

As a consequence
$$
\lim_{N\to\infty} \sum_{i=1}^9 I_N^i(t) = 0
$$

uniformly in $t \in [0.T]$.

Acknowledgments

It is a pleasure to acknowledge fruitful discussions with Karl Oelschläger. Thanks also to Matteo Ortisi.

References

1. Ash R. B. & Gardner M. F., Topics in Stochastic Processes, Academic Press, London; 1975.
2. Banasiak J., Positivity in Natural Sciences. This Volume.
3. Billingsley P., Convergence of Probability Measures, Wiley, New York; 1968.
4. Boi, S., Capasso V., Morale, D. "Modeling the aggregative behavior of ants of the species Polyergus rufescens.", Nonlinear Analysis: Real World Applications, 1, 2000, p. 163-176.
5. Bodnar, M., J.J.L. Velazquez, An integro-differential equation arising as a limit of individual cell-based models, J. Diff. Eqs., 222, 2, (2006), 341380.
6. Burger M., Capasso V., Morale D., On an aggregation model with long and short range interactions, Nonlinear Anal. Real World Appl., 2006. In Press.
7. Capasso V., Mathematical Structures of Epidemic Systems, Springer-Verlag, Heidelberg; 1993.
8. Capasso V., Bakstein D. An Introduction to Continuous-Time Stochastic Processes-Theory, Models and Applications to Finance, Biology and Medicine. Birkhäuser, Boston, 2004.

9. Carrillo J., Entropy solutions for nonlinear degenerate problems, Arch. Rat. Mech. Anal., 269–361, 147; 1999.
10. Diekmann, O., M. Gyllenberg, H.R. Thieme, Lack of uniqueness in transport equations with a nonlocal nonlinearity, Math. Models and Meth. in Appl. Sciences **10** (2000), 581-592.
11. Dudley, R.M., Convergence of Baire measures, Studia Math. 27 (1966), 251-268.
12. Dudley, R.M., Real Analysis and Probability, Cambridge Studies in Advanced Mathematics 74, Cambridge University Press, Cambridge, 2002.
13. Durrett, R., Levin, S.A., "The importance of being discrete (and spatial)." Theor. Pop. Biol., **46**, 1994, 363-394.
14. Dynkin E. B., Markov Processes, Springer-Verlag, Berlin, Vols. 1–2; 1965.
15. Ethier S. N. & Kurtz T. G., Markov Processes, Characterization and Convergence, Wiley, New York; 1986.
16. Feller W., An Introduction to Probability Theory and Its Applications, Wiley, New York; 1971.
17. Gihman I. I. & Skorohod A. V., The Theory of Random Processes, Springer-Verlag, Berlin; 1974.
18. Grünbaum, D., Okubo, A. "Modelling social animal aggregations" In "Frontiers of Theoretical Biology" (S. Levin Ed.), Lectures Notes in Biomathematics, 100, Springer Verlag, New York, 1994, 296-325.
19. Has'minski, R.Z. Stochastic Stability of Differential Equations. Sijthoff & Noordhoff, Alphen aan den Rijn, The Netherlands and Rockville, Maryland, USA, 1980.
20. Karlin S., Taylor H. M., A First Course in Stochastic Processes, Academic Press, New York; 1975.
21. Lachowicz M., Links Between Microscopic and Macroscopic Descriptions. This Volume.
22. Malrieu, F. Convergence to equilibrium for granular media equations and their Euler schemes. The Annals of Applied Probability, **13**, 540-560, 2003.
23. Métivier M., Notions Fondamentales de la Théorie des Probabilités, Dunod, Paris; 1968.
24. Mogilner A., L. Edelstein-Keshet, A non-local model for a swarm, J. Math. Bio. **38**, 534-549, 1999.
25. Morale, D., V. Capasso, K. Ölschlaeger, An interacting particle system modelling aggregation behavior: from individuals to populations, J. Math. Bio. **50**, 49-66(2005).
26. Morale D., Capasso V., & Oelschläger K., A rigorous derivation of the mean-field nonlinear integro-differential equation for a population of aggregating individuals subject to stochastic fluctuations, Preprint 98–38 (SFB 359), IWR, Universität Heidelberg, Juni; 1998.
27. Nagai T. & Mimura M., Some nonlinear degenerate diffusion equations related to population dynamics, J. Math. Soc. Japan, 539–561, 35; 1983.
28. Oelschläger K. A law of large numbers for moderately interacting diffusion processes. Z. Wahrscheinlichkeitstheorie verw. Gabiete 69, 279-322; 1985.
29. Oelschläger K., Large systems of interacting particles and the porous medium equation, J. Differential Equations, 294–346, 88; 1990.
30. A. Okubo, S. Levin, Diffusion and Ecological Problems : Modern Perspectives (Springer, Berlin), 2002.
31. Ortisi, M. *Limiting Behavior of an Interacting Particle Systems*, Ph.D., University of Milano, Italy, 2007.

32. Rogers L. C. G. & Williams D., Diffusions, Markov Processes and Martingales, Vol. 1, Wiley, New York; 1994.
33. Shiryaev A. N., Probability, Springer-Verlag, New York; 1995.
34. van der Vaart, Aad W. and Wellner, Jon A., Weak Convergence and Empirical Processes, with Applications to Statistics, Springer, New York, 1996.
35. Veretennikov, A.Y. On polynomial mixing bounds for stochastic differential equations. Stochastic Processes and their Applications, **70**, 115-127, 1997.
36. Veretennikov, A.Y. On polynomial mixing and convergence rate for for stochastic differential equations. Theory Probab. Appl., **44**, 361-374, 1999.
37. Veretennikov, A.Y. On subexponential mixing rate for Markov processes. Theory Probab. Appl., **49**, 110-122, 2005.
38. Conradt, L. and Roper, T.J., Group decision-making in animals, Nature, **42**, 155-158, 2003.
39. Conradt, L. and Roper, T.J., Consensus decision-making in animals, Trends Ecol. Evolut., **20**, 449-456, 2005.
40. Meunier, H., Leca, J. B., Deneubourg, J.L., and Petit,O., Group movement in capuchin monkeys: the utility of an experimental study and a mathematical model to explore the relationship between individual and collective behaviours. Preprint, 2006.

Modelling Aspects of Cancer Growth: Insight from Mathematical and Numerical Analysis and Computational Simulation

Mark A.J. Chaplain

Division of Mathematics, University of Dundee, Dundee DD1 4HN, Scotland
chaplain@maths.dundee.ac.uk

Summary. In this chapter we present a variety of reaction-diffusion-taxis (i.e. macroscopic) models of several key stages of solid tumour growth – avascular growth, the immune response to solid tumours and invasive growth. The basis for all of the models is deriving the continuum PDE using a conservation of mass argument. In the model for avascular growth we examine the potential role pre-pattern theory (diffusion-driven instability à la Turing) may play in the generation of spatio-temporal heterogeneity of mitotic activity on the surface of multicell spheroids. In the model for the immune response to cancer, working from an initial "microscale" cell interaction scheme, we derive a system of PDEs which are used to predict the capacity of the immune system to eradicate cancer (or not). In the final model of cancer invasion, once again we initially focus on "microscale" activity of matrix degrading enzymes and their binding to cell-surface receptors to derive a PDE model of the process of cancer invasion of the local tissue. For each system, we carry out mathematical and numerical analyses of the model and perform computational simulations and attempt to draw relevant biological conclusions and make experimentally-testable predictions from the results.

1 Introduction

As is stated in the Preface of this book: *"Cancer may be regarded as a paradigmatic microcosm for all of biology - i.e. as an observable system where mutation and evolution take place."* As such, cancer provides a broad canvas on which to apply the brush strokes of mathematical modelling. In this chapter, using an artistic analogy, we will adopt the approach of Monet and the Impressionists rather than Seurat and the Pointillists, and will focus on macroscopic modelling of certain aspects of cancer growth. Of course, just as there is a link between the individual dots of a painting by Seurat and the "big picture" (really a question of "scale" or perspective), so too in mathematical modelling, as we have seen in the chapter by M. Lachowicz, do there exist links between the microscopic and the macroscopic. Although we will not deal with discrete systems in this chapter, the seminal work of Othmer and Stevens [58] provides a practical foundation for the exploration of discrete and continuum approaches to modelling biological phenomena. A specific example in the area of cancer modelling is provided by the work of Anderson and Chaplain [3] in modelling angiogenesis - the growth of new blood vessels. The remainder of this chapter is devoted then to the formulation, analysis and simulation of continuum macroscopic models of cancer growth and development, where systems of nonlinear partial differential equations play the main role.

1.1 Macroscopic Modelling

Deterministic reaction-diffusion equations have been used to model the spatio-temporal growth and spread of tumours both at an early stage in its growth [73] and at the later invasive stage [57] [30] [63]. Under the continuum hypothesis, the spatio-temporal state of a system of cells and/or chemical interactions is described by partial differential equations (PDEs) derived from considerations of conservation of matter. Suppose we have a fixed but arbitrary volume V enclosed by a smooth surface S and we consider the flux of cells across S. By the conservation of mass, the rate at which the number of cells changes within V must equal the net flux of cells across the S, plus the number of cells created or lost within V through mitosis and/or death. Thus we can write:

$$\frac{d}{dt} \int_V n(\mathbf{x}, t)\, d\mathbf{x} = \int_S -\mathbf{J}(\mathbf{x}, t) \cdot d\mathbf{S} + \int_V F(n, c_i)\, d\mathbf{V}, \qquad (1)$$

where $n(\mathbf{x}, t)$ is the density of cells at position $\mathbf{x}$ and time t; $\mathbf{J}$ is the flux of cells across S, $(= \partial V)$, per unit volume per unit time; $F(n, \mathbf{c})$ describes the net rate of mitotic, proliferating, and/or degrading cells and is generally described by a polynomial or fractional function in n and $\mathbf{c} = (c_1, c_2, \ldots, c_k)$, representing the concentrations of all chemical species that may be present. Using the divergence theorem ($\int_V \nabla \cdot \mathbf{J}\, d\mathbf{V} = \int_S \mathbf{J} \cdot d\mathbf{S}$), (1) may be written

$$\frac{d}{dt} \int_V n(\mathbf{x}, t)\, d\mathbf{V} = \int_V (-\nabla \cdot \mathbf{J} + F(n, \mathbf{c}))\, d\mathbf{V}. \qquad (2)$$

Assuming the domain is fixed in time, we may differentiate beneath the integral and then using the fact that the choice of V was arbitrary, we have that at every point $(\mathbf{x}, t)$ the following conservation equation holds:

$$\frac{\partial n}{\partial t} = -\nabla \cdot \mathbf{J} + F(n, \mathbf{c}) \qquad (3)$$

In a similar way, we may derive a partial differential equation governing the spatio-temporal evolution of the concentration of each chemical species c_i reacting with n. In the vast majority of biological situations, chemical species simply diffuse, and so the generic reaction-diffusion equation for each c_i is of the form:

$$\frac{\partial c_i}{\partial t} = D_{c_i} \nabla^2 c_i + G(\mathbf{c}, \mathbf{n}) \tag{4}$$

Systems of the above form have been used to model a wide variety of biological phenomena and a number of examples can be found in the books by [54], [24] and [65]. A more formal derivation of the equation can be found in [56].

In this chapter we examine and analyse three mathematical models of the above general form concerning three important aspects or phases of solid tumour growth: (i) avascular tumour growth; (ii) the response of the immune system to a solid tumour, and (iii) invasive cancer spread through tissue. For each model, we give a brief discussion of the underlying biology, we derive and formulate the model and then undertake a mathematical and numerical analysis. Full details of the biological systems associated with each model, as well as other details concerning the models such as parameter values and estimation, can be found in the papers of [13] [47] [48] [14] [15], while technical details of some of the numerical techniques adopted may be found in [31]. In the next section, we give a brief overview and description of how cancer arises and develops, before presenting the three mathematical models in the subsequent sections.

1.2 Cancer Growth and Development

The development of a primary solid tumour (e.g., a carcinoma) begins with a single normal cell becoming transformed as a result of mutations in certain key genes. This transformed cell differs from a normal one in several ways, one of the most notable being its escape from the body's homeostatic mechanisms, leading to inappropriate proliferation. An individual tumour cell has the potential, over successive divisions, to develop into a cluster (or nodule) of tumour cells. Further growth and proliferation leads to the development of an avascular tumour consisting of approximately 10^6 cells. The avascular tumour cannot grow any further, owing to its dependence on diffusion as the only means of receiving nutrients and removing waste products. If the development of the solid tumour were to remain in this avascular state, little or no damage would be done to the host since avascular tumours are relatively small and remain localised in the host tissue and do not spread. However two crucial and inter-linked processes permit the avascular tumour to grow further - tumour-induced angiogenesis (the recruitment of blood vessels) and tissue invasion by the tumour cells. The tumour cells first secrete angiogenic factors which in turn induce endothelial cells in a neighbouring blood vessel to degrade their basal lamina and begin to migrate towards the tumour. Endothelial cell migration through the extracellular matrix is driven by a chemotactic response to the angiogenic factors and a haptotactic response to components in the matrix such as fibronectin and collagen. The migration is facilitated by the local degradation of the tissue by the endothelial cells. As it migrates, the endothelium begins to form sprouts which can then form loops and branches through which blood circulates. From these branches more sprouts form and the whole process repeats forming a capillary network which eventually connects with the tumour, completing angiogenesis and supplying the tumour with the nutrients it needs to grow further. There is now also the possibility of tumour cells finding their way into the circulation and being deposited in distant sites in the body, resulting in metastasis. The complete process of metastasis involves several sequential steps, each of which must be successfully completed by cells of the primary tumour before a secondary tumour (a metastasis) is formed. A summary of the key stages of the metastatic cascade is as follows:

- growth of the initial avascular primary tumour;
- recruitment of new blood vessels (angiogenesis) and vascularisation of the primary tumour;
- escape of cancer cells from the primary tumour;
- local degradation of the surrounding tissue by cancer cells and continued migration;
- cancer cells enter the lymphatic or blood circulation system (*intravasation*);
- cancer cells must then survive their journey in the circulation system;
- cancer cells must escape from the blood circulation (*extravasation*);
- cancer cells (from the primary tumour) must then establish a new colony in distant organs;
- the new colony of cells must then begin to grow to form a new, secondary tumour in the new organ.

In the following sections of the chapter, we present and analyse our macroscopic mathematical models.

2 Modelling Avascular Solid Tumour Growth

2.1 Introduction

In this section, we apply reaction-diffusion pre-pattern theory to a specific problem on a spherical domain - that of a growing avascular solid tumour. We also suggest actual chemicals known to be produced by tumours (autocrine growth factors) which could give rise to the pre-patterns and examine their relevance in the light of clinical and experimental observations. From a general mathematical perspective, the model we present in this section is concerned with examining reaction-diffusion systems on the surface of the unit sphere $S = \{\mathbf{f} \in \mathbb{R}^3 : |\mathbf{f}| = 1\}$. The generic reaction-diffusion system which we will analyse (numerically) in this section may be written:

$$\mathbf{u_t} = \mathbf{D}\Delta_*\mathbf{u} + \mathbf{f(u)} \tag{5}$$

on the space-time domain $(\mathbf{f}, t) \in S \times [0, \infty)$, where $(u_1, \dots, u_s)^T = \mathbf{u} = \mathbf{u}(\mathbf{f}, \mathbf{t})$ is a vector (for example, of chemical concentrations such as growth activating and inhibiting factors), $D = \text{diag}\{d_1, \dots, d_s\}$ is a diagonal matrix of positive diffusion coefficients, Δ_* is the Laplace-Beltrami operator:

$$\Delta_* u = \frac{1}{\sin\theta} \left\{ \frac{\partial}{\partial\theta} \left(\sin\theta \frac{\partial u}{\partial\theta} \right) + \frac{1}{\sin\theta} \frac{\partial^2 u}{\partial\phi^2} \right\},$$

and $\mathbf{f} : \mathbb{R}^s \to \mathbb{R}^s$ is a (nonlinear) autonomous vector-valued function representing the reaction kinetics.

For our purposes here we restrict to the special two-species case

$$u_t = \Delta_* u + \gamma f(u, v),$$
$$v_t = d\Delta_* v + \gamma g(u, v),$$

with d, γ given positive parameters and f, g given functions, but in fact the method applies equally well to any number of chemical species. In general system (5) arises naturally in studies of pre-pattern formation in biological systems [33, 49, 53, 54, 79]. There it is of interest to study the stability of spatially homogeneous steady states of (5) with respect to the diffusion represented by $D\Delta_*\mathbf{u}$, and in particular to identify spatial patterns which evolve in practice from unstable (in the Turing sense) homogeneous steady states. In the next section we describe the application of pre-pattern theory to solid tumour growth and present some simulations of our system. A more detailed description of the background biology, modelling and numerical technique used in this section can be found in [13].

2.2 Linearised Stability Theory

The reaction kinetics governing a general reaction-diffusion system may either be a pure or cross activator-inhibitor mechanism [53]. The distinction between these two types of kinetics lies in whether the self-activating chemical (cf. [88]) either activates (pure) or inhibits (cross) the second species. There is experimental evidence which suggests both mechanisms may be applicable in the case of growth factors secreted by tumour cells (cf. [19], [55]). For illustrative purposes only, here we assume that the kinetics are a cross activator-inhibitor mechanism, and we denote the concentration of the two chemical species by u and v.

For (6) in general, a homogeneous steady state is defined to be a pair $(u_0, v_0) \in \mathbb{R}^2$ such that

$$f(u_0, v_0) = g(u_0, v_0) = 0. \tag{6}$$

In linear stability analysis, (6) is said to exhibit "diffusion driven instability" if (u_0, v_0) is a linearly stable solution of (6) when diffusion is neglected, but unstable otherwise. The ranges of values of the parameters γ and d for which such a phenomenon can arise (the so called "Turing space") can be found in [54]. In the particular case under consideration here, due to the spherical geometry, the eigenfunctions of the diffusion operator Δ_* can be chosen as the spherical harmonics:

$$Y_n^m(\mathbf{f}) = c_n^m P_n^{|m|}(\cos\theta)\exp(im\phi), \tag{7}$$
$$(\theta,\phi) \in [0,\pi] \times [0,2\pi], \quad n = 0,1,2,\dots, |m| \le n,$$

where $\mathbf{f} = \sigma(\theta,\phi) := (\sin\theta\cos\phi, \sin\theta\sin\phi, \cos\theta) \in S$, P_n^m are the associated Legendre functions and c_n^m is chosen to be the normalising factor: $c_n^m = \sqrt{\frac{2n+1}{4\pi}\frac{(n-|m|)!}{(n+|m|)!}}$. It is well known that

$$\Delta_* Y_n^m = -k^2 Y_n^m, \tag{8}$$

where $k^2 = n(n+1)$. These properties of Δ_* can be combined with the general stability theory in [54] where, in particular, it is shown that diffusion-driven instability can occur in (6), provided the values of the partial derivatives of f and g evaluated at (u_0, v_0) satisfy the inequalities

$$f_u + g_v < 0, \quad |A| > 0,$$
$$df_u + g_v > 0, \quad (df_u + g_v)^2 - 4d|A| > 0, \tag{9}$$

with A denoting the Jacobian matrix, that is $A = \begin{bmatrix} f_u & f_v \\ g_u & g_v \end{bmatrix}$. These inequalities give a range of values of d for which instability may occur and the unstable modes are the spherical harmonics (7) with $|m| \le n$, provided n lies in the range

$$\gamma L(f_u, f_v, g_u, g_v, d) < k^2 = n(n+1) < \gamma M(f_u, f_v, g_u, g_v, d) \tag{10}$$

where

$$L = \frac{[df_u + g_v] - \{[df_u + g_v]^2 - 4d|A|\}^{1/2}}{2d}, \tag{11}$$

and

$$M = \frac{[df_u + g_v] + \{[df_u + g_v]^2 - 4d|A|\}^{1/2}}{2d}. \tag{12}$$

So, if there exists at least one n satisfying (10), then there is a possibility that a trajectory starting from a random perturbation of (u_0, v_0) will evolve into a spatially heterogeneous pattern generated by the spherical harmonics $Y_n^m, |m| \le n$.

The above arguments are the main source of many results concerning spatial pattern formation, see, for example, [12], [13] [54]. For the purpose of testing this theory it is useful to note that we can isolate a specific mode for "excitation" by choosing the parameters γ and d so that the width of the interval $[\gamma L, \gamma M]$ is sufficiently narrow so that (10) is satisfied for a unique n. Then the corresponding spherical harmonics are isolated unstable modes.

Following [12] [13] we can apply this theory to case of the interaction of two chemicals on the surface of a solid spherical tumour. As stated previously, we assume that the reaction kinetics are of a cross activator-inhibitor mechanism, that is to say, regions of high concentration of the activator (or growth promoting factor, GPF) correspond to regions of low concentration of the inhibitor (or growth inhibitory factor, GIF) and vice versa [22]. Under the assumption that the growth factors are produced only by the live cells at the surface of the tumour, then the problem is essentially a 2-dimensional one on the spherical surface. The eigenfunctions are simply the surface harmonics Y_n^m and the wavenumbers in this case are given by $k^2 = n(n+1)$. Illustrations of the application of this linear stability analysis are given in [12] and [13].

Although this theory is not rigorous, it is widely known to produce results which are consistent with applications ([12, 53, 54]). Its main disadvantage is that it depends on analytic techniques and thus is restricted to reasonably simple domains and tractable kinetics. Numerical simulations, on the other hand, are much more generally applicable and so we now introduce our numerical method for simulating (5).

2.3 The Role of Pre-Pattern Theory in Solid Tumour Growth and Invasion

Application to a Spherical Tumour

Solid tumours are known to progress through two distinct phases of growth - the avascular phase and the vascular phase. During the former growth phase the tumour remains in a diffusion-limited, dormant state while during the latter growth phase, invasion and metastasis may take place. The initial avascular growth phase can be studied in the laboratory by culturing cancer cells in the form of three-dimensional *multicell spheroids* ([52, 75] and references therein). It is well known that these spheroids, whether grown from established tumour cell lines or actual *in vivo* tumour specimens, possess growth kinetics which are very similar to *in vivo* tumours. Typically, these avascular nodules may grow to a few millimetres in diameter depending on the cell types and the culture conditions used, although carcinoma *in vivo* may reach dormancy at a smaller size of between $250 - 500\mu$m. Cells towards the centre of the spheroid, being deprived of vital nutrients, die and give rise to a necrotic core. Proliferating cells can be found in the outer three to five cell layers, that is, *essentially on the surface of the tumour*. Lying between these two regions is a layer of quiescent cells, a proportion of which can be recruited into the outer layer of proliferating cells. Much experimental data has been gathered on the internal architecture of spheroids, and studies regarding the distribution of vital nutrients (for example, oxygen) and metabolites within the spheroids have been carried out [29, 86].

The transition from the dormant avascular state to the vascular state, wherein the tumour possesses the ability to invade surrounding tissue and metastasise to distant parts of the body, depends upon its ability to induce new blood vessels from the surrounding tissue to grow and eventually connect with the tumour. This permits vascular growth to take place. It is during this stage of growth that the insidious process of invasion takes place. In certain types of cancer, for example, carcinomas arising within an organ, this process typically consists of columns of cells projecting from the central mass of cells and extending into the surrounding tissue area and the local spread of these carcinomas often assume an irregular jagged shape.

Prior to successful completion of angiogenesis, the avascular tumour, although dormant (or quasi-dormant) with regard to its growth, is still very much in a "dynamic state of equilibrium", with cell birth and proliferation in balance with cell loss and death. The cancer cells are also known to produce and secrete a variety of growth-activating and growth-inhibiting chemicals [39, 66].

In formulating our mathematical model we take into account certain important experimental/biological observations from multicell spheroid studies and make other reasonable mathematical assumptions, namely:

- We assume that the tumour is perfectly spherical in shape and that it has grown in a radially symmetric manner.
- We assume that the tumour has reached its diffusion-limited avascular maximum size and consists of a large internal necrotic core surrounded by a thin layer of proliferating cells at the surface. The thin layer of live cells essentially defines the surface of the solid tumour.
- Experimental results have demonstrated that tumour cells secrete both growth-inhibiting and growth-activating chemicals in an autocrine manner [66] and that the balance and interaction between these factors play an important role in the development and progression of tumours.
- Transforming growth factor betas (TGF-βs) constitute a family of local mediators that regulate the proliferation and functions of many cell types. Indeed TGF-βs have an identified effect of specifically suppressing tumour cell proliferation in many types of cancers [41, 50, 68, 89], including carcinomas.
- TGF-βs also are known to induce apoptosis (cell death) in carcinoma cells [89] and can stimulate the synthesis of the extracellular matrix and equally importantly the tumour stroma. They have therefore been implicated in controlling cancer invasion [1].
- There is also much evidence to demonstrate that many types of tumour cells (including carcinoma cells) also secrete a variety of growth-activating factors. For example, epidermal growth factor (EGF) and transforming growth factor-α (TGF-α) [91]; basic fibroblast growth factor (bFGF) [78]; platelet-derived growth factor (PDGF) [85]; insulin-like growth factor (IGF) [67]; interleukin-1α (IL-1α) [38] and granulocyte colony-stimulating factor (G-CSF) [51].

- We assume that the production of the growth activating and growth in-hibitory factors is restricted to the thin layer of live, proliferating cells at the tumour surface.
- Not only can we identify specific growth inhibitors and activators (as op-posed to generic chemicals), but there is direct experimental evidence that in tumour cell lines these chemicals interact and modulate the effect of each other [36, 46].

In addition to the above experimental observations, it is also well known that the timescale of a growing tumour is very much slower than the diffusion timescale of chemicals. Any chemical which is produced by the tumour cells will therefore diffuse and reach a steady-state distribution within its domain on a much faster timescale than the growing tumour itself. We therefore consider the possibility of the development of a genuine heterogeneous chemical pre-pattern on the surface of a solid tumour which takes place prior to successful angiogenesis. This chemical pre-pattern predisposes cells in certain regions on the surface of the tumour (that is, in regions where the concentration of the growth-activating factor is high) to invasion and subsequently facilitates the vascular, invasive growth. Such cellular heterogeneity in tumours is well documented [7, 40, 64, 72].

The mathematical model we propose consists of a system of reaction-diffusion equations on the surface of a sphere (that is, the tumour surface), modelling the interaction of the growth-activating (u) and growth-inhibiting (v) chemicals which are produced by the tumour cells. The specific system we consider is given by:

$$
\begin{aligned}
u_t &= \Delta_* u + \gamma(a - u + u^2 v), \\
v_t &= d\Delta_* v + \gamma(b - u^2 v),
\end{aligned}
\tag{13}
$$

where, as before, Δ_* is the Laplace-Beltrami operator, d, γ, a, b are positive constants. Using the numerical scheme developed in [13], we can solve the above system on the surface of a sphere with a set of parameter values (see legend in Fig. 1) which satisfy the conditions for Turing-instability and we can obtain spatially heterogeneous steady-state distributions of the two chemicals on the surface of the tumour (cf. [12, 13]). These can be seen in Fig. 1.

It is a well-known feature of solid tumours such as carcinomas that they invade the surrounding local tissue with columns of cells projecting outward from the central mass. We suggest that while a solid tumour is in its avascular, dormant state, a steady-state chemical pre-pattern, is set up. This is biolog-ically feasible given the difference in timescales between the tumour growth rate and the diffusion rate of the chemicals ([12]). Once angiogenesis takes place and the tumour becomes vascularized, tumour cells which are located on the surface in regions of high concentrations of the growth promoting factor will be stimulated into proliferating faster and begin to invade the local tis-sue through increased migration. A chemical pre-pattern of this type is also consistent with the observation that tumours can directly manipulate their

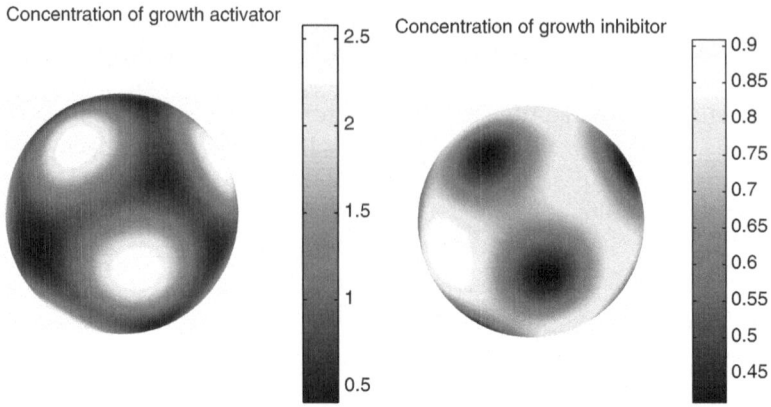

Fig. 1: Steady-state, spatially heterogeneous distributions of the chemical concentration profiles u (growth-activator) and v (growth-inhibitor) over the surface of a multicell spheroid. Initial conditions were taken as small perturbations around the spatially-homogeneous steady-state. Parameter values: $d = 25$, $\gamma = 100$, $a = 0.2$, $b = 1$

local environment by secretion of the growth factors. Thus this chemical pre-pattern will not only predispose the tumour cells to higher proliferation and increased mobility but will also directly affect the local surrounding tissue as well, thus facilitating invasion of the tissue by the cells [1, 41].

2.4 Model Extension: Application to a Growing Spherical Tumour

The results of the previous section were obtained by considering the reaction-diffusion system on a domain of fixed size, that is, the surface of the *unit sphere*. The fact that a tumour grows on a much slower timescale than the diffusion of the chemicals enabled a genuine chemical pre-pattern to form. The model, as described, is therefore most applicable when applied to a solid tumour which has already reached its diffusion-limited avascular size. However, in the case of smaller tumours which are still growing, growth promoting and growth inhibiting chemicals will still be produced by the tumour cells. These chemicals will reach a steady-state distribution (on a faster timescale than the tumour growth rate) and a pre-pattern will be formed. If the tumour is not at the stage of its growth where invasion of the tissue occurs, then it will continue to grow, the chemicals will form a new pre-pattern (on a faster timescale) and so on. Thus a more appropriate and realistic way to model the distribution of the chemicals on the surface of a growing tumour would be to consider the reaction-diffusion system on a growing, time-dependent domain.

   We therefore now consider the application of the results of previous section to the case of the growing domain described above. The reaction-diffusion system is therefore considered on the domain $S(t)$, the surface of the sphere of radius $R(t)$, that is, $S(t) = \{\mathbf{f} \in \mathbb{R}^3 : |\mathbf{f}| = R(t)\}$. The reaction-diffusion system on $S(t)$ (cf. [13]) is then

$$u_t = \frac{1}{[R(t)]^2} \Delta_* u + \gamma(a - u + u^2 v) , \tag{14}$$

$$v_t = \frac{d}{[R(t)]^2} \Delta_* v + \gamma(b - u^2 v) , \tag{15}$$

which is to be solved for functions u, v of θ, ϕ and t (cf. the formulation of [17]).

It is possible to prescribe in detail the specific growth law of an avascular tumour and then couple the ODE modelling this to (14), (15). However, since we are interested only in qualitative results, it is sufficient to consider monotonically increasing functions of time for $R(t)$, and here we restrict to the case $R(t) = 1 + \alpha t$, $\alpha > 0$, representing linear growth. We solved (14) and (15) using our numerical scheme with $\alpha = 0.1$. The actual spatio-temporal distributions of the growth activating chemical on the surface of the tumour are given in Fig. 2. Clearly one can see that the spatial pattern generated is heterogeneous and changes with time (a similar result is seen for the growth-inhibiting chemical). The results of these numerical simulations give a predictive insight into the "dynamic activity" which occurs during the growth of solid tumours and are consistent with the experimentally and clinically observed proliferative heterogeneity of cancer cells in solid tumours [7, 29, 40, 64, 72].

2.5 Discussion and Conclusions

In this section we have studied a system of reaction-diffusion equations on the surface of a sphere. We have applied the pre-pattern theory (Turing-type models) of reaction-diffusion systems to a novel biological (pathological) problem – that of the growth of solid tumours, for example, carcinomas – and, moreover, have suggested a number of specific chemicals which may be involved in this process. Finally, we have studied the system of reaction-diffusion equations on a growing domain using a moving-boundary formulation. This formulation models the dynamic process of tumour growth more realistically.

We have also shown that the spatially heterogeneous chemical pre-patterns which arise on the surface of a sphere may be an important process occurring in solid tumour growth and may help to explain certain clinically and experimentally observed phenomena in carcinoma and multicell spheroids, that is, the heterogeneous distributions of proliferating cells in carcinoma and multicell spheroids and the characteristic invasive patterns of these cancers. Of course there are many other factors and processes which are involved in tumour growth, for example, the distribution of nutrient supply to the cancer cells. These are also very important and we certainly do not claim that the results of the model provide a complete answer to the problem of cancer growth

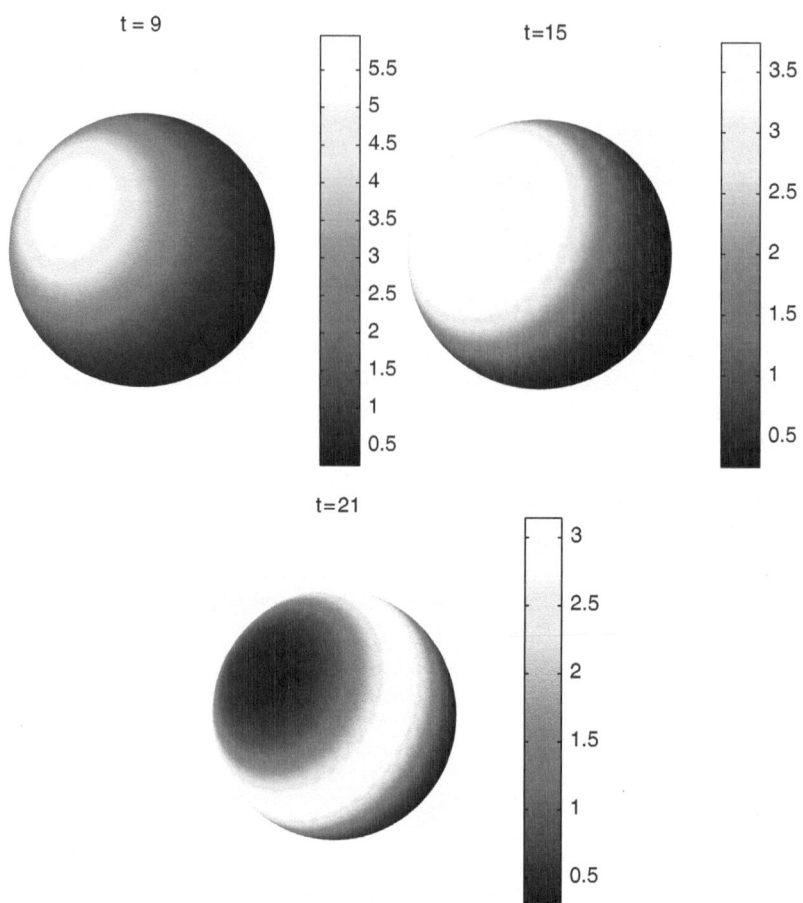

Fig. 2: Plots of the concentration profiles of the growth activating chemical, u, over the surface of the multicell spheroid at times $t = 9, 15$ and 21 in the situation of a growing tumour. As time increases the spatially heterogeneous pattern evolves with the changing chemical concentration profiles on the surface of the tumour. Parameter values $a = 0.2, b = 1, \gamma = 5, d = 100$

and invasion but rather may be an important part of the complex overall mechanisms governing solid tumour growth (cf. [66, 67, 69, 77]).

The application of our system of reaction-diffusion equations to a growing, spherical domain has enabled us to model more realistically an actual growing solid tumour and we believe that the results of the numerical simulations of Sect. 2.4 are highly consistent with in vitro experimentally observed proliferative heterogeneity of cancer cells in solid tumours at all stages of their development [7, 29, 40, 64, 72]. This aspect of reaction-diffusion theory (applications on growing domains) is attracting a good deal of interest and is

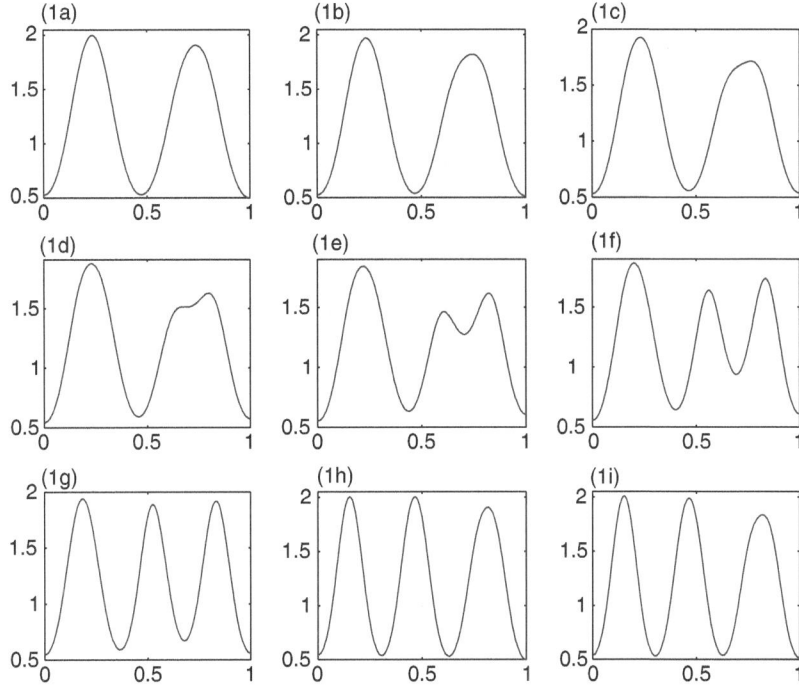

Fig. 3: Numerical simulation of (13) on a growing domain. The results show the dynamic evolution of a pattern in a 1-dimensional growing domain $x = [0, s(t)]$ where growth occurs only at the right hand boundary $x = s(t)$. The initial pattern shows "two stripes" with a third "stripe" being added once the domain has grown large enough. As the domain continues to grow further stripes are added one-by-one at the right hand boundary. See [44] for full details

providing a better explanation for many hitherto unexplained aspects of pattern formation (see, for example, [17, 18, 44]). Figure 3 shows how a pattern emerges dynamically in a 1-dimensional growing domain $x = [0, s(t)]$, which is growing only at the boundary $x = s(t)$ ([44]).

Finally, the results of the model suggest that some degree of control or regulation of cancer invasion may be possible through manipulation of the levels of growth factors as has already been suggested experimentally by [1,41], that is, it may be possible to intervene with the growth factor kinetics in such a way as to ensure that, even if one cannot halt the growth of a cancer, one may be able to prevent the highly heterogeneous distributions of proliferating cells from occurring [25, 51, 70]. Given that a solid tumour can be detected at an early enough stage in its development (for example, the avascular stage), this fact alone may prevent the irregular spread of columns of cancer cells into the surrounding tissue and may reduce the likelihood of the secondary spread of the disease.

3 Mathematical Modelling of T-Lymphocyte Response to a Solid Tumour

3.1 Introduction

"Cancer dormancy" is a term used to describe the phenomenon of a prolonged quiescent state in which tumour cells are present, but tumour progression is not clinically apparent [71, 81, 90]. As a condition, cancer dormancy is often observed in breast cancers, neuroblastomas, melanomas, osteogenic sarcomas, and in several types of lymphomas, and is often found "accidentally" in tissue samples of healthy individuals who have died suddenly. In some cases, cancer dormancy has been found in cancer patients after several years of front-line therapy and clinical remission. The presence of these cancer cells in the body determines, finally, the outcome of the disease. In particular, age, stress factors, infections, act of treatment itself or other alterations in the host can provoke the initiation of uncontrolled growth of initially dormant cancer cells and subsequent waves of metastases [81]. Recently, some molecular targets for the induction of cancer dormancy and the re-growth of a dormant tumour have been identified [32, 80]. However, the precise nature of the phenomenon remains poorly understood.

One of the main factors (but not the only one) contributing to the induction and maintenance of cancer dormancy is the reaction of the host immune system to the tumour cells [71, 81]. Indeed, tumour-associated antigens can be expressed on tumour cells at very early stages of tumour progression [16] and, as a consequence, during the avascular stage, tumour development can be effectively controlled by *tumour-infiltrating cytotoxic lymphocytes* (TICLs) [43]. The TICLs may be cytotoxic lymphocytes (CD8+ CTLs), natural killer-like (NK-like) cells and/or lymphokine activated killer (LAK) cells [21, 27, 45, 87].

In [47] we developed a mathematical model for the spatio-temporal response of cytotoxic T-lymphocytes to a solid tumour. For a particular choice of parameters the model was able to simulate the phenomenon of cancer dormancy by depicting spatially unstable and heterogeneous tumour cell distributions that were nonetheless characterized by a relatively *small total number* of tumour cells. This behaviour was consistent with several immunomorphological investigations. However, the alteration of certain parameters of the model was enough to induce bifurcations into the system, which in turn resulted in the existence of travelling-wave-like solutions in the numerical simulations. These travelling waves were of great importance because when they existed, the tumour invaded the healthy tissue at its full potential escaping the host's immune surveillance.

It is worth mentioning that the cancer dormancy solutions were characterized by an irregular evolution, which according to several objective indications, was an actual manifestation of spatio-temporal chaos. In particular, a bifurcation analysis of the ODE kinetics of our system has revealed the existence of oscillatory solutions for the ODE system emerging through a Hopf bifurcation. We have presented these results in [47] and have indicated several connections with spatio-temporal chaotic systems that couple oscillatory kinetics with diffusion (see for example the excellent work on $\lambda - \omega$ systems presented in [74]). Furthermore, we have been able to correlate numerically the existence of the stable limit cycle that emerged through the Hopf bifurcation with the irregular spatio-temporal evolution of the PDE system and the onset of cancer dormancy.

In recent years several papers have begun to investigate the mathematical modelling of the various aspects of the immune system response to cancer. The development of models which reflect several spatial and temporal aspects of tumour immunology can be regarded as the first step towards an effective computational approach in investigating the conditions under which tumour recurrence takes place and in the optimising of both spatial and temporal aspects of the application of various immunotherapies. Key papers in this area include [2, 6, 8–10, 20], which focus on the modelling of tumour progression and immune competition by generalized kinetic (Boltzmann) models and [59–61], which focus on the development of tumour heterogeneities as a result of tumour cell and macrophage interactions. Moreover, [83] is concerned with receptor-ligand (Fas-FasL) dynamics, [42] investigates the process of macrophage infiltration into avascular tumours, [47] focus on the dynamics of tumour cell-TICL interactions, and finally [28] and [76] analyze various immune system and immunotherapy models in the context of cancer dynamics.

In this section we present a travelling wave analysis of a sub-system of the model presented in [47]. The full system involved some spatially non-uniform kinetic terms through the introduction of a Heaviside function modelling some aspects of the geometry over which the system was solved. This spatial non-uniformity in the kinetics complicates the travelling wave analysis and for the sake of mathematical simplicity it is not treated here (see also the comments in [47] concerning the effect that the Heaviside function in the kinetics has on the spatio-temporal simulations). Furthermore, we also do not consider the chemotaxis aspect of the full system. Thus the model under consideration will be a nonlinear reaction-diffusion system of partial differential equations.

3.2 The Mathematical Model

For the sake of completeness we first of all introduce the complete model as it has been discussed in [47]. Let us consider a simplified process of a small, growing, avascular tumour which elicits a response from the host immune system and attracts a population of lymphocytes. The growing tumour is directly attacked by TICLs which, in turn, secrete soluble diffusible factors (chemokines). These factors enable the TICLs to respond in a chemotactic manner (in addition to random motility) and migrate towards the tumour cells. Our model will therefore consist of six dependent variables denoted E, T, C, E^*, T^* and α, which are the local densities/concentrations of TICLs, tumour cells, TICL-tumour cell complexes, inactivated TICLs, 'lethally hit' (or 'programmed-for-lysis') tumour cells, and a single (generic) chemokine respectively.

The local interactions between the TICLs and tumour cells may be described by the simplified kinetic scheme given in Figure 1 (see [47] for full details). The parameters k_1, k_{-1} and k_2 are non-negative kinetic constants: k_1 and k_{-1} describe the rate of binding of TICLs to tumour cells and detachment of TICLs from tumour cells $without$ damaging cells; k_2 is the rate of detachment of TICLs from tumour cells, resulting in an irreversible programming of the tumour cells for lysis (i.e. death) with probability p or inactivating/killing TICLs with probability $(1 - p)$. Using the law of mass action the above kinetic scheme can be "translated" into a system of ordinary differential equations. Furthermore, we consider other kinetic interaction terms between the variables and examine migration mechanisms for the TICLs, tumour cells and also consider diffusion of the chemokines. We assume that there is no "nonlinear" migration of cells and no nonlinear diffusion of chemokine i.e. all random motility, chemotaxis and diffusion coefficients are assumed constant.

We assume that the TICLs have an element of random motility and also respond chemotactically to the chemokines. There is a source term modelling the underlying TICL production by the host immune system, a linear decay (death) term and an additional TICL proliferation term in response to the presence of the tumour cells. Combining these assumptions with the local kinetics (derived from Figure 4) we have the following PDE for TICLs:

$$\frac{\partial E}{\partial t} = \overbrace{D_1 \nabla^2 E}^{\text{random motility}} \overbrace{- \chi \nabla \cdot (E \nabla \alpha)}^{\text{chemotaxis}} + \overbrace{s \cdot h(\mathbf{x})}^{\text{supply}} + \overbrace{\frac{fC}{g+T}}^{\text{proliferation}}$$

$$\overbrace{- d_1 E}^{\text{decay}} \overbrace{- k_1 ET + (k_{-1} + k_2 p)C}^{\text{local kinetics}}, \tag{16}$$

where D_1, χ, s, f, g, d_1, k_1, k_{-1}, k_2, p are all positive constants. D_1 is the random motility coefficient of the TICLs and χ is the chemotaxis coefficient. The parameter s represents the "normal" rate of flow of mature lymphocytes into the tissue (non-enhanced by the presence of tumour cells). The function $h(\mathbf{x})$ is a Heaviside function, which aims to model the existence of a subregion of the domain of interest where initially there are only tumour cells and where

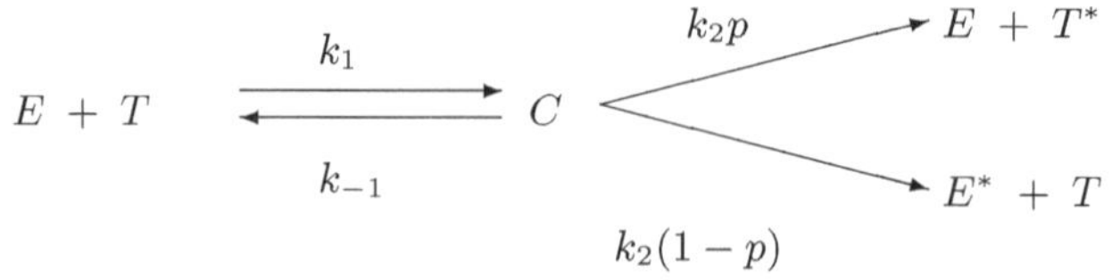

Fig. 4: Schematic diagram of local lymphocyte-cancer cell interactions

lymphocytes do not reside. This region of the domain is penetrated by effector cells subsequently through the processes of diffusion and chemotaxis only (see [47] for a full discussion regarding this assumption). The proliferation term $fC/(g+T)$ represents the experimentally observed enhanced proliferation of TICLs in response to the tumour and has been derived through data fitting [47]. This functional form is consistent with a model in which one assumes that the enhanced proliferation of TICLs is due to signals, such as released interleukins, generated by effector cells in tumour cell-TICL complexes. We note that the growth factors that are secreted by lymphocytes in complexes (e.g IL-2) act mainly in an autocrine fashion. That is to say they act on the cell from which they have been secreted and thus, in our spatial setting, their action can be adequately described by a "local" kinetic term only, without the need to incorporate any additional information concerning diffusivity.

We assume that the chemokines are produced when lymphocytes are activated by tumour cell-TICL interactions. Thus we define chemokine production to be proportional to tumour cell-TICL complex density C. Once produced the chemokines are assumed to diffuse throughout the tissue and to decay in a simple manner with linear decay kinetics. Therefore the PDE for the chemokine concentration is:

$$\frac{\partial \alpha}{\partial t} = \overbrace{D_2 \nabla^2 \alpha}^{\text{diffusion}} + \overbrace{k_3 C}^{\text{production}} - \overbrace{d_4 \alpha}^{\text{decay}}, \tag{17}$$

where D_2, k_3, d_4 are positive parameters.

We assume that migration of the tumour cells may be described by simple random motility and that on the kinetic level the growth dynamics of a solid tumour may be described adequately by a logistic term (see [47] for a full discussion concerning the validity of these assumptions). Hence the PDE governing the evolution of tumour cell density is:

$$\frac{\partial T}{\partial t} = \overbrace{D_3 \nabla^2 T}^{\text{random motility}} + \overbrace{b_1(1 - b_2 T)T}^{\text{logistic growth}} \overbrace{- k_1 ET + (k_{-1} + k_2(1-p))C}^{\text{local kinetics}}, \tag{18}$$

where D_3 is the random motility coefficient of the tumour cells, b_1, b_2, k_1, k_{-1}, k_2, p are positive parameters.

We assume that there is no diffusion of the complexes, only interactions governed by the local kinetics derived from Figure 4. The absence of a diffusion term is justified by the fact that formation and dissociation of complexes occurs on a time scale of tens of minutes, whereas the random motility of the tumour cells, for example, occurs on a time scale of tens of hours. Thus, the cell-cell complexes do not have time to move. Therefore the equation for the complexes is given by

$$\frac{\partial C}{\partial t} = \overbrace{k_1 ET - (k_{-1} + k_2)C}^{\text{local kinetics}}. \tag{19}$$

We assume that inactivated and 'lethally hit' cells (i.e. cells which will die) are quickly eliminated from the tissue (for example, by macrophages) and do not substantially influence the immune processes being analyzed. Inactivated cells also do not migrate and therefore we have:

$$\frac{\partial E^*}{\partial t} = \overbrace{k_2(1-p)C}^{\text{local kinetics}} - \overbrace{d_2 E^*}^{\text{decay}}, \tag{20}$$

$$\frac{\partial T^*}{\partial t} = \overbrace{k_2 pC}^{\text{local kinetics}} - \overbrace{d_3 T^*}^{\text{decay}}. \tag{21}$$

It is easy to see that equations (20) and (21) are only coupled to the full system through the complexes C and that neither E^* nor T^* have any effect on the variable C. Thus, equations (16), (17), (18) and (19) essentially dictate the behaviour of the complete system.

The system of equations (16), (17), (18) and (19) is closed by applying appropriate boundary and initial conditions. In the one-dimensional case, we define the spatial domain to be the interval $[0, x_0]$ and we assume that there are two distinct regions in this interval – one region entirely occupied by tumour cells, the other entirely occupied by the immune cells. We propose that an initial interval of tumour localization is $[0, l]$, where $l = 0.2x_0$. In this framework the function $h(x)$ (cf. equation 16) is defined by:

$$h(x) = \begin{cases} 0, & \text{if } x - l \leq 0, \\ 1, & \text{if } x - l > 0. \end{cases}$$

and the initial conditions are given by:

$$E(x,0) = \begin{cases} 0 & \text{if } 0 \leq x \leq l, \\ E_0(1 - \exp(-1000(x-l)^2)) & \text{if } l < x \leq x_0, \end{cases}$$

$$T(x,0) = \begin{cases} T_0(1 - \exp(-1000(x-l)^2)) & \text{if } 0 \leq x \leq l, \\ 0 & \text{if } l < x \leq x_0, \end{cases} \tag{22}$$

$$C(x,0) = \begin{cases} 0 & \text{if } x \notin [l-\epsilon, l+\epsilon], \\ C_0 \exp(-1000(x-l)^2) & \text{if } x \in [l-\epsilon, l+\epsilon], \end{cases}$$

$$\alpha(x,0) = 0, \forall x \in [0, x_0],$$

where

$$E_0 = \frac{s}{d_1}, \quad T_0 = \frac{1}{b_2}, \quad C_0 = \min(E_0, T_0), \quad 0 < \epsilon \ll 1. \tag{23}$$

In addition, zero-flux boundary conditions are imposed on the variables E, α and T. A full discussion of the biological interpretation of the particular initial and boundary conditions can be found in [47].

The closed system is non-dimensionalized by choosing an order-of-magnitude scale for the E, T and C cell densities, of E_0, T_0 and C_0 respectively, as suggested by the initial conditions. The chemokine concentration α is normalised through some reference concentration α_0 discussed in [47]. Time is scaled relative to the diffusion rate of the TICLs i.e. $t_0 = x_0^2 D_1^{-1}$ and the space variable x is scaled relative to the length of the region under consideration i.e. $x_0 = 1$ cm.

An estimation of the parameters of the system based on experimental data has been obtained in [47]. The experimental data used were concerned with *dormant* murine B cell lymphomas [81]. The corresponding numerical simulations of the non-dimensionalized system with the estimated values for the parameters under discussion were able to reproduce several characteristics of a tumour in its dormant state. Figures 5(a), 6(a) and 7(a) show the initial conditions for the TICL, tumour cell and TICL-tumour cell densities respectively. Figures 5(b)–(d) show the evolution of the (non-dimensionalized) spatial distribution of TICL density within the tissue at times corresponding to 700, 1000 and 1300 days respectively. The time instances depicted in the figures show the formation of an unstable, heterogeneous spatial distribution of TICL density throughout the tissue. Figures 6(b)–(d) show the spatial distribution of tumour cell density within the tissue at times corresponding to 700, 1000 and 1300 days. The figures show a train of solitary-like waves invading the tissue and subsequently creating a spatially heterogeneous distribution of tumour cell density throughout. Figures 7(b)–(d) show time instances of the corresponding TICL-tumour cell complex distribution at 700, 1000 and 1300 days respectively.

In addition to observing the above spatio-temporal distributions of each cell type within the tissue, the temporal dynamics of the overall populations of each cell type (i.e total cell number) was examined. This was achieved by calculating the total number of each cell type within the whole tissue space using numerical quadrature. Figure 8(a) shows the variation in the number of TICLs within the tissue over time (approximately 80 years, an estimated average lifespan). Initially, the total number of TICLs within the tissue increases and then subsequently oscillates around some stationary level (approximately 5.9×10^6 cells). Long-time numerical calculations indicated that this behaviour will persist for all time. A similar scenario is observed for the tumour cell population. From figure 8(b), we observe that initially, the tumour cell population decreases in number before subsequently oscillating around some stationary value (approximately 10^7 cells) for all time. Figure 8(c) gives the corresponding temporal dynamics of the complexes.

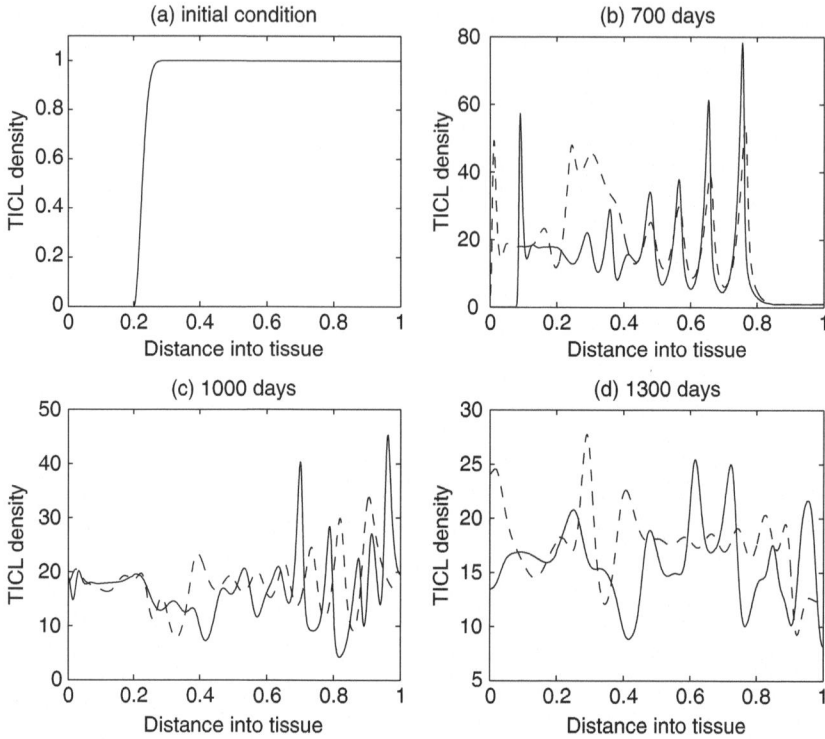

Fig. 5: Figures showing plots of the TICL density over time. As time evolves the complicated spatio-temporal dynamics can be observed. The solid line curves show results when chemotaxis of the TICLs is incorporated into the system. The dashed line curves show results when there is no chemotaxis (i.e. $\chi = 0$), only random motility of TICLs

In [47] we were able to correlate numerically the irregular spatio-temporal evolution of the system (16)–(19) depicted in the above simulations with the existence of a stable limit cycle that emerged through a Hopf bifurcation in the spatially homogeneous ODE kinetics. This can be verified via a bifurcation analysis of the non-dimensionalized homogeneous ODE kinetics of equations (16)–(19) with the Heaviside function omitted (i.e. $h(x) \equiv 1$) and with respect to parameter k_1, which is crucial in revealing the existence of the Hopf bifurcation under discussion. The bifurcation diagrams presented here have been generated with the version of the AUTO routine that is implemented within the XPP software package. Figure 9 shows part of the bifurcation diagram of TICL density E versus parameter k_1. In the case of our system AUTO was able to detect a (super-critical) Hopf bifurcation at $k_1 = 8.421 \times 10^{-8}$ day^{-1}cells^{-1}cm. The solid dots represent the maximum and minimum values of the periodic solutions that emerge when k_1 lies in a par-

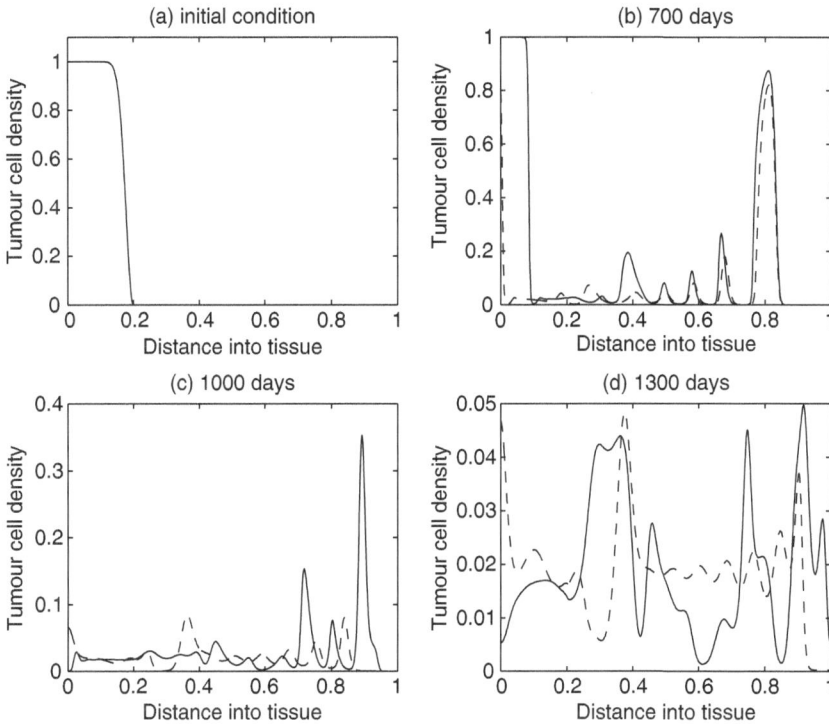

Fig. 6: Figures showing plots of the tumour cell density over time. As time evolves the complicated spatio-temporal dynamics can be observed. The solid line curves show results when chemotaxis of the TICLs is incorporated into the system. The dashed line curves show results when there is no chemotaxis (i.e. $\chi = 0$), only random motility of TICLs

ticular interval. Most of the limit cycles that emerge through this bifurcation, including the one that is generated for $k_1 = 1.3 \times 10^{-7}$ day^{-1}cells^{-1}cm (the parameter value that is associated with the irregular spatio-temporal simulations presented here), have been characterized by AUTO as stable. However there is a region around $k_1 = 1.92 \times 10^{-7}$ day^{-1}cells^{-1}cm where unstable limit cycles exist. Figure 10 shows a detailed view of this part of the bifurcation diagram. Here the solid dots represent stable limit cycle solutions, whereas the open circles represent unstable limit cycle solutions. As can be seen we have co-existence of stable and unstable limit cycles. Figure 11, which has been generated with the MATLAB continuation toolbox MATCONT, reveals the structure of the projections of the limit cycles emerging at the co-existence region to the (E, T) phase-plane. The existence of a fold or limit-point cycle (LPC) is evident.

The above spatio-temporal simulations appear to indicate that eventually the tumour cells develop very small-amplitude oscillations about a 'dormant'

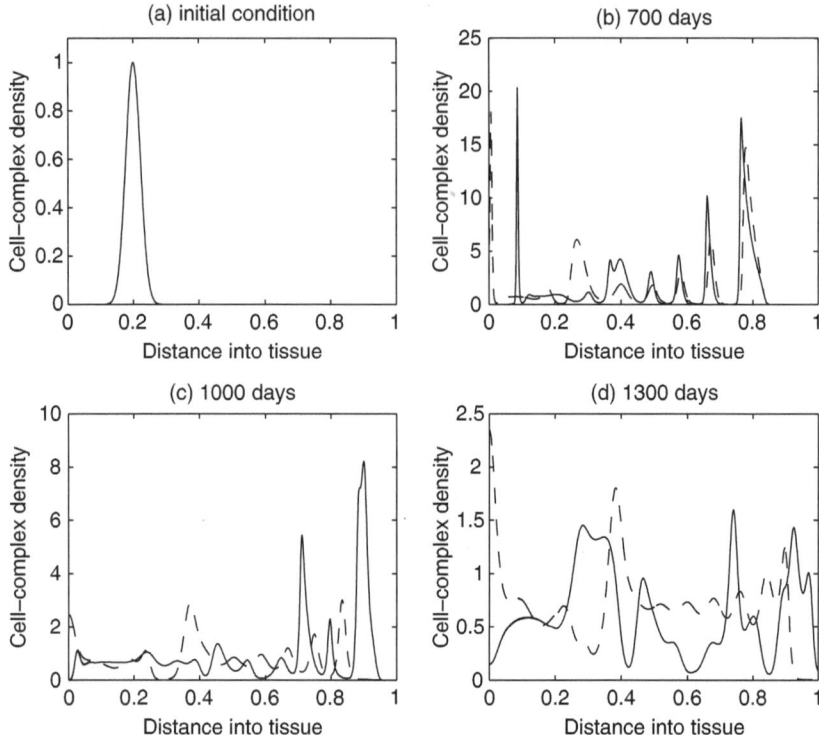

Fig. 7: Figures showing plots of the TICL-tumour-cell complex density over time. As time evolves the complicated spatio-temporal dynamics can be observed. The solid line curves show results when chemotaxis of the TICLs is incorporated into the system. The dashed line curves show results when there is no chemotaxis (i.e. $\chi = 0$), only random motility of TICLs

state, indicating that the TICLs have successfully managed to keep the tumour under control. The numerical simulations demonstrate the existence of cell distributions that are quasi-stationary in time but unstable and heterogeneous in space. However, one would expect that by reducing the probability p of tumour cells being killed by lymphocytes, travelling-wave-like solutions of more canonical nature should emerge. Indeed, reducing the parameter p in the simulations results in the emergence of solutions of a composite type consisting of steady-state and travelling-wave components both in the full system and in the system without chemotaxis (i.e the system of equations (16), (18) and (19) with $\chi = 0$). Figures 12–14 show the evolution of these composite solutions from the initial conditions for the full system (with chemotaxis incorporated).

We now investigate the travelling-wave components of the composite solutions that arise for a particular range of values of parameter p (see also the bifurcation analysis undertaken in [47]). For the sake of mathematical

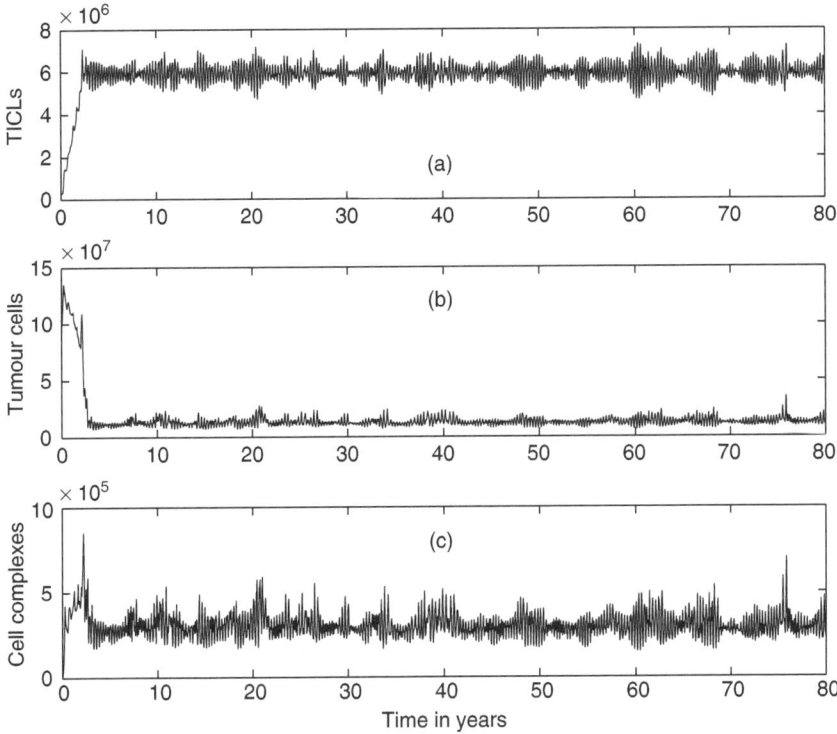

Fig. 8: Total number of (a) lymphocytes, (b) tumour cells, and (c) tumour cell-TICL complexes within tissue over a period of 80 years

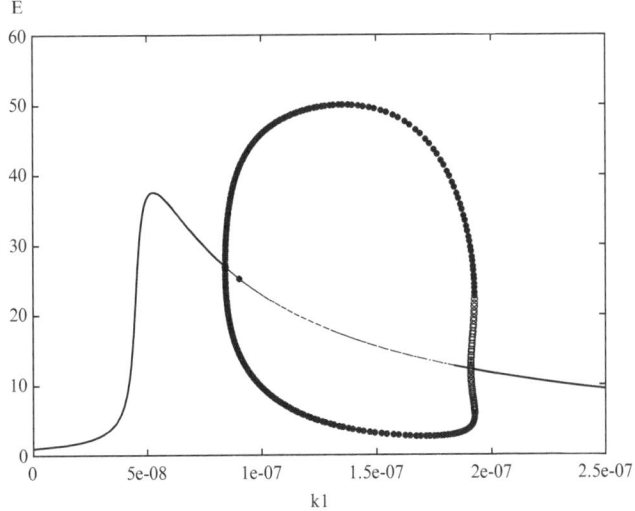

Fig. 9: Bifurcation diagram of TICL density E versus the parameter k_1

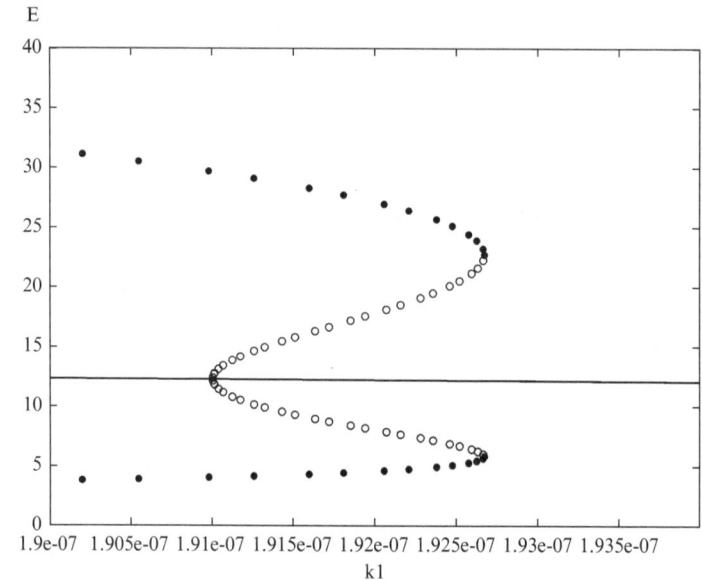

Fig. 10: Detailed view of the co-existence region of Figure 9

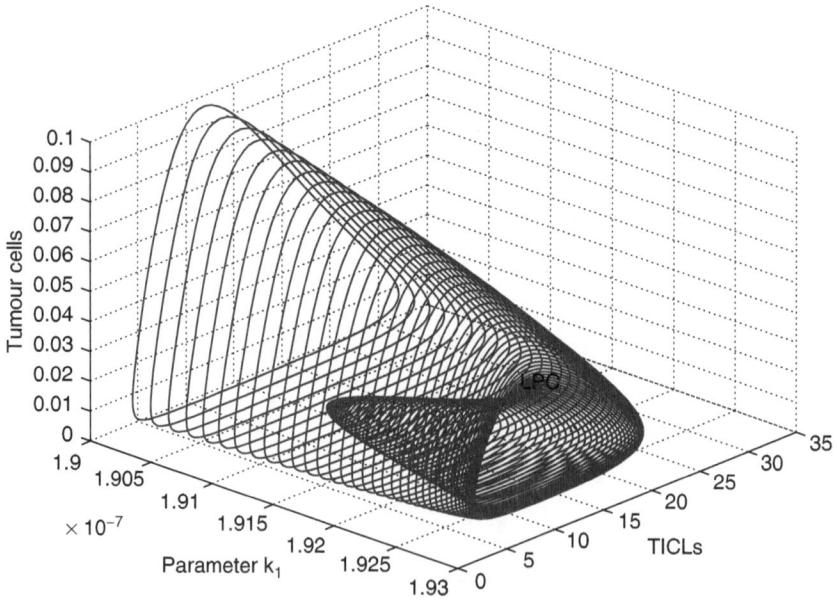

Fig. 11: Limit-cycle-projection continuation in the co-existence region

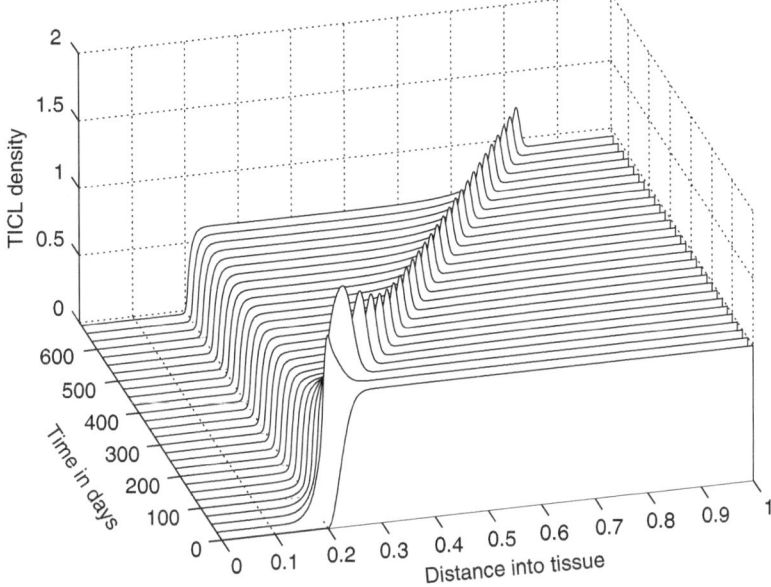

Fig. 12: The evolution from the initial conditions of a "composite" solution concerning TICL density E

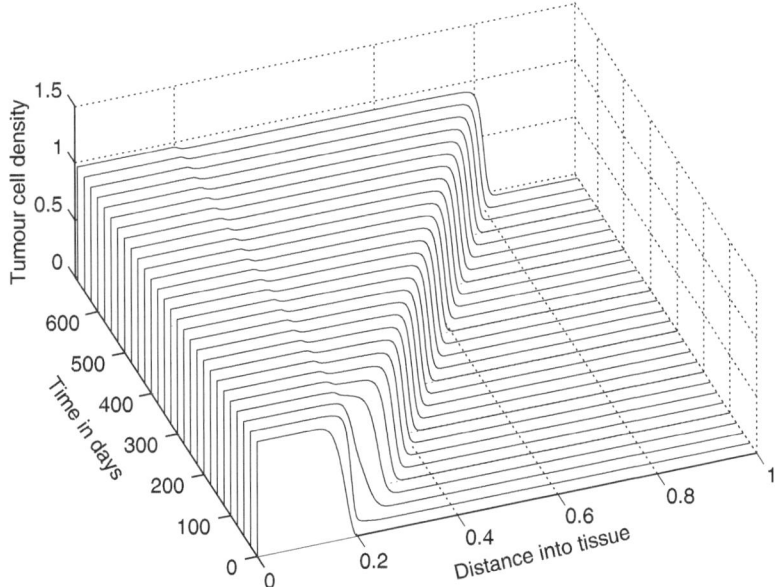

Fig. 13: The evolution from the initial conditions of a "composite" solution concerning tumour cell density T

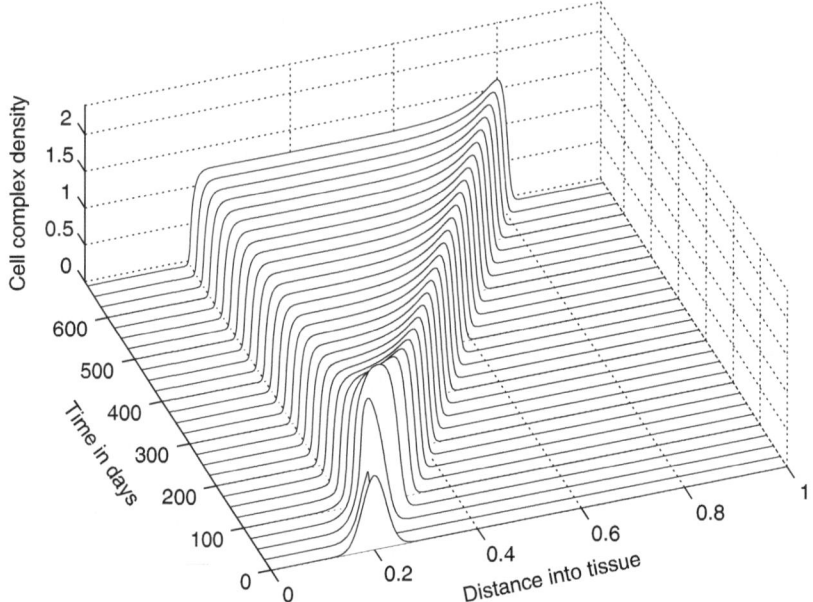

Fig. 14: The evolution from the initial conditions of a "composite" solution concerning cell-complex density C

simplicity we will not consider the effect of chemotaxis. That is to say we will investigate the solutions of the system of equations (16), (18) and (19) with $\chi = 0$. We would like to note here that this is a reasonable simplifying assumption since, according to the numerical simulations that follow, the formation of the travelling-wave components is not affected by the chemotaxis term. Furthermore, we omit the Heaviside function i.e. we set $h(x) \equiv 1$. We note that the Heaviside function is responsible for the formation of the steady-state components of the composite solutions (see also the relevant discussion in [47] [48] concerning the effect that the Heaviside function has on the spatio-temporal simulations) and thus by omitting it we choose to focus on the travelling-wave components. Specifically then, we will focus on the following non-dimensionalized reaction-diffusion system:

$$\frac{\partial E}{\partial t} = \nabla^2 E + \sigma + \frac{\rho C}{\eta + T} - \sigma E - \mu ET + \varepsilon C, \qquad (24)$$

$$\frac{\partial T}{\partial t} = \omega \nabla^2 T + \beta_1 (1 - \beta_2 T)T - \phi ET + \lambda C, \qquad (25)$$

$$\frac{\partial C}{\partial t} = \mu ET - \psi C, \qquad (26)$$

Table 1: Non-dimensionalized parameter values

$\sigma = 41200$	$\rho = 59760$	$\eta = 0.0404$	$\mu = 6.5 \times 10^7$
$\varepsilon = 31128000.01$	$\omega = 1$	$\beta_1 = 1.8 \times 10^5$	$\beta_2 = 1$
$\phi = 42912.6214$	$\lambda = 15892.19418$	$\psi = 3.12 \times 10^7$	

where

$$\sigma = \frac{st_0}{E_0} = d_1 t_0, \qquad \rho = \frac{ft_0 C_0}{E_0 T_0}, \qquad \mu = \frac{k_1 t_0 T_0 E_0}{C_0} = k_1 t_0 T_0,$$

$$\eta = \frac{g}{T_0}, \qquad \varepsilon = \frac{t_0 C_0 (k_{-1} + k_2 p)}{E_0}, \, \omega = \frac{D_3 t_0}{x_0^2} = D_3 D_1^{-1},$$

$$\beta_1 = b_1 t_0, \qquad \beta_2 = b_2 T_0, \qquad \phi = k_1 t_0 E_0,$$

$$\lambda = \frac{t_0 C_0 (k_{-1} + k_2 (1 - p))}{T_0}, \, \psi = t_0 (k_{-1} + k_2),$$

and E_0, T_0 and C_0 are the order-of-magnitude scales defined by (23). The parameter p affects the non-dimensionalized parameters ε and λ and thus the reduction of p leads to different values of ε and λ than the ones used in the tumour-dormancy simulations. In what follows we employ the parameter values given in Table 1. These are obtained from the estimated dimensional parameters by reducing the parameter p. In particular parameter p is here set at 0.99, whereas in the simulation results depicted in Figures 5–8, $p = 0.9997$.

The system of equations (24), (25) and (26) has been solved numerically over the interval $[0, 1]$ with zero-flux boundary conditions imposed and the initial conditions given by the non-dimensionalization of equations (22). Figures 15, 16 and 17 show the results of the numerical simulations, which clearly depict the evolution of standard travelling waves from the initial conditions. We note that these travelling waves – and the corresponding composite solutions of the full system – are of great biological importance because, when they exist, the tumour invades the healthy tissue at its full potential. In the next section we undertake a travelling wave analysis of system (24)–(26).

3.3 Travelling Wave Analysis

The numerical simulations of the previous section indicate that the system of equations (24), (25) and (26) exhibits travelling wave solutions for some choice of parameters. Two of the main approaches for establishing travelling wave solutions for systems of PDEs are (a) the geometric treatment of an appropriate phase-space, where one essentially is interested in intersections between unstable and stable manifolds and (b) the Leray-Schauder (degree-theoretic) method, which employs homotopy techniques (see e.g. [11, 82]).

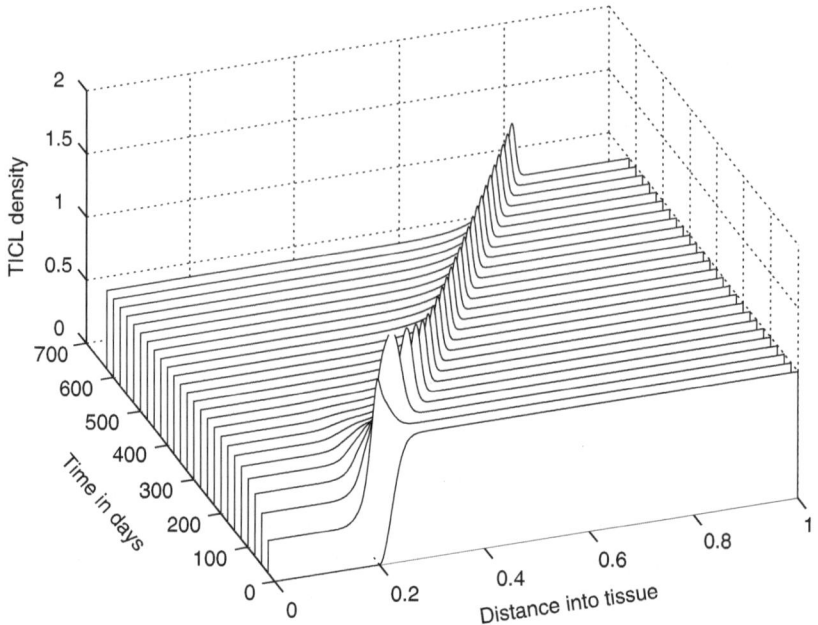

Fig. 15: The evolution from the initial conditions of the travelling wave of effector cell density

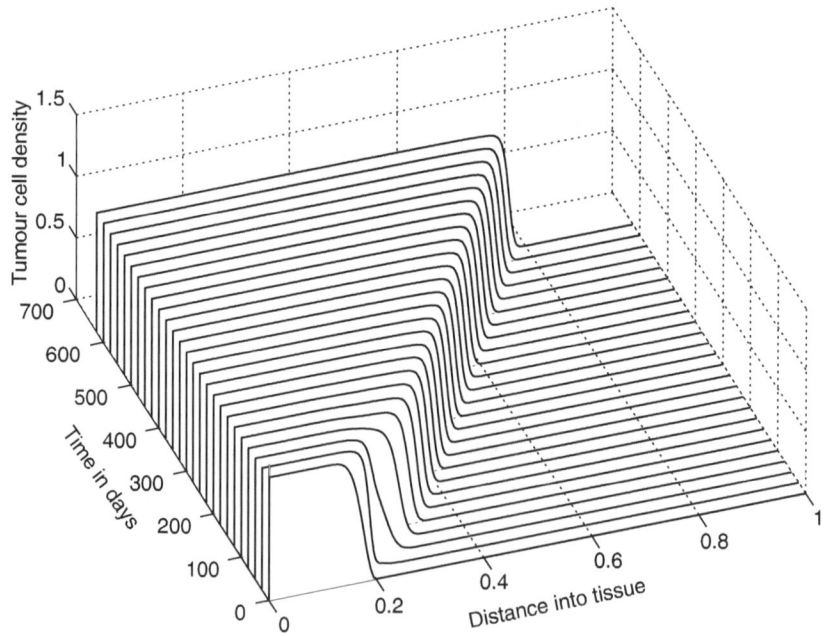

Fig. 16: The evolution from the initial conditions of the "invasive" travelling wave of tumour cell density

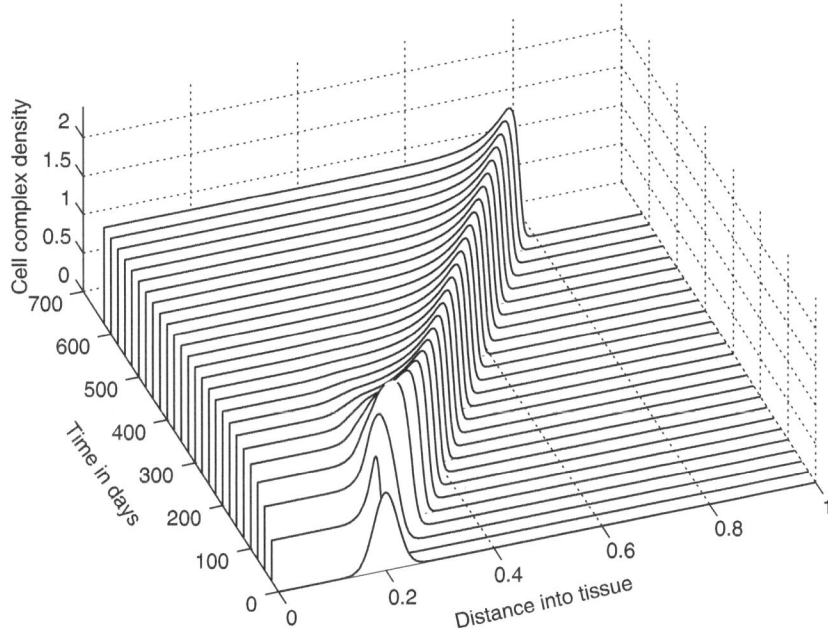

Fig. 17: The evolution from the initial conditions of the travelling wave of cell-complex density

From a numerical analysis point of view the former approach is used either in conjuction with a shooting method over a truncated domain or by trying to identify a "trivial" heteroclinic connection for some choice of parameters and then follow its deformation as the parameters are changing using numerical continuation.

In all cases the main purpose is to establish the existence of a travelling wave solution *without* any available information concerning its nature. Our approach, however, is going to be "computer-assisted" in the sense that we are going to make use of the information that the numerics of the previous section can provide us.

Since we are interested in waves travelling from the left part of the domain to the right, we specify a travelling coordinate $z = x - ct$, where $c > 0$ and we let

$$\widetilde{E}(z) = E(x,t), \ \widetilde{T}(z) = T(x,t), \text{ and } \widetilde{C}(z) = C(x,t).$$

We note that we assign the same wave velocity c to each variable, as suggested by the numerical simulations. By substituting $\widetilde{E}$, $\widetilde{T}$ and $\widetilde{C}$ into the system of equations (24), (25) and (26) and omitting the tildes for the sake of clarity we get:

$$-c\frac{dE}{dz} = \frac{d^2E}{dz^2} + \sigma + \frac{\rho C}{\eta + T} - \sigma E - \mu ET + \varepsilon C, \tag{27}$$

$$-c\frac{dT}{dz} = \omega\frac{d^2T}{dz^2} + \beta_1(1 - \beta_2 T)T - \phi ET + \lambda C, \tag{28}$$

$$-c\frac{dC}{dz} = \mu ET - \psi C. \tag{29}$$

Our intention is to take advantage of phase-space techniques and thus we formulate the system of equations (27), (28) and (29) as a dynamical system in $\mathbb{R}^5$. In particular, by defining the new variables

$$E_1 = \frac{dE}{dz} \quad \text{and} \quad T_1 = \frac{dT}{dz},$$

the system of equations (27), (28) and (29) can be formulated as

$$\frac{d\mathbf{x}}{dz} = \mathbf{f}(\mathbf{x}), \quad \text{where} \quad \mathbf{x} = \begin{pmatrix} E_1 \\ E \\ T_1 \\ T \\ C \end{pmatrix} \in \mathbb{R}^5 \tag{30}$$

and

$$\mathbf{f}(\mathbf{x}) = \begin{pmatrix} -cE_1 - \sigma - \dfrac{\rho C}{\eta + T} + \sigma E + \mu ET - \varepsilon C \\ E_1 \\ -\dfrac{c}{\omega}T_1 - \dfrac{\beta_1}{\omega}(1 - \beta_2 T)T + \dfrac{\phi}{\omega}ET - \dfrac{\lambda}{\omega}C \\ T_1 \\ -\dfrac{1}{c}\mu ET + \dfrac{1}{c}\psi C \end{pmatrix}. \tag{31}$$

Since the wave velocity c is unknown, system (30) can be regarded as a nonlinear eigenvalue problem. Several analytical methods have been developed for estimating c in this framework. However, the numerical solutions of equations (24), (25) and (26) readily yield a value of $c \approx 850$. In the analysis which follows, we therefore use this numerical estimate for c to fix the wavespeed at the constant (non-dimensional) value of 850 and hence take c as a fixed parameter.

The steady states of system (30) can be found by solving the (nonlinear) equation $\mathbf{f}(\mathbf{x}) = \mathbf{0}$. Several numerical optimization methods can be employed for this task. However, for the purposes of the travelling wave analysis, the numerical simulations of the previous section indicate that we should identify a heteroclinic connection between $\mathbf{x}^0$ and $\mathbf{x}^1$, where

$$\mathbf{x}^0 \approx \begin{pmatrix} 0 \\ 0.62 \\ 0 \\ 0.97 \\ 1.24 \end{pmatrix} \quad \text{and} \quad \mathbf{x}^1 = \begin{pmatrix} 0 \\ 1 \\ 0 \\ 0 \\ 0 \end{pmatrix} \tag{32}$$

One can improve the estimate for $\mathbf{x}^0$ by using the above value as an initial condition in an optimization algorithm. This would also confirm that $\mathbf{x}^0$ is indeed a steady state of system (30). The fact that $\mathbf{x}^1$ is also a steady state of (30) is trivial.

We are interested in the existence of an orbit $\mathbf{x}_{con}(z)$ of (30) that satisfies

$$\lim_{z \to -\infty} \mathbf{x}_{con}(z) = \mathbf{x}^0 \quad \text{and} \quad \lim_{z \to \infty} \mathbf{x}_{con}(z) = \mathbf{x}^1. \tag{33}$$

We consider the linearizations

$$\frac{d\mathbf{x}}{dz} = Df(\mathbf{x}^0)\mathbf{x} \quad \text{and} \quad \frac{d\mathbf{x}}{dz} = Df(\mathbf{x}^1)\mathbf{x} \tag{34}$$

of the vector field $\mathbf{f}$ at equilibria $\mathbf{x}^0$ and $\mathbf{x}^1$ respectively. It is a straightforward task to determine the spectrum of the Jacobian matrices $Df(\mathbf{x}^0)$ and $Df(\mathbf{x}^1)$. Indeed, there are five real eigenvalues of $Df(\mathbf{x}^0)$, three positive and two negative, with the positive ones implying the existence of a three dimensional *unstable* manifold $W^u(\mathbf{x}^0)$. Furthermore, there are five real eigenvalues of $Df(\mathbf{x}^1)$, two positive and three negative, with the negative ones implying the existence of a three dimensional *stable* manifold $W^s(\mathbf{x}^1)$. We note that

$$\dim(W^u(\mathbf{x}^0)) + \dim(W^s(\mathbf{x}^1)) = \dim \mathbb{R}^5 + 1. \tag{35}$$

Equation (35) suggests that $W^u(\mathbf{x}^0)$ and $W^s(\mathbf{x}^1)$ probably intersect transversally along an one-dimensional curve in the five-dimensional phase-space [34], [35]. If this is the case then this curve would define a (generic) heteroclinic connection.

The values of the parameters of the system under discussion suggest that an approximation of the connecting orbit by perturbing the system of equations (27), (28) and (29) can be feasible. In particular we perturb equations (27) and (28) by ignoring the effect of the second derivatives on the system (see also [23]). That is to say, we consider the perturbed system:

$$-c\frac{dE}{dz} = \sigma + \frac{\rho C}{\eta + T} - \sigma E - \mu ET + \varepsilon C, \tag{36}$$

$$-c\frac{dT}{dz} = \beta_1(1 - \beta_2 T)T - \phi ET + \lambda C, \tag{37}$$

$$-c\frac{dC}{dz} = \mu ET - \psi C. \tag{38}$$

We note here that by ignoring the second derivatives in effect we choose to focus on a *first order* approximation to equations (27) and (28).

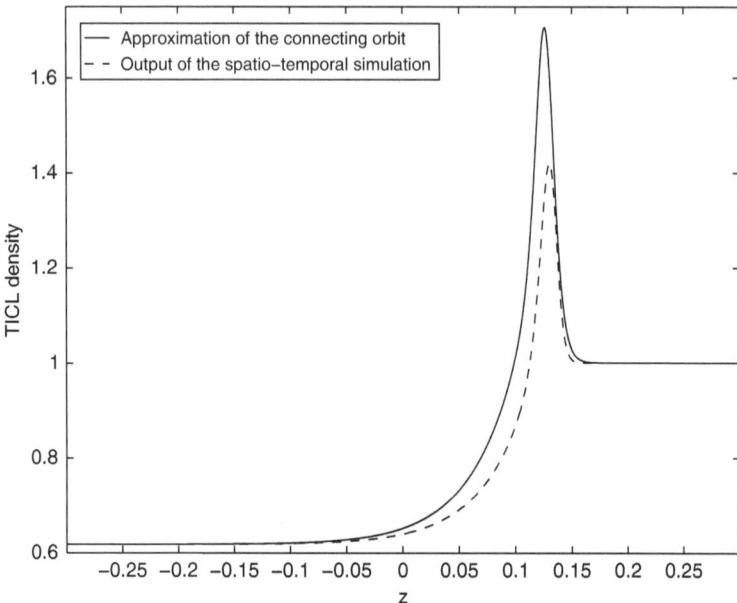

Fig. 18: Figure showing the approximation of the connecting orbit in the (E, z)-plane from the travelling wave analysis (solid line). The orbit was computed over the truncated domain $[-0.3, 0.3]$

Let $\Pi(\mathbf{x}^0)$ and $\Pi(\mathbf{x}^1)$ be the projections of $\mathbf{x}^0$ and $\mathbf{x}^1$ onto the phase-space defined by equations (36), (37) and (38). It is obvious that $\Pi(\mathbf{x}^0)$ and $\Pi(\mathbf{x}^1)$ are steady states of the perturbed system. There is a three-dimensional unstable manifold $W^u(\Pi(\mathbf{x}^0))$ associated with $\Pi(\mathbf{x}^0)$ and a one-dimensional stable manifold $W^s(\Pi(\mathbf{x}^1))$ associated with $\Pi(\mathbf{x}^1)$. We have used the computational package XPP to investigate numerically the phase-space of the system of equations (36), (37) and (38). XPP provides the implementation of numerical algorithms for tracking one-dimensional invariant manifolds and in the case of $\Pi(\mathbf{x}^1)$ it was able to confirm that $W^s(\Pi(\mathbf{x}^1))$ defines a heteroclinic connection between $\Pi(\mathbf{x}^0)$ and $\Pi(\mathbf{x}^1)$. Figures 18, 19 and 20 show approximations to the connecting orbit defined by $W^s(\Pi(\mathbf{x}^1))$ in the (E, z), (T, z) and (C, z) planes respectively. These compare very well with the results of the spatio-temporal simulations of the full PDE system.

3.4 Discussion

We have undertaken a travelling wave analysis of a mathematical model developed in [47], describing the growth of solid tumours in the presence of an immune system response. For a particular choice of parameters the model is able to simulate the phenomenon of cancer dormancy; a clinical condition

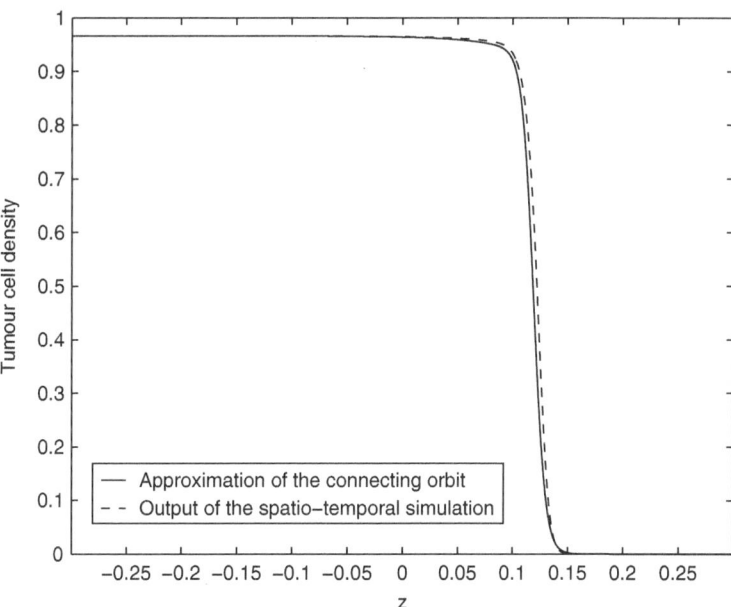

Fig. 19: Figure showing the approximation of the connecting orbit in the (T, z)-plane (solid line). The orbit was computed over the truncated domain $[-0.3, 0.3]$

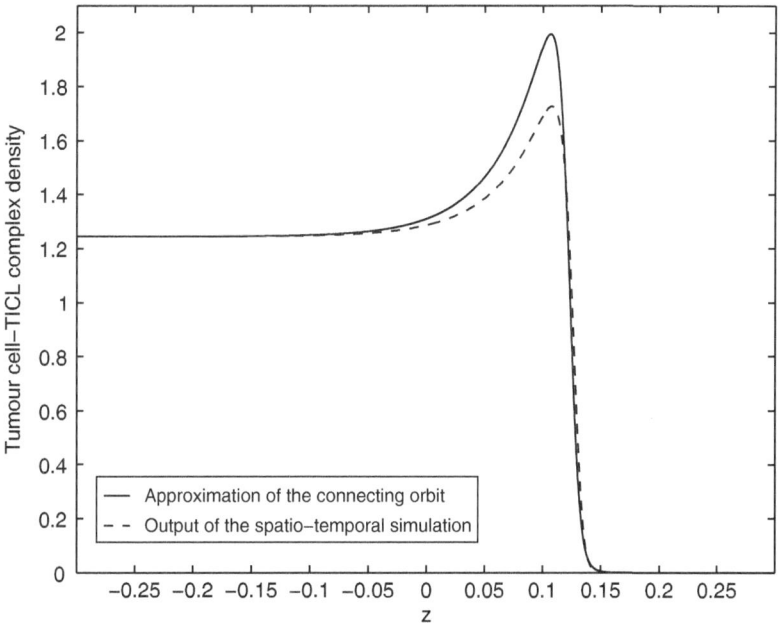

Fig. 20: Figure showing the approximation of the connecting orbit in the (C, z)-plane (solid line). The orbit was computed over the truncated domain $[-0.3, 0.3]$

that has been observed in breast cancers, neuroblastomas, melanomas, osteogenic sarcomas, and in several types of lymphomas. The behaviour of the cancer dormancy simulations can be described as highly irregular, depicting unstable and heterogeneous tumour cell distributions that are nonetheless characterized by a reletively low total number of tumour cells. This behaviour is consistent with several immunomorphological investigations with tumour spheroids infiltrated by TICLs.

However, the alteration of certain parameters of the model is enough to induce bifurcations into the system, which in turn result in the evolution of travelling-wave-like solutions in the numerical simulations. These travelling waves are of great importance because, when they exist, the tumour invades the healthy tissue at its full potential. The existence of these travelling waves for a particular choice of parameters was established for a reduced system, which nonetheless captures the essential elements of the full model presented in [47].

We would like to note here that the first order approximation employed in deriving equations (36)–(38) could be further investigated in the context of geometric perturbation theory. More precisely, we believe that Fenichel's invariant manifold theorem [26, 34] has a special role to play here and, as a matter of fact, it seems that one could try to analyse system (30) by employing the relevant techniques discussed in [34]. Nonetheless, several problems arise in this direction with perhaps the most prominent of them being the unavailability of simple analytic expressions for the steady states of the system under discussion.

The work presented here provides an adequate framework for studying and identifying critical parameters of the process in which cancer cells are present in a tissue but do not clinically occur for a long period of time, but can begin to grow progressively at a later date. Thus our modelling and analysis offers the potential for quantitative analysis of mechanisms of tumour-cell-host-cell interactions and for the optimization of tumour immunotherapy and genetically engineered anti-tumour vaccines.

4 Mathematical Modelling of Cancer Invasion

4.1 Introduction

As has already been described in this chapter, *in vivo* cancer growth is a complicated phenomenon involving many inter-related processes at many spatial and temporal scales. Solid tumour growth occurs in two distinct phases, the initial growth being characterised as the (relatively harmless) avascular phase (see section 2), the later growth as the vascular phase. Also, as we have seen in section 3, during the early avascular stage of solid tumour growth there may also be an immune response to the cancer from the host, with cells of the immune system (most notably T-lymphocytes) responding to and attacking the cancer cells. However, unfortunately solid tumours do not always remain avascular. The transition from avascular growth to vascular growth depends upon the crucial process of angiogenesis and is necessary for the tumour to attain nutrients and dispose of waste products. To achieve vascularization, tumour cells secrete a diffusible substance known as tumour angiogenesis factor (TAF) into the surrounding tissue. This has the effect of stimulating nearby capillary blood vessels to grow towards and penetrate the tumour, re-supplying the tumour with vital nutrient. Invasion and metastasis can now take place. By the time a tumour has grown to a size whereby it can be detected by, in the case of breast-cancer, simple self-examination, there is a strong likelihood that it has already reached the vascular growth phase. The primary aim of screening and the associated image enhancement technologies is therefore to detect cancers prior to this stage.

In this section we present and analyse a mathematical model of the urokinase plasminogen activation system, its role in tissue invasion, metastasis, tumour heterogeneity, and investigate its clinical implications. The interactions of cancer cells and the various chemical species are described by a system of taxis-diffusion-reaction equations of the form

$$\partial_t n = \varepsilon \Delta n - \nabla \cdot \left(n \sum_{j=1}^{l} p_j(\mathbf{c}) \nabla c_j \right) + f_0(n, \mathbf{c}), \tag{39}$$

$$\partial_t \mathbf{c} = \mathbf{D}\Delta \mathbf{c} + \mathbf{g}_0(n, \mathbf{c}), \tag{40}$$

for $(t, \mathbf{x}) \in (0, T] \times \Omega$ and where we denote the time and space dependent concentrations of the chemical species by the vector valued function

$$\mathbf{c} : [0, T] \times \bar{\Omega} \to \mathbb{R}^l.$$

and we denote the density of the cancer cells by

$$n : [0, T] \times \bar{\Omega} \to \mathbb{R}.$$

Here, $\Omega \subset \mathbb{R}^d$, $d = 1$ or $d = 2$, is a bounded domain and $[0, T]$ is the time interval of interest. Furthermore, the cell random motility coefficient ε and the diagonal matrix $\mathbf{D}$ of chemical diffusion coefficients, the taxis functions associated with each chemical c_j, denoted by $p_j : \mathbb{R}^l \to \mathbb{R}, j = 1, \ldots, l$, as well as the reaction terms $f_0 : \mathbb{R} \times \mathbb{R}^l \to \mathbb{R}$ and $\mathbf{g}_0 : \mathbb{R} \times \mathbb{R}^l \to \mathbb{R}^l$ are given. The temporal derivative is denoted by ∂_t and spatial gradient operator by ∇. The PDE system 39 is supplied with appropriate initial and boundary conditions.

In Sect. 4.2, we give a brief overview of the invasion of tissue by cancer cells and of the formation of metastasis. Particular emphasis is placed on the processes of proteolysis and extracellular matrix degradation. In Sect. 4.4, we describe the mathematical model of the urokinase plasminogen activation system. In Sect. 4.6, we undertake a linear stability analysis of a spatially uniform steady state of the model. This analysis gives rise to the observation that the model allows for taxis-driven instability of that steady state. In Sect. 4.9, we present simulations of the model reinforcing the results obtained from the linear stability analysis. Here we also comment briefly on appropriate numerical schemes for the simulation of the model equations. Finally, we conclude this section with a discussion of implications of the results obtained and a summary in Sect. 4.12.

4.2 Cancer Invasion of Tissue and Metastasis

As has been noted in the previous section, during its early growth stage a solid tumour is relatively harmless and is still *avascular*, i.e. it lacks its own network of blood vessels for supplying nutrients, including oxygen, and for removing wastes. The critical event that converts the small, localised mass of cancer cells into a rapidly growing malignancy comes when the tumour becomes *vascularized*, whereby the tumour acquires its own blood supply and microcirculation. A vascularized tumour has two distinct advantages over an avascular tumour – it receives a direct supply of nutrients which results in a rapid increase in growth (tumour mass); it can shed cells directly into the bloodstream. In addition to blood-borne spread via angiogenesis, cancers also possess the ability to actively invade the local tissue. Invasion and metastasis are the most insidious and life-threatening aspects of cancer.

Whether physiological or malignant invasion, the regulation for its necessary events involves spatial and temporal coordination, as well as certain cyclic "on-off" processes, at the level of individual cells. Motility, coupled with regulated, intermittent adhesion to the extracellular matrix and degradation of matrix molecules, allows an invading cell to move through the three-dimensional tissue matrix. At the leading edge of the motile cell, receptor-ligand and proteolytic-antiproteolytic complexes coordinate sensing, protrusion, burrowing and traction of the cell. The most significant turning point in the disease (cancer), however, is the establishment of metastasis. The metastatic spread of tumour cells is the predominant cause of cancer deaths, and with few exceptions, all cancers can metastasize. Metastases can appear shortly after surgery but can also remain undetected for more than a decade before manifesting themselves clinically. This indicates that disseminated cancer cells can persist in a dormant state, unable to form a progressively increasing tumor mass. Such heterogeneity of outcome indicates that the fate of tumour cells that disseminate to distant organs before surgery must be regulated by either inherent cancer cell properties or the milieu of the target organs, or both. Identifying the mechanisms that keep metastases in their dormant, occult state is one of the most challenging and important avenues of cancer research.

4.3 Proteolysis and Extracellular Matrix Degradation

The prognosis of a cancer is primarily dependent on its ability to invade and metastasize, and a crucial component of these processes is the degradation of extracellular matrix. Many steps that occur during tumour invasion and metastasis (as well as in a number of distinct physiological events in the healthy organism) require the regulated turnover of extracellular matrix (ECM) macromolecules. A more localized degradation of matrix components is required when cells migrate through a basal lamina. It is now widely believed that the breakdown of these barriers is catalyzed by proteolytic enzymes (proteases) released from the invading cancer cells. Most of these proteases belong to one of two general classes: matrix metalloproteases [62], or serine proteases [4] [5]. Proteases give cancers their defining deadly characteristic – the ability of malignant cells to break out of tissue compartments.

The enzymatic system we will focus on modelling in this section is the urokinase plasminogen activator system which consists of the urokinase plasminogen activator (uPA), the urokinase plasminogen activator receptor (uPAR), the extracellular matrix protein vitronectin (VN), and the plasminogen activator inhibitors type-1 (PAI-1). Full details of this system and the mathematical model can be found in Chaplain and Lolas (2005). We give a brief overview here.

uPA is an extracellular serine protease produced by cells as a single-chain proenzyme pro-uPA. Two major functional domains make up the uPA molecule: the protease domain and the growth factor domain (not discussed here). The protease moiety activates plasminogen which in turn generates plasmin, a serine protease capable of digesting basement membrane and extracellular matrix proteins. Plasmin itself is a broadly acting enzyme that not only catalyzes the breakdown of many of the known ECM and basement membrane molecules, such as vitronectin, fibrin, laminin and collagens, but also may activate metalloproteinases. As we have described in the previous sections, plasmin is a protease which is generated at the cell surface from its inactive precursor, plasminogen, via the proteolytic activity of urokinase plasminogen activator (uPA). In addition to fibrin, plasmin cleaves many extracellular matrix proteins, including fibronectin, laminin, vitronectin and thrombospondin and can activate many of the matrix metalloproteinases which degrade still other matrix constituents. Plasmin also can affect the activity of cytokines and growth factors, notably TGF-beta, which influences the composition of the extracellular milieu. In this regard, the unrestrained generation of plasminogen activator (uPA) is potentially hazardous to cells. Therefore, to maintain tissue *homeostasis* and avoid unrestrained tissue damage, plasmin activity must be tightly controlled. Such regulation is achieved at multiple "checkpoints" within the plasminogen system. A primary role in plasmin regulation is played by the availability of the plasminogen activators and their corresponding inhibitors.

Thus, the unrestrained generation of plasmin from plasminogen by the action of the uPA is potentially hazardous to cells. In this regard, the process of plasminogen activation in a healthy organism is strictly controlled through the availability of uPAs, localized activation, and interaction with specific inhibitors (PAIs). One of these inhibitors, PAI-1, which is believed to be the most abundant, fast-acting inhibitor of uPA *in vivo*. In other words, for cells to protect themselves they must secrete a surplus of inhibitors to guarantee restraint of pericellular proteolysis. Indeed secreted uPA is often associated with plasminogen activator inhibitor-1 (PAI-1) and remains inactive.

This linkage suggests to us that four molecules: uPA, PAI-1, uPAR and vitronectin, constitute the core of an integrated dynamical system which allows spatial and temporal rearrangements of its components at cell surfaces during cell migration and invasion. Moreover, it has become clear that the system has a multi-functional role in tumour biology. The system seems to function not only in cancer cell migration and invasion, but also in remodelling of the tissue surrounding the cancer cells, which may contribute decisively to the overall process of metastasis.

4.4 The Mathematical Model of Proteolysis and Cancer Cell Invasion of Tissue

In this section we present the mathematical model of cancer cell invasion of tissue, based on an explicit consideration of plasmin production, and we investigate how interactions between cancer cells, urokinase plasminogen activator (uPA), plasminogen activator inhibitor-1 (PAI-1), plasmin, and the extracellular matrix substrate (ECM, or a component of ECM such as vitronectin) may regulate tumour invasion and metastasis. In the model we make the basic assumption that the concentration of urokinase plasminogen activator receptor (uPAR) is proportional to the cancer cells density and therefore do not explicitly model the evolution of uPAR.

In what follows, we denote cancer cell density by $n(t, \mathbf{x})$, urokinase plasminogen activator (uPA) concentration by $u(t, \mathbf{x})$, plasminogen activator inhibitor-1 (PAI-1) concentration by $p(t, \mathbf{x})$, plasmin concentration by $m(t, \mathbf{x})$, and the extracellular matrix substrate (ECM, vitronectin) density by $v(t, \mathbf{x})$. We consider the following fundamental biological interactions that are known to occur during the invasion process.

Cancer Cells

When considering the spatio-temporal evolution of the cancer cell density $n(t, \mathbf{x})$, the dominant factors governing the cancer cell locomotion are random motion, chemotaxis due to urokinase plasminogen activator (uPA) and plasminogen activator inhibitor-1 (PAI-1) as well as haptotaxis due to vitronectin (VN) and other ECM components. Besides locomotion, we also include cancer cell proliferation in the model and assume a logistic growth law.

Representing by D_n the cell random motility, χ_n uPA-mediated chemotaxis, ζ_n PAI-1-mediated chemotaxis, and ξ_n VN-mediated haptotaxis coefficients, as well as by μ_1 the cancer cell proliferation rate and by n_0 the maximum sustainable cell density for cancer cells, the cancer cell equation takes the following mathematical form:

$$\partial_t n = \underbrace{D_n \Delta n}_{\text{random motion}} - \nabla \cdot (\underbrace{\chi_n n \nabla u}_{\text{uPA-chemo}} + \underbrace{\zeta_n n \nabla p}_{\text{PAI-1-chemo}} + \underbrace{\xi_n n \nabla v}_{\text{VN-hapto}}) + \underbrace{\mu_1 n (1 - n n_0^{-1})}_{\text{proliferation}} .$$

(41)

Extracellular Matrix

It is known that ECM does not diffuse and therefore we omit any diffusion term (or other "migration" terms) from its model equation. Furthermore, based on the experimental evidence that uPA activates plasminogen to produce the cancer cell-surface associated protein plasmin which in turn catalyzes the breakdown of many of the known ECM and basement membrane molecules we assume that plasmin degrades the extracellular matrix upon contact. Moreover, several studies suggest that normal cells, as well as cancer cells such as gliomas, have the ability to produce numerous ECM components. On the other hand, as has previously been mentioned, the major role of PAI-1 is to neutralise uPA production and thus we assume that PAI-1 binding to uPA results indirectly in the production of VN (vitronectin). Finally, we include a logistic reaction term in the model in order to account for the remodelling of the extracellular matrix by other cells, e.g. fibroblasts, in the tissue.

Denoting by δ the degradation rate, μ_2 the proliferation rate, v_0 the maximum sustainable density for the extracellular matrix, ϕ_{21} the production rate of PAI-1/uPA binding, and ϕ_{22} the neutralisation of PAI-1 due to binding to VN, we arrive for the ECM density $v(t, \mathbf{x})$ at the equation:

$$\partial_t v = - \underbrace{\delta v m}_{\text{degradation}} + \underbrace{\phi_{21} u p}_{\text{uPA/PAI-1}} - \underbrace{\phi_{22} v p}_{\text{PAI-1/VN}} + \underbrace{\mu_2 v (1 - v v_0^{-1})}_{\text{proliferation}} . \qquad (42)$$

Urokinase Plasminogen Activator

The evolution of the concentration of the urokinase plasminogen activator (uPA) $u(t, \mathbf{x})$, which is secreted by the cancer cells and acts as a chemoattractant, is assumed to occur through diffusion driving its motion, cancer cells acting as sources, while its binding to PAI-1 as well as to cancer cell surface receptors (uPAR) dominates its removal from the system.

If we denote by D_u the uPA's diffusion coefficient, α_{31} its rate of production by cancer cells, ϕ_{31} its rate of neutralisation by PAI-1 inhibition, and ϕ_{33} its rate of binding to uPAR cell-surface receptors, we have:

$$\partial_t u = \underbrace{D_u \Delta u}_{\text{diffusion}} - \underbrace{\phi_{31} p u}_{\text{uPA/PAI-1}} - \underbrace{\phi_{33} n u}_{\text{uPA/uPAR}} + \underbrace{\alpha_{31} n}_{\text{production}} . \qquad (43)$$

Plasminogen Activator Inhibitor-1

The conservation equation for the concentration of the plasminogen activator inhibitor-1 (PAI-1) $p(t, \mathbf{x})$ is similar to that of uPA considered above. Thus, diffusion drives its motion, while its production is a result of plasmin activation and PAI-1's neutralisation in the system occurs by its binding to VN or to uPA.

Denoting by D_p the PAI-1 diffusion coefficient, α_{41} the rate of production as a result of plasmin formation, ϕ_{41} the neutralisation rate due to uPA binding and by ϕ_{42} the neutralisation rate due to VN binding we have:

$$\partial_t p = \underbrace{D_p \Delta p}_{\text{diffusion}} - \underbrace{\phi_{41}\, p\, u}_{\text{PAI-1/uPA}} - \underbrace{\phi_{42}\, p\, v}_{\text{PAI-1/VN}} + \underbrace{\alpha_{41}\, m}_{\text{production}} . \tag{44}$$

Plasmin

In examining the conservation of mass regarding plasmin concentration $m(t, \mathbf{x})$ we note that diffusion drives its motion. Furthermore, we assume that binding of uPA to uPAR provides the cell surface with a potential proteolytic activity via activation and cell-surface co-localization of plasminogen and thus leads to plasmin formation. Additionally, we assume that the binding of PAI-1 to VN indirectly results in the binding of uPA to uPAR and therefore in enhanced plasmin formation. However, the activity of uPA, and therefore the formation of plasmin, is inhibited by the binding of the serine protease inhibitor-1 (PAI-1) to uPA. The inhibiting effect on that process due to PAI-1/uPA binding is already taking into account in the uPA equation as the neutralisation term $-\phi_{31} p u$ for uPA and hence leads to a reduction of the uPA concentration. The enhancing effect of PAI-1/VN binding on the process of plasmin production is taken into account by modifying the production rate ϕ_{53} by an appropriately chosen factor to account for this effect. Finally, we also include a term representing natural decay of plasmin at rate ϕ_{54}. Hence we arrive at the following equation

$$\partial_t m = \underbrace{D_m \Delta m}_{\text{diffusion}} + \underbrace{(1 + \phi'_{53}\, p\, v)}_{\text{factor}} \underbrace{\phi_{53}\, u\, n}_{\text{uPA/uPAR}} - \underbrace{\phi_{54} m}_{\text{decay}}, \tag{45}$$

where we denote by D_m the plasmin diffusion coefficient, ϕ_{53} its rate of production due to uPA/uPAR binding, ϕ'_{53} is a constant describing the strength of the enhancement of the plasmin production due to PAI-1/VN binding, ϕ_{52} its rate of production due to PAI-1/VN binding, ϕ_{51} its inactivation rate due to uPA inhibition by PAI-1/uPA binding, and ϕ_{54} is a decay constant.

The complete system of five equations describing the interactions of cancer cells, ECM, uPA, PAI-1 and plasmin is given by equations (41, 42, 43, 44, 45).

We consider the system to hold on a bounded spatial domain $\Omega \subset \mathbb{R}^d$, for $d = 1$ or $d = 2$, representing a region of tissue. In Sect. 4.9 we present simulation results for different cases of domains, each parametrised by a positive parameter M: the one-dimensional domain $\Omega_1 := (0, M) \subset \mathbb{R}$, and the two-dimensional circular domain $\Omega_3 := \{\mathbf{x} \in \mathbb{R}^2 : |\mathbf{x}| < M\}$.

The system must be closed by appropriate boundary and initial conditions for each of the dependent variables. We assume that cancer cells, and as a consequence uPA, PAI-1 and plasmin, remain within the domain of tissue Ω under consideration and therefore zero-flux boundary conditions are imposed on $\partial\Omega$, the boundary of Ω. For the ECM density $v(t, \mathbf{x})$ no boundary conditions can be prescribed.

As for the initial conditions, we assume that initially there is a cluster of cancer cells already present at $\mathbf{x} = \mathbf{0}$ and that they have penetrated a short distance into the extracellular matrix, while the remaining space is occupied by the matrix alone. Additionally, we assume that the uPA protease as well as the PAI-1 inhibitor initial concentration are proportional to the initial cancer cell density while the plasmin protease is not yet produced by the cancer cells. We give precise functional forms of the initial data in the end of the following section on the nondimensionalisation of the model equation systems, see 51.

4.5 Nondimensionalisation of the Model Equations

We recast the PDE system in terms of dimensionless variables, rescaling distance with the maximum distance of the cancer cells at this early stage of invasion $L = 0.1$ cm, time with $\tau = L^2 D^{-1}$, where D represents a chemical diffusion coefficient $\sim 10^{-6} \mathrm{cm}^2 \mathrm{s}^{-1}$, and the dependent variables n, v, u, p, m with appropriate reference density and concentration values n_0, v_0, u_0, p_0, m_0, i.e.

$$\tilde{t} = \frac{t}{\tau}, \ \tilde{\mathbf{x}} = \frac{\mathbf{x}}{L}, \ \tilde{n} = \frac{n}{n_0}, \ \tilde{v} = \frac{v}{v_0}, \ \tilde{u} = \frac{u}{u_0}, \ \tilde{p} = \frac{p}{p_0}, \ \tilde{n} = \frac{m}{m_0} \ .$$

After the corresponding rescaling of the models parameters, we obtain, after dropping the tildes for notational convenience, the non-dimensional system of equations:

$$\partial_t n = D_n \Delta n - \nabla \cdot (\chi_n n \nabla u + \zeta_n n \nabla p + \xi_n n \nabla v) + \mu_1 \, n \, (1 - n) \,, \quad (46)$$

$$\partial_t v = -\delta \, v \, m + \phi_{21} \, u \, p - \phi_{22} \, v \, p + \mu_2 \, v \, (1 - v) \,, \quad (47)$$

$$\partial_t u = D_u \Delta u - \phi_{31} \, p \, u - \phi_{33} \, n \, u + \alpha_{31} \, n \,, \quad (48)$$

$$\partial_t p = D_p \Delta p - \phi_{41} \, p \, u - \phi_{42} \, p \, v + \alpha_{41} \, m \,, \quad (49)$$

$$\partial_t m = D_m \Delta m + (1 + \phi'_{53} \, p \, v)\phi_{53} \, u \, n - \phi_{54} m. \quad (50)$$

The boundary conditions (all no-flux) are not affected by the rescaling and so hold also for the non-dimensionalised systems. As for the initial conditions, we use

$$n(0, \mathbf{x}) = \exp(-|\mathbf{x}| \, \epsilon^{-1}) \,, \quad (51)$$

$$v(0, \mathbf{x}) = 1 - \frac{1}{2} \, \exp(-|\mathbf{x}| \, \epsilon^{-1}) \,, \quad (52)$$

$$u(0, \mathbf{x}) = \frac{1}{2} \, \exp(-|\mathbf{x}| \, \epsilon^{-1}) \,, \quad (53)$$

$$p(0, \mathbf{x}) = \frac{1}{20} \, \exp(-|\mathbf{x}| \, \epsilon^{-1}) \,, \quad (54)$$

$$m(0, \mathbf{x}) = 0 \,, \quad (55)$$

for $\mathbf{x} \in \bar{\Omega}$ and where we have taken $\epsilon = 0.01$.

The spatial domain Ω on which these systems hold changes accordingly: the parameter M is simply rescaled to M/L. The parameter values used throughout the remainder of this section are as follows: $D_n = 10^{-4}, D_u = 2.5 \times 10^{-3}, D_p = 3.5 \times 10^{-3}, D_m = 4.9 \times 10^{-3}, \chi_n = 1.5 \times 10^{-2}, \xi_n = 2.85 \times 10^{-2}, \zeta_n = 3.5 \times 10^{-2}, \delta = 7.5, \alpha_{41} = \alpha_{31} = 0.5, \phi_{33}, \phi_{21}, \phi_{31}, \phi_{41} = 0.15, \phi_{22} = \phi_{42} = 0.12, \phi_{53}, \phi'_{53} = 0.1, \phi_{54} = 0.03, \mu_1 = 0.2, \mu_2 = 0.85$. Full details can be found in [14].

4.6 Model Analysis

We collect the five individual solution components of the model system into a vector $\mathbf{w}(t, \mathbf{x})$

$$\mathbf{w}(t, \mathbf{x}) := (n(t, \mathbf{x}), v(t, \mathbf{x}), u(t, \mathbf{x}), p(t, \mathbf{x}), m(t, \mathbf{x})).$$

4.7 Spatially Uniform Steady States

There is only one positive spatially uniform steady state for the above choice of parameter values, given by

$$\mathbf{w}^* := (n^*, v^*, u^*, p^*, m^*) = (1, , , ,). \tag{56}$$

This steady state is linearly stable which can be seen by evaluating the Jacobian matrix $J_R(\mathbf{w})$ of the reaction terms, given by

$$\begin{bmatrix} \mu_1(1-2n) & 0 & 0 & 0 & 0 \\ 0 & -\delta m - \phi_{22}p + \mu_2(1-2v) & \phi_{21}p & \phi_{21}u - \phi_{22}v & -\delta v \\ -\phi_{33}u + \alpha_{31} & 0 & -\phi_{31}p - \phi_{33}n & -\phi_{31}u & 0 \\ 0 & -\phi_{42}p & -\phi_{41}p & -\phi_{41}u - \phi_{42}v & \alpha_{41} \\ (1 + \phi'_{53}pv)\phi_{53}u & \phi_{53}\phi'_{53}pun & (1 + \phi'_{53}pv)\phi_{53}n & \phi_{53}\phi'_{53}vun & -\phi_{54} \end{bmatrix} \tag{57}$$

at $\mathbf{w}^*$ and observing that its eigenvalues have all a negative real part.

4.8 Taxis-Driven Instability and Dispersion Curves

Based on the grounds that perturbed solutions

$$\mathbf{w}(t, \mathbf{x}) := (n(t, \mathbf{x}), v(t, \mathbf{x}), u(t, \mathbf{x}), p(t, \mathbf{x}), m(t, \mathbf{x})) \equiv \mathbf{w}^* + \tilde{\mathbf{w}}(t, \mathbf{x})$$

of the model system around its unique positive, spatially uniform steady state $\mathbf{w}^* := (n^*, v^*, u^*, p^*, m^*)$, cf. 56, have perturbations $\tilde{\mathbf{w}}(t, \mathbf{x})$ of the form

$$\tilde{\mathbf{w}}(t, \mathbf{x}) = \sum_{\lambda, \mathbf{k}} \mathbf{a}_{\lambda, \mathbf{k}} \exp(\lambda t + i\mathbf{k} \cdot \mathbf{x})$$

with coefficient vectors $\mathbf{a}_{\lambda,\mathbf{k}} \in \mathbb{R}^5$, we now study the stability of the steady state $\mathbf{w}^*$ for the full model including diffusion and taxis. We do this on an infinite spatial domain ignoring the boundary conditions.

We linearise the model equations around the steady state $\mathbf{w}^*$ and obtain upon dropping higher order terms

$$\frac{\partial \tilde{\mathbf{w}}}{\partial t} = J_T(\mathbf{w}^*)\frac{\partial^2 \tilde{\mathbf{w}}}{\partial x^2} + J_R(\mathbf{w}^*)\tilde{\mathbf{w}} \,. \tag{58}$$

Here, the reaction Jacobian $J_R(\mathbf{w})$ is given by 4.7 and the transport Jacobian, accounting for diffusion and taxis, is

$$J_T(\mathbf{w}) = \begin{bmatrix} D_n & -\xi_n n & -\chi_n n & -\zeta_n n & 0 \\ 0 & 0 & 0 & 0 & 0 \\ 0 & 0 & D_u & 0 & 0 \\ 0 & 0 & 0 & D_p & 0 \\ 0 & 0 & 0 & 0 & D_m \end{bmatrix} \tag{59}$$

Inserting a single mode $\tilde{\mathbf{w}}(t,\mathbf{x}) = \mathbf{1}\exp(\lambda t + i\mathbf{k}\cdot\mathbf{x})$ in the linearised problem 58, we arrive at

$$(\lambda I + k^2 J_T(\mathbf{w}^*) - J_R(\mathbf{w}^*)) \cdot \tilde{\mathbf{w}}(t,\mathbf{x}) = \mathbf{0} \,. \tag{60}$$

This relation is satisfied only (for all $(t,\mathbf{x})$) provided λ is an eigenvalue of the matrix $J(\mathbf{w}^*) := J_R(\mathbf{w}^*) - k^2 J_T(\mathbf{w}^*)$. This defines which perturbation modes are admissible, i.e. which growth factors λ are allowed with a given wave number $\mathbf{k}$.

We are interested in non-trivial perturbation modes $\mathbf{1}\exp(\lambda t + i\mathbf{k}\cdot\mathbf{x})$ which grow with time, i.e. have $\Re\lambda > 0$, and hence destabilise the linearly stable fixed point $\mathbf{w}^*$ of the reaction system.

In order to see whether such modes exists, we consider the maximum real part of the spectrum $\sigma(\cdot)$ of the matrix $J(\mathbf{w}^*)$,

$$\max\{\Re\,\sigma(J(\mathbf{w}^*))\}\,.$$

This value is plotted against k^2 to give the so-called *dispersion relation*. For the choice of parameter values given above, it can be shown that there do indeed exist perturbation modes which grow with time.

4.9 Numerical Results

In this section we show numerical simulation results for the model given in Sect. 4.4. These firstly represent the typical behaviour of the models solutions in 1D and 2D space and, secondly, illustrate the theoretical results on taxis-driven instability as obtained in Sect. 4.6. We start off with a few comments on the numerical methods employed in obtaining the simulation results.

4.10 Numerical Technique

NAG Routine

For simulation with a spatially one-dimensional domain and a non-zero diffusion coefficient the subroutine D03PCF of the NAG library is used. Routine D03PCF is a method of lines (MOL) discretisation for parabolic PDEs in one space variable using standard central finite differences for the discretisation in space and BDF formulas for the time-stepping. This method is not capable of successfully simulating the system in the case of zero diffusion. The reason for that are travelling waves and spikes with steep fronts which are not resolved correcly but rather non-physical oscillations are introduced locally by the central discretisation leading to over- and undershoots in the numerical solutions. In particular, negative solution values are caused by undershoot and these negative values feed back into the reaction kinetics which becomes unstable and we observe a subsequent blow-up of the numerical solution within a short time. A slight amount of diffusion can counterbalance these effects by smoothing out local small-scale oscillations before they harm the global system behaviour. However, the precise amount of diffusion required in a simulation run is not known a priori and simply choosing a value which works also defeats the purpose of running simulations with zero diffusion. Alternatively, negative solution components can be set to zero at the end of a time step in the MOL. This is also not desirable since it interferes with the systems mass balance (time- and space-dependent sources are introduced) and hence might have unwanted and uncontrolled side effects.

Custom Finite Volume Code for TDR Systems

The simulation results obtained with the NAG routine have been backed up by simulations with a custom made code for the simulation of taxis-diffusion-reaction systems, see [31] and the references cited there. Furthermore, this code is used for the simulation of the model with zero-diffusion in 1D and for simulation runs (with and without diffusion) in 2D.

The code is also based on the method of lines but in contrast to the NAG routine no central differences for the discretisation of the taxis terms in space are used. Instead the approach makes use of a finite volumes discretisation in space which employs for the taxis term a higher-order, upwind-biased discretisation with nonlinear limiter function. This dedicated treatmant of the taxis terms ensures an, in general, second order of accuracy of the spatial discretisation and leads to a large positive ODE system. Here *positive* means that the exact and unique solution of the ODE system with arbitrary non-negative initial condition remains non-negative for all later times. Conditions which ensure this property of an ODE system are, for instance, given in [37]. So the approach taken here does not lead to negative solution values and all the problems which go with them, see above. This is achieved by the discretisation automatically by locally introducing just the amount of *numerical diffusion* which is required to ensure the property of positivity. No user intervention is required. With these measures taken, the code also allows for computations in the zero-diffusion case.

After the discretisation in space a large ODE system is obtained which needs to be solved numerically. In particular, simulations of 2D problems lead to a very large dimension of the ODE system and suitable numerical schemes have to be employed. Due to the stiffness of the equations, the method selected must be implcit for computational efficiency. We have opted here for using the linearly-implicit Runge–Kutta method ROWMAP [84]. This method requires only the a subroutine for the evaluation of the right-hand side of the ODE. No explicit subroutine for the evaluation of the Jacobian matrix is required. The linear systems in the stages of the method are then solved using a multiple Arnoldi process [84]. An automatic selection and control of the time-step size further increase the efficiency and reliability of the method.

Comsol Multiphysics

Comsol Multiphysics is a Finite Element (FE) modelling and simulation software. It solves a wide range of PDE problems from various application fields. The flexibility of the FE method allows for complicated spatial geometries. We use the software tool for simulations of the model in two spatial dimensions: on a square, as is also done with the finite volumes code above) and on a circle. We use second-order Lagrangian elements on a triangulation of the spatial domain. The simulations run well for the model with diffusion but again, as for the NAG subroutine, in the zero-diffusion case we experience difficulties with oscillations and negative solutions. These difficulties also show up if the simple scalar linear advection equation is solved with Comsol Multiphysics. One way to avoid these problems is to add a small amount of diffusion to the equation, an another, more elaborate one, is to use a streamline-diffusion FE method. Nevertheless, in the latter case a parameter needs to be provided which is not readily available. For this reason we use Comsol Multiphysics only for simulations of the model with diffusion.

4.11 Computational Simulation Results

4.12 Discussion and Conclusions

In this section we have presented a mathematical model of the invasion of tissue by cancer cells through the secretion of matrix degrading enzymes. The model focused specifically on the role of the urokinase plasminogen activation system. The main achievement of this model is that fairly simple mathematical models representing the binding interactions of the components of the plasminogen activation system coupled with cell migration were able to capture the main characteristic effects of the system in cancer progression and invasion. The results of the computational simulations of the model show a

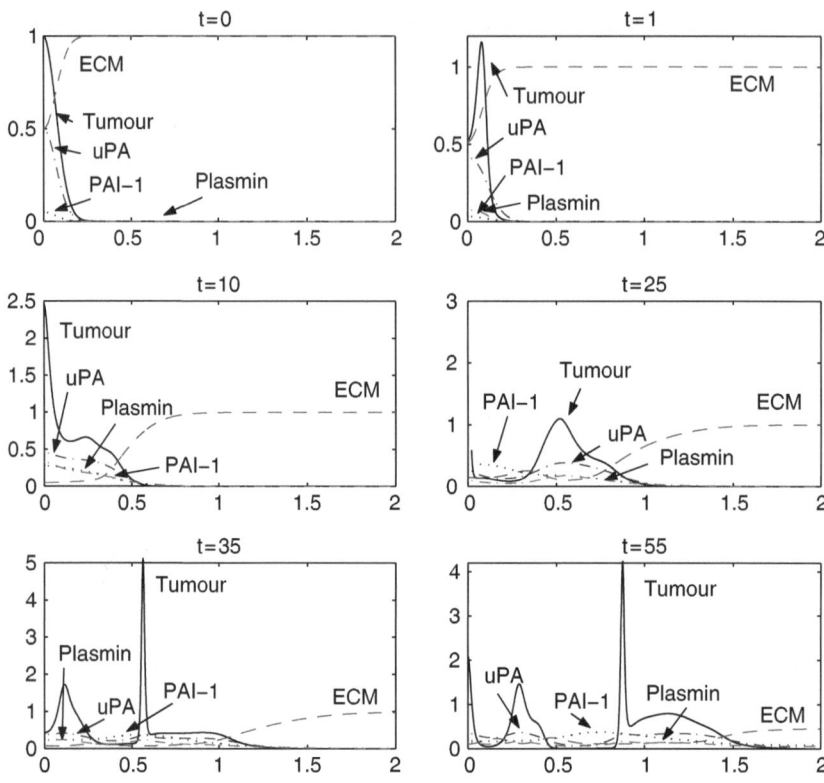

Fig. 21: Sequence of profiles showing the evolution of the tumour cell density $n(x,t)$ (solid black line), the uPA protease concentration $u(x,t)$ (dot-dashed blue line), the ECM density $v(x,t)$ (dashed red line), the PAI-1 concentration $v(x,t)$ (dotted black line) and the plasmin concentration $m(x,t)$ (dot-dashed magenta line) from $t = 0 - 55$. Parameter values as per the text

very rich dynamic spatio-temporal behaviour. Figures 21 and 22 show the 1-dimensional results, while Figure 23 shows the 2-dimensional results. The observed spatio-temporal heterogeneities in the solution profiles arise from the complex interplay between proliferative effects - cancer cell proliferation and matrix remodelling - and gradient-driven migration (chemotaxis and haptotaxis). Cancer cells initially degrade the matrix through the uPA system (ultimately via plasmin activation) and can initially move via taxis into the degraded matrix, coupled with cell proliferation. However, as the matrix remodels, this provides the cancer cells with the opportunity to "re-degrade" this re-modelled matrix with the consequence that multiple clusters of invading cancer cells appear throughout the domain. While high levels of the inhibitor PAI-1 halt invasion at the beginning, as time evolves clusters of cancer cells in distinct spatial locations form.

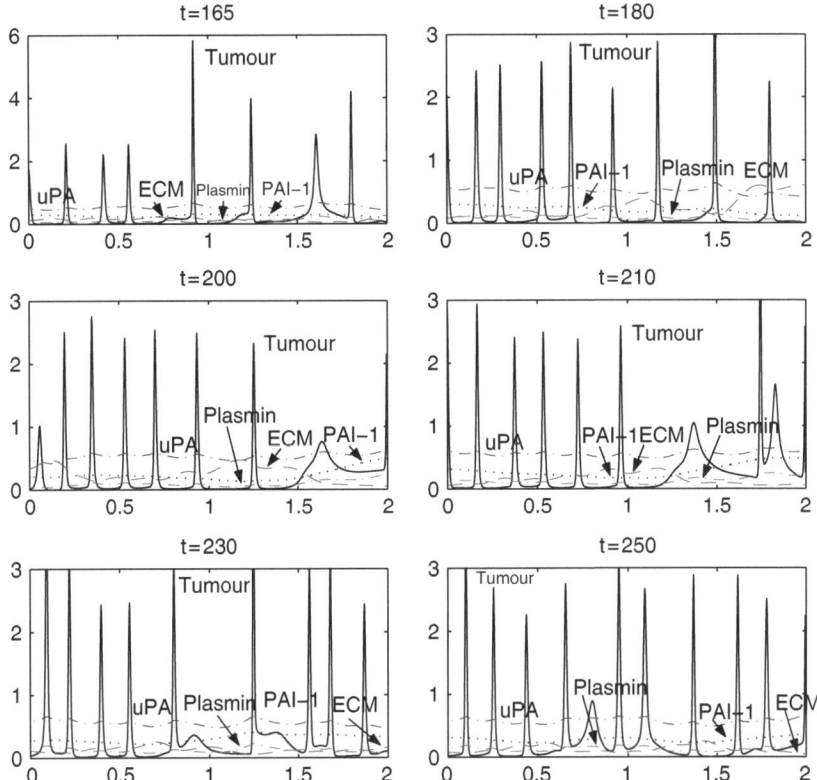

Fig. 22: Sequence of profiles showing the evolution of the tumour cell density $n(x, t)$ (solid black line), the uPA protease concentration $u(x, t)$ (dot-dashed blue line), the ECM density $v(x, t)$ (dashed red line), the PAI-1 concentration $v(x, t)$ (dotted black line) and the plasmin concentration $m(x, t)$ (dot-dashed magenta line) from $t = 165 - 250$. Parameter values as per the text

The results of the model have shown that the spatially heterogeneous distributions of cancer cells which arise as a consequence of simple binding reactions and gradient-driven migration may help to explain certain clinically and experimentally observed phenomena in carcinoma and multicellular spheroids, *i.e.* the heterogeneous spatial distribution of proliferating cancer cells and tissue. The applications of our complete plasmin taxis-reaction-diffusion equations to cancer invasion enable us to model more realistically solid tumour invasion of tissue and we believe that the results of the numerical simulations are highly consistent with *in vitro* as well as *in vivo* experimentally observed proliferative heterogeneity of cancer cells in solid tumours at their invasive stage. The results of our models are in line with recent experimental results, that show that when breast cells become malignant, plasmin is

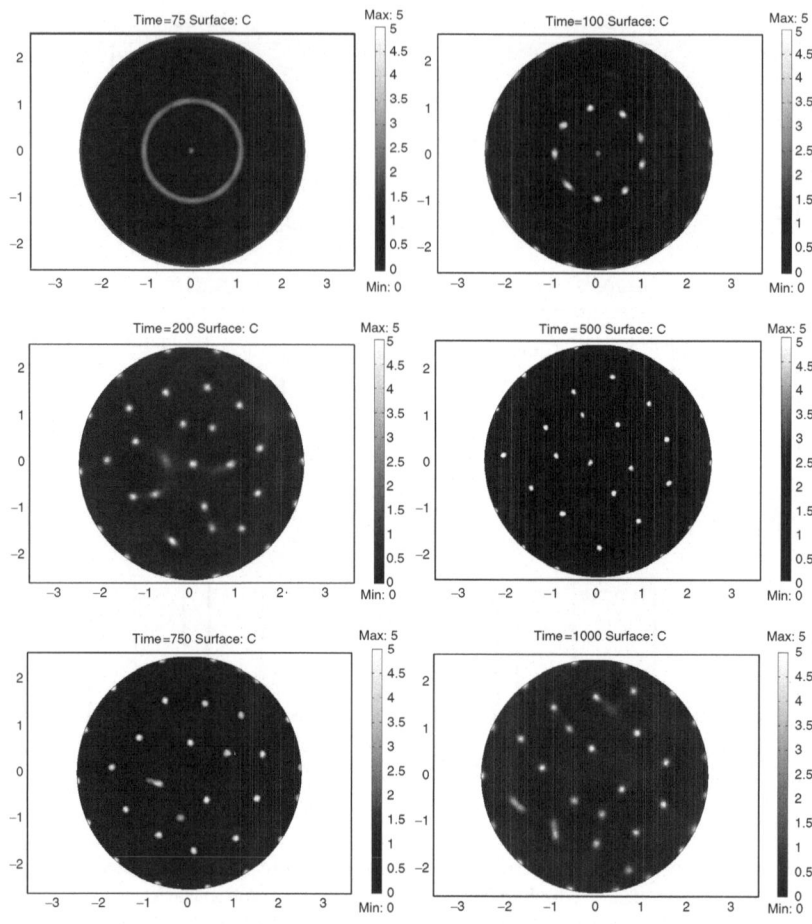

Fig. 23: Sequence of profiles showing the evolution of the tumour cell density $n(x,t)$ in a two-dimensional circular domain from $t = 75 - 1000$. The plots show that the dynamic heterogeneity persists and that no heterogeneous steady state is reached

activated on their membrane and their morphology is changed from sheet-like structures to multicellular heterogeneous masses. Considering these results, one may speculate that through the activation of plasmin on their membrane, micrometastatic tumour cells form multicellular clusters and thus manage to shield themselves from the action of chemotherapeutic drugs, thereby impeding chemotherapies and raising therapeutic drug doses to prohibitively high levels. In this regard, compromising the protective shield or even targeting plasmin with its α_2-antiplasmin inhibitor may be a reasonable target of new therapeutic designs for cancer therapies.

5 Summary

In this chapter we have formulated and analysed three macroscopic mathematical models of important aspects or phases of solid tumour growth: (i) avascular tumour growth; (ii) the response of the immune system to a solid tumour, and (iii) invasive cancer spread through tissue. Using a combination of analytical and computational approaches, we have shown that it is possible to begin to make quantitative predictions regarding the spatio-temporal dynamics of the system variables. The challenges for the future lie in adopting a "holistic" or "systems biology" approach where so-called multi-scale models are formulated. These multi-scale models will be composed of several "sub-models" each operating at a different spatial or temporal scale, with interactions between two or more of the scales. Developing appropriate analytical and numerical methods to analyse such models will require using a variety of mathematical techniques. We believe that the apparently diverse range of techniques described in the other chapters in this book will in the near future be brought together in order to formulate and analyse the emerging multi-scale models. With this in mind, it is perhaps fitting to close this chapter with a few prescient words of wisdom from D. Hilbert:

"He who seeks for methods without having a definite problem in mind seeks in the most part in vain. The further a mathematical theory is developed, the more harmoniously and uniformly does its construction proceed, and unsuspected relations are disclosed between hitherto separated branches of the science."

References

1. Albo, D., Berger, D.H., Wang, T.N., Xu, X.L., Rothman, V. and Tuszynski, G.P.: Thrombospondin-1 and transforming-growth-factor-beta1 promote breast tumor cell invasion through up-regulation of the plasminogen/plasmin system. *Surgery*, **122**, 493–499 (1997)
2. Ambrosi, D., Bellomo, N., Preziosi, L.: Modelling tumor progression, heterogeneity, and immune competition. *J. Theor. Medicine*, **4**, 51–65 (2002)
3. Anderson. A.R.A., Chaplain, M.A.J.: Continuous and discrete mathematical models of tumour-induced angiogenesis. *Bull. Math. Biol.*, **60**, 857–899 (1998)
4. Andreasen P.A., Kjøller L., Christensen L. and Duffy M.J.: The urokinase-type plasminogen activator system in cancer metastasis: A review. *Int. J. Cancer*, **72**, 1–22 (1997)
5. Andreasen P.A., Egelund R. and Petersen H.H.: The plasminogen activation system in tumor growth, invasion, and metastasis. *Cell. Mol. Life Sci.*, **57**, 25–40 (2000)
6. Arlotti, L., Gamba, A., Lachowicz, M.: A kinetic model of tumor/immune system cellular interactions. *J. Theor. Medicine*, **4**, 39–50 (2002)
7. Becciolini, A., Balzi, M., Barbarisi, M., Faraoni, P., Biggeri, A. and Potten, C.S.: 3H-thymidine labelling index (TLI) as a marker of tumour growth heterogeneity: evaluation in human solid carcinomas. *Cell Prolif.* **30**, 117–126 (1997)

8. Bellomo, N., Firmani, B., Guerri, L.: Bifurcation analysis for a nonlinear system of integro-differential equations modelling tumor-immune cells competition. *Appl. Math. Letters*, **12**, 39–44 (1999)
9. Bellomo, N., Preziosi, L.: Modelling and mathematical problems related to tumor evolution and its interaction with the immune system. *Math. Comp. Modelling*, **32**, 413–452 (2000)
10. Bellomo, N., Bellouquid, A., De Angelis, E.: The modelling of the immune competition by generalized kinetic (Boltzmann) models: Review and research perspectives. *Math. Comp. Modelling*, **37**, 65–86 (2003)
11. Berestycki, H., Larrouturou, B., Lions, P.L.: Multi-dimensional travelling wave solutions of a flame propagation model. *Arch. Rational Mech. Anal.*, **111**, 33–49 (1990)
12. Chaplain, M.A.J.: Reaction-diffusion prepatterning and its potential role in tumor invasion. *J. Bio. Sys.*, **3**, 929–936 (1995)
13. Chaplain, M.A.J., Ganesh, M. and Graham, I.G.: Spatio-temporal pattern formation on spherical surfaces: numerical simulation and application to solid tumour growth. *J. Math. Biol.*, **42**, 387–423 (2001)
14. Chaplain, M.A.J., Lolas, G.: Mathematical modelling of cancer cell invasion of tissue: The role of the urokinase plasminogen activation system. *Math. Modell. Methods. Appl. Sci.*, **15**, 1685–1734 (2005)
15. Chaplain, M.A.J., Lolas, G.: Mathematical modelling of cancer invasion of tissue: Dynamic heterogeneity. *Net. Hetero. Med.* **1**, 399–439 (2006)
16. Coulie, P.G.: Human tumor antigens recognized by T-cells: new perspectives for anti-cancer vaccines? *Mol. Med. Today*, **3**, 261–268 (1997)
17. Crampin, E.J., Gaffney, E.A. and Maini, P.K.: Reaction and diffusion on growing domains: Scenarios for robust pattern formation. *Bull. Math. Biol.*, **61**, 1093–1120 (1999)
18. Crampin, E.J., Hackborn, W.W. and Maini, P.K.: Pattern formation in reaction-diffusion models with nonuniform domain growth. *Bull. Math. Biol.*, **64**, 747–769 (2002)
19. Danforth, D.N., Sgagias, M.K.: Tumor-necrosis-factor-alpha enhances secretion of transforming-growth-factor-beta (2) in MCF-7 breast-cancer cells, *Clin. Canc. Res.* 2:827-835, 1996.
20. De Angelis, E., Delitala, M., Marasco, A., Romano, A.: Bifurcation analysis for a mean field modelling of tumor and immune system competition. *Math. Comp. Modelling*, **37**, 1131–1142 (2003)
21. Deweger, R.A., Wilbrink, B., Moberts, R.M.P., Mans, D., Oskam, R., den Otten, W.: Immune reactivity in SL2 lymphoma-bearing mice compared with SL2-immunized mice. *Cancer Immun. Immunotherapy*, **24**, 1191–1192 (1987)
22. Dillon, R., Maini, P.K., Othmer, H.G.: Pattern formation in generalized Turing systems. I. Steady-state patterns in systems with mixed boundary conditions. *J. Math. Biol.* 32:345-393, 1994.
23. Dunbar, S.R.: Travelling wave solutions of diffusive Lotka-Volterra equations. *J. Math. Biology*, **17**, 11–32 (1983)
24. Edelstein-Keshet L.: Mathematical Models in Biology. Random House, New York (1988)
25. Ethier, S.P.: Growth factor synthesis and human breast cancer progression. *J. Natl. Cancer Inst.*, **87**, 964–973 (1995)
26. Fenichel, N.: Geometric singular perturbation theory for ordinary differential equations. *J. Diff. Eqns.*, **31**, 53–98 (1979)

27. Forni, G., Parmiani, G., Guarini, A., Foa, R.: Gene transfer in tumour therapy. *Annals Oncol.*, **5**, 789–794 (1994)
28. Foryś, U.: Marchuk's model of immune system dynamics with application to tumour growth. *J. Theor. Medicine*, **4**, 85–93 (2002)
29. Freyer, J.P. and Sutherland, R.M.: Proliferative and clonogenic heterogeneity of cells from EMT6/Ro multicellular spheroids induced by the glucose and oxygen supply. *Cancer Res.*, **46**, 3513–3520 (1986)
30. Gatenby R.A., Gawlinski, E.: A reaction-diffusion model of cancer invasion. *Cancer Research*, **56**, 5745–5753 (1996)
31. Gerisch, A., Chaplain, M.A.J.: Robust Numerical Methods for Taxis–Diffusion–Reaction Systems: Applications to Biomedical Problems. *Math. Comput. Modell.*, **43**, 49–75 (2006)
32. Ghiso, J.A.: Inhibition of FAK signaling activated by urokinase receptor induces dormancy in human carcinoma cells in vivo. *Oncogene*, **21**, 2513–2524 (2002)
33. Gierer, A. and Meinhardt, H.: A theory of biological pattern formation. *Kybernetik*, **12**, 30–39 (1972)
34. Ashwin, P.B., Bartuccelli, M.V., Bridges, T.J., Gourley, S.A.: Travelling fronts for the KPP equation with spatio-temporal delay *Z. Angew. Math. Phys.*, **53**, 103–122 (2002)
35. Guckenheimer, J., Holmes. P.: Nonlinear Oscillations, Dynamical Systems, and Bifurcations of Vector Fields. Springer, New York (1983)
36. Hata, A., Shi, Y.G. and Massagué, J.: TGF-β signaling and cancer: structural and functional consequences of mutations in Smads. *Molecular Medicine Today*, **4**, 257–262 (1998)
37. Horváth, Z.: Positivity of Runge–Kutta and diagonally split Runge–Kutta methods. *Appl. Numer. Math.*, **28**, 309–326 (1998)
38. Ito, R., Kitadai, Y., Kyo, E., Yokozaki, H., Yasui, W., Yamashita, U., Nikai, H. and Tahara, E.: Interleukin 1α acts as an autocrine growth stimulator for human gastric carcinoma cells. *Cancer Res.*, **53**, 4102–4106 (1993)
39. Iversen, O.H.: The hunt for endogenous growth-inhibitory and or tumor suppression factors - their role in physiological and pathological growth-regulation. *Adv. Cancer Res.*, **57**, 413–453 (1991)
40. Jannink, I., Risberg, B., Vandiest, P.J., and Baak, J.P.A.: Heterogeneity of mitotic-activity in breast-cancer. *Histopathol.*, **29**, 421–428 (1996)
41. Keski-Oja, J., Postlethwaite, A.E. and Moses, H.L.: Transforming growth factors and the regulation of malignant cell growth and invasion. *Cancer Invest.*, **6**, 705–724 (1988)
42. Kelly, C.E., Leek, R.D., Byrne, H.M., Cox, S.M., Harris, A.L., Lewis, C.E.: Modelling macrophage infiltration into avascular tumours. *J. Theor. Med.*, **4**, 21–38 (2002)
43. Loeffler, D., Ratner, S.: In vivo localization of lymphocytes labeled with low concentrations of HOECHST-33342. *J. Immunol. Meth.*, **119**, 95–101 (1989)
44. Lolas, G.: Spatio-temporal pattern formation and reaction diffusion equations. MSc Thesis, University of Dundee, Dundee (1999)
45. Lord, E.M., Burkhardt, G.: Assessment of in situ host immunity to syngeneic tumours utilizing the multicellular spheroid model. *Cell. Immunol.*, **85**, 340–350 (1984)
46. Massagué, J.: TGFβ signal transduction. *Annu. Rev. Biochem.*, **67**, 753–791 (1998)

47. Matzavinos, A., Chaplain, M.A.J., Kuznetsov, V.: Mathematical modelling of the spatio-temporal response of cytotoxic T-lymphocytes to a solid tumour. *Math. Med. Biol.*, **21**, 1–34 (2004)

48. Matzavinos, A., Chaplain, M.: Travelling wave analysis of a model of the immune response to cancer. *C. R. Biologies*, **327**, 995–1008 (2004)

49. Meinhardt, H.: *Models of Biological Pattern Formation*. Academic Press, London, (1982)

50. Moses, M.L., Yang, E.Y. and Pietenpol, J.A.: TGF-β stimulation and inhibition of cell proliferation: new mechanistic insights. *Cell*, **63**, 245–247 (1990)

51. Mueller, M.M., Herold-Mende, C.C., Riede, D., Lange, M., Steiner, H.-H. and Fusenig, N.E.: Autocrine growth regulation by granulocyte colony-stimulating factor and granulocyte macrophage colony-stimulating factor in human gliomas with tumor progression. *Am. J. Pathol.*, **155**, 1557–1567 (1999)

52. Mueller-Klieser, W.: Multicellular spheroids: A review on cellular aggregates in cancer research. *J. Cancer Res. Clin. Oncol.*, **113**, 101–122 (1987)

53. Murray, J.D.: Parameter space for Turing instability in reaction diffusion mechanisms: a comparison of models. *J. theor. Biol.*. **98**, 143–163 (1982)

54. Murray, J.D.: Mathematical Biology (Second Edition). Springer-Verlag, London (1993)

55. Nagy N., Vanky, F.: Transforming-growth-factor-beta, (TGF beta), secreted by the immunogenic ex vivo human carcinoma cells, counteracts the activation and inhibits the function of autologous cytotoxic lymphocytes. Pretreatment with interferon gamma and tumor-necrosis-factor-alpha reduces the production of active TGF beta. *Canc. Immunol. Immunother.* **45**, 306–312, (1998)

56. Okubo, A.: Diffusion and Ecological Problems: Mathematical Models. Springer-Verlag, New York (1980)

57. Orme M.E., Chaplain M.A.J.: A mathematical model of vascular tumour growth and invasion, *Math. Comp. Model.*, **23**, 43–60 (1996)

58. Othmer, H., Stevens, A.: Aggregation, blowup and collapse: The ABCs of taxis and reinforced random walks. *SIAM J. Appl. Math.*, **57**, 1044–1081 (1997)

59. Owen, M.R., Sherratt, J.A.: Pattern formation and spatio-temporal irregularity in a model for macrophage-tumour interactions. *J. Theor. Biol.*, **189**, 63–80 (1997)

60. Owen, M.R., Sherratt, J.A.: Modelling the Macrophage Invasion of Tumours: Effects on Growth and Composition. *IMA J. Math. Appl. Med. Biol.*, **15**, 165–185 (1998)

61. Owen, M.R., Sherratt, J.A.: Mathematical modelling of macrophage dynamics in tumours. *Math. Models Meth. Appl. Sci.*, **9**, 513–539 (1999)

62. Parsons S.L., Watson S.A., Brown P.D., Collins H.M., Steele R.J.C.: Matrix metalloproteinases. *British Journal of Surgery*, **84**, 160 – 166 (1997).

63. Perumpanani A.J., Sherratt J.A., Norbury J., Byrne H.M.: Biological inferences from a mathematical model for malignant invasion. *Invasion & Metastasis*, **16**, 209–221, (1996)

64. Palmqvist, R., Oberg, A., Bergstrom, C., Rutegard, J.N., Zackrisson, B. and Stenling, R.: Systematic heterogeneity and prognostic significance of cell proliferation in colorectal cancer. *Br. J. Cancer*, **77**, 917–925 (1998)

65. Preziosi, L. (Ed.): Cancer Modelling and Simulation. Chapman & Hall/CRC Press (2003)

66. Pusztai, L., Lewis, C.E. and Yap, E. (eds.). *Cell Proliferation in Cancer: Regulatory Mechanisms of Neoplastic Cell Growth.* Oxford University Press, Oxford, (1996)

67. Quinn, K.A., Treston, A.M., Unsworth, E.J., Miller, M.-J., Vos, M., Grimley, C., Battey, J., Mulshine, J.L. and Cuttitta, F.: Insulin-like growth factor expression in human cancer cell lines. *J. Biol. Chem.,* **271**, 11477–11483 (1996)

68. Rahimi, N., Tremblay, E., McAdam, L., Roberts, A. and Elliott, B.: Autocrine secretion of TGF-beta 1 and TGF-beta 2 by pre-adipocytes and adipocytes: A potent negative regulator of adipocyte differentiation and proliferation of mammary carcinoma cells. *In Vitro Cell. Dev. Biol. Animal,* **34**, 412–420 (1998)

69. Rosfjord, E.C. and Dickson, R.B.: Growth factors, apoptosis and survival of mammary epithelial cells. *J. Mammary Gland Biol. Neoplasia,* **4**, 229–237 (1999)

70. Rozengurt, E.: Autocrine loops, signal transduction and cell cycle abnormalities in the molecular biology of lung cancer. *Curr. Opin. Oncol.,* **11**, 116–122 (1999)

71. Schirrmacher, V.: T-cell immunity in the induction and maintenance of a tumour dormant state. *Semin. Cancer Biol.,* **11**, 285–295 (2001)

72. Sessa, F., Bonato, M., Bisoni, D., Bosi, F. and Capella, C.: Evidence of a wide heterogeneity in cancer cell population in gallbladder adenocarcinomas. *Lab. Invest.,* **76**, 860 (1997)

73. Sherratt J.A., Nowak M.A.: Oncogenes, anti-oncogenes and the immune response to cancer: A mathematical model. *Proc. Roy. Soc. Series B,* **248**, 261–271 (1992)

74. Sherratt, J.A., Lewis, M.A., Fowler, A.C.: Ecological chaos in the wake of invasion. *Proc. Natl. Acad. Sci. USA,* **92**, 2524–2528 (1995)

75. Sutherland, R.M.: Cell and environment interactions in tumor microregions: the multicell spheroid model. *Science* **240**, 177–184 (1988)

76. Szymańska, Z.: Analysis of Immunotherapy Models in the Context of Cancer Dynamics. *Appl. Math. Comp. Sci.,* **13**, 407–418 (2003)

77. Tahara, E., Yasui W. and Yokozaki, H.: Abnormal growth factor networks in neoplasia, chapter 6, pp. 133-153, in: L. Pusztai, C.E. Lewis and E. Yap (eds.). *Cell Proliferation in Cancer: Regulatory Mechanisms of Neoplastic Cell Growth.* Oxford University Press, Oxford, (1996)

78. Takahashi, J.A., Mori, H., Fukumoto, M., Igarashi, K., Jaye, M., Oda, Y., Kikuchi, H. and Hatanaka, M.: Gene expression of fibroblast growth factors in human gliomas and meningiomas: demonstration of cellular source of basic fibroblast growth factor mRNA and peptide in tumor tissues. *Proc. Natl. Acad. Sci. USA,* **87**, 5710–5714 (1990)

79. Turing, A.M.: The chemical basis of morphogenesis. *Phil. Trans. Roy. Soc. Lond.,* **B237**, 37–72 (1952)

80. Udagawa, T., Fernandez, A., Achilles, E.G., Folkman, J., D'Amato, R.J.: Persistence of microscopic human cancers in mice: alterations in the angiogenic balance accompanies loss of tumor dormancy. *FASEB J.,* **16**, 1361–1370 (2002)

81. Uhr, J.W., Marches, R.: Dormancy in a model of murine B-cell lymphoma. *Semin. Cancer Biol.,* **11**, 277–283 (2001)

82. Volpert, A.I., Volpert, V.A., Volpert, V.A.: Traveling Wave Solutions of Parabolic Systems. American Mathematical Society, Translations of Mathematical Monographs, **140** (2000)

83. Webb, S.D., Sherratt, J.A., Fish, R.G.: Cells behaving badly: a theoretical model for the Fas/FasL system in tumour immunology. *Math. Biosci.,* **179**, 113–129 (2002)

84. Weiner, R., Schmitt, B.A., Podhaisky, H.: ROWMAP — a ROW-code with Krylov techniques for large stiff ODEs. *Appl. Numer. Math.*, **25**, 303–319 (1997)
85. Westermark, B. and Heldin, C.-H.: Platelet-derived growth factor in autocrine transformation. *Cancer Res.*, **51**, 5087–5092 (1991)
86. Wibe, E., Lindmo, T. and Kaalhus, O.: Cell kinetic characteristics in different parts of multicellular spheroids of human origin. *Cell Tissue Kinet.*, **14**, 639–651 (1981)
87. Wilson, K.M., Lord, E.M.: Specific (EMT6) and non-specific (WEHI-164) cytolytic activity by host cells infiltrating tumour spheroids. *Brit. J. Cancer*, **55**, 141–146 (1987)
88. Wu, S., Boyer, C.M., Whitaker, R.S., Berchuck, A., Wiener, J.R., Wienberg, J.B., Bast, R.C.: Tumor-necrosis-factor-alpha as an autocrine and paracrine growth-factor for ovarian cancer - monokine induction of tumor-cell proliferation and tumor-necrosis-factor-alpha expression. *Cancer Res.* 53:1939-1944, 1993.
89. Yanagihara, K. and Tsumuraya, M.: Transforming growth factor $\beta1$ induces apoptotic cell death in cultured human gastric carcinoma cells. *Cancer Res.*, **52**, 4042–4045 (1992)
90. Yefenof, E.: Cancer dormancy: from observation to investigation and onto clinical intervention. *Semin. Cancer Biol.*, **11**, 277–283 (2001)
91. Yoshida, K., Kyo, E., Tsujino, T., Sano, T., Niimoto, M., and Tahara, E.: Expression of epidermal growth factor, transforming growth factor-α and their receptor genes in human carcinomas: implication for autocrine growth. *Cancer Res.*, **81**, 43–51 (1990)

Lins Between Microscopic and Macroscopic Descriptions

Mirosław Lachowicz

Institute of Applied Mathematics and Mechanics, Faculty of Mathematics, Informatics and Mechanics, ul. Banacha, 2, 02-097 Warsaw, Poland
`lachowic@mimuw.edu.pl`

La blave e sta tai nui

Summary. The problem of relationships between various scales of description seems to be one of the most important problems of the mathematical modeling of complex systems (including e.g. the modeling of solid tumor growth). In the lecture we provide the theoretical framework for modeling at the microscopic scale in such a way that the corresponding models at the macro–, meso– and micro– scales are asymptotically equivalent, i.e. the solutions are close to each other in a properly chosen norm.

In mathematical terms we state the rigorous links between the following mathematical structures:

1. Continuous (linear) semigroups of Markov operators: the *micro*–scale of stochastically interacting entities (cells, individuals,...);
2. Continuous nonlinear semigroups related to the solutions of bilinear Boltzmann–type nonlocal kinetic equations: the *meso*–scale of statistical entities;
3. Dynamical systems related to bilinear reaction–diffusion–chemotaxis equations: the *macro*–scale of densities of interacting entities.

1 Introduction

The unity of knowledge and *the unity of science* have been widely discussed issues both within the philosophy of science and within several scientific fields (cf. [3,44]). The unity of knowledge is one of the ideas constantly present in human thinking. Many philosophers have affirmed that "to know" means "to bring to unity". In consequence the highest form of knowledge should involve including all phenomena *within one system* ([44]).

The system would become closer to perfection the smaller the number of rules it is based on. Looking for such a system can be realized, for example, by postulating on the superiority of some groups of phenomena compared to others, and in consequence concentrating effort on the *"compact"* unification of the former groups with the convention that the latter do not have essential effect, or can be deduced from the former ([44]).

Typical examples are *"systems of the world"* such as those raised by Laplace which, describing the phenomena of mechanical type, was treated, by enthusiastic commentators, the general models of the whole nature.

Some physicists with a philosophical inclination suggest returning to the *Pitagorian-Platonian basis* and searching for *the unity of science* by bringing all that happens in nature into the laws of physics, and then into mathematical structures. One may mention here W. Heisenberg and C.F. von Weizsäcker.

It is interesting that von Weizsäcker ([105]) claims that the verification of a theory (model) by an experiment is not philosophically decisive because every experiment has to be preceded by the acceptance of another theory. Therefore the experiments should not judge the acceptance or the rejection of a theory, but rather the theory should be reliable by its structure.

Such a programme seems to be rather *idealistic*. However the similar programme was proposed within the framework of *reductionism*, treating physics and particularly *particle physics*, as the basis of different domains of science (cf. [3]).

Usually there are a number of theories and models describing the behavior of a phenomenon. Particularly it is visible in physics, where there are many different models that relate to basic phenomena, with some definitions of their applicability. The choice of the model proper for the phenomenon usually is only intuitive, and it is a play between the simplicity of description and the possibility of including important aspects of the phenomenon.

The first step towards the unity of science can rely upon the affirmation of specific relationships among different theories and models. Knowledge about nature will be more reliable when relationships between different descriptions are more visible.

The mathematical contribution to this process would be seen in the search for mathematical relationships between the solutions of various models that describe the same reality. Usually such relationships have an asymptotic nature.

The relationships between two consistent models (theories) can be investigated in a way that in one of these models one or more characteristic (dimensionless) parameters are chosen such that they have a clear physical meaning and the limit of these parameters to given critical values means a transition from one model to another.

One may consider the following basic examples.

Example 1. Classical Mechanics and Relativistic Mechanics

In Relativistic Mechanics one may consider the (dimensionless) parameter

$$\gamma = frac{v_{\mathrm{char}}}{c} > 0, \tag{1}$$

where v_{char} is the characteristic speed related to the phenomenon in question and c is the speed of light. The limit

$$\gamma \to 0 \tag{2}$$

would mean the transition from Relativistic Mechanics to Classical Mechanics in the sense of convergence of solutions to the corresponding models.

Example 2. Classical Mechanics and Quantum Mechanics

In Quantum Mechanics one may consider the (dimensionless) parameter

$$H = \frac{\hbar}{\mathcal{M}_{\text{char}}} > 0, \tag{3}$$

where $\hbar$ is Planck's constant and $\mathcal{M}_{\text{char}}$ is the characteristic value of the moment of momentum (or spin). The limit

$$H \to 0 \tag{4}$$

would mean the transition from Quantum Mechanics to Classical Mechanics.

Example 3. Theory of Inviscid Fluids (TIF) and Theory of Viscid Fluids (TVF)

In continuum theory of viscid fluids (TVF) one may consider the Reynolds number $\mathfrak{R} > 0$ — the dimensionless parameter such that $\frac{1}{\mathfrak{R}}$ describes the "size" of viscosity. The limit

$$\mathfrak{R} \to \infty \tag{5}$$

would mean the transition from TVF to TIF.

Example 4. The Macroscopic Description in the framework of Theory of Continuum and the Mesoscopic Description in the framework of Kinetic Theory

In Kinetic Theory one may consider the Knudsen number $\mathfrak{K} > 0$, i.e. the dimensionless parameter describing the "size" of the mean free path. The limit

$$\mathfrak{K} \to 0 \tag{6}$$

would mean the transition from Mesoscopic to Macroscopic Description (*Hydrodynamic Limit.*

Example 5. The macroscopic model of interacting populations (MMIP) and the model of N interacting entities of various populations (MIE)

The limit

$$N \to \infty \tag{7}$$

would mean the transition from the MIE to the MMIP (cf. Section 2).

The asymptotic analysis plays a significant role in the mathematical analysis of equations modeling natural phenomena. Mathematical models of processes that take place in nature contain a number of parameters that represent physical quantities essential for the processes in question. These parameters, defined as dimensionless are elements either of $\mathbb{N}$ or of $\mathbb{R}_+$, according to their physical sense. The essence of the asymptotic analysis is to examine the behavior of the solutions to model equations when the selected parameters tend to their critical values. From the physical point of view, such a limit may signify the transition from the application range of one theory to another one, as we have seen before.

One of the most important problems of applied mathematics is the understanding relationships between different models developed within the framework of different levels of description. One can distinguish three possible scales of description as follows

(Mi) at the level of interaction of entities (the microscopic scale),

(Me) at the level of a statistical description of a test–entity (the mesoscopic scale),

(Ma) at the level of densities of subpopulations (the macroscopic scale).

In the context of the theory of rarefied gases (cf. [28, 30, 63, 64, 77, 91]) the above three levels are formulated in the following way

(Mi) is the level of particle dynamics (Newton's laws),

(Me) is the level of Boltzmann's description

(Ma) is the level of continuum description.

The hydrodynamic limits (Example 4) refer to the transition **(Me)** $\rightarrow$ **(Ma)**.

In the case of theory of rarefied gases the most important, but still unsolved, problem is the derivation of kinetic equations, and then of hydrodynamic equations, from (deterministic) particle dynamics. The problem refers to the first part of Hilbert's 6th Problem: the axiomatization of physics — cf. [106].

There is a huge bibliography on the transition "**(Mi)** $\rightarrow$ **(Me)**" — see [30,91] — and on the transition "**(Me)** $\rightarrow$ **(Ma)**" — see e.g. [47,63,64,66], but the relationships between the different models are still not fully understood. In particular, the theory "**(Mi)** $\rightarrow$ **(Me)**" does not fit that of "**(Me)** $\rightarrow$ **(Ma)**". The most difficult step is the rigorous derivation of the mesoscopic description **(Me)** from the Newtonian laws of motion **(Mi)**. The situation is easier if one starts from stochastic particle systems (i.e. assuming stochastic interactions between particles) — see references in [69] and Section 3.4 in [6].

The mathematical links between various levels of description of the theory of rarefied gases may be considered as a prototype of a theory for complex systems like those considered in Sections 5 – 8.

An alternative approach can be found in [26].

2 Microscopic (Stochastic) Systems

Usually the description of biological populations is carried out on a macroscopic level of interacting subpopulations. The mathematical structures are deterministic reaction — diffusion equations [85, 96, 97]. They describe the (deterministic) evolution of densities of subpopulations rather than the interactions between their individual entities. However, in many cases the description on a micro–scale of interacting entities (e.g. cells) seems to be more appropriate.

The problem of relationships between the various scales of description seems to be one of the most important problems of the mathematical modeling of complex systems, including the modeling of solid tumor growth. The following strategy can be applied. One starts with the deterministic macroscopic model for which the identification of parameters by an experiment is easier. Then one provides the theoretical framework for modeling at the microscopic scale in such a way that the corresponding models at the macro– and micro– scales are asymptotically equivalent, i.e. the solutions are close to each other in a properly chosen norm. Then, if the microscopic model is chosen suitably, one may hope that it covers not only the macroscopic behavior of the system in question, but also some of its microscopic features. The microscopic model by its nature is richer and it may describe a larger variety of phenomena.

In mathematical terms we are interested in the links between the following mathematical structures:

(**Mi**) the *micro*–scale of stochastically interacting entities (cells, individuals,...), in terms of continuous (linear) semigroups of Markov operators — continuous stochastic semigroups — cf. [13, 79];

(**Me**) the *meso*–scale of statistical entities, in terms of continuous nonlinear semigroups related to the solutions of bilinear Boltzmann–type nonlocal kinetic equations — cf. [76];

(**Ma**) the *macro*–scale of densities of interacting entities (in terms of dynamical systems related to bilinear reaction–diffusion–chemotaxis equations — cf. [85]).

In Refs. [66–72] such a conceptual framework was developed for various situations of biological interest. In particular, in Ref. [69] a general class of microscopic stochastic systems was proposed and the limit (**Mi**) → (**Me**) was rigorously studied. Papers [67, 70–72] consider the links between (**Mi**), (**Me**) and (**Ma**) for a class of competitive systems, a general class of population systems of the Lotka-Volterra–type without or with (weak) diffusion, coagulation equations, and reaction–diffusion–chemotaxis equations, respectively.

The idea of approximating the Boltzmann equation by stochastic particle systems appeared in the 1930–ies in a paper by Leontovich [80] but had tended to be ignored by mathematicians. In the spatially homogeneous case the derivation of a simplified Boltzmann-type equation (the so-called carica-ture of a Maxwellian gas) from a stochastic model was proposed by Kac [58] (cf. [82]). Nowadays there is a huge amount of literature on various stochastic approaches in kinetic theory — see for example [49, 62, 74, 93, 95, 102, 103] and references therein.

The idea of [74] can be continued ([69]) in a more general context. Consider a system composed of N interacting entities. Each entity n ($n \in \{1, 2, \dots, N\}$) is characterized by the pair (j_n, u_n), where $j_n \in \mathbb{J}$ characterizes the population of the n–entity and $u_n \in \mathbb{U}$ — its (physical or biological) state (e.g. position, velocity, activation state, domination, fitness), $\mathbb{J} \subset \mathbb{N}_{-1} = \{-1, 0, 1, 2, \dots\}$, and $\mathbb{U}$ is a domain in $\mathbb{R}^d$, $d \geq 1$. The evolution is determined by the inter-actions between pairs of entities. Analogously to kinetic theory only binary interactions are taken into account, but generalization into multiple interac-tions is possible. The n–entity interacts with the m–entity and the interaction take place at random time. During the interaction both entities may change their population or/and their state.

The rate of interaction of the individuals of the j–th population ($j \in \mathbb{J}$) with state u ($u \in \mathbb{U}$) with the individual of the k–th population ($k \in \mathbb{J}$) with state v ($v \in \mathbb{U}$) is given by the function

$$a = a(j, u; k, v), \qquad a : \left(\mathbb{J} \times \mathbb{U}\right)^2 \to \mathbb{R}_+ . \tag{8}$$

The transition into the j–th population ($j \in \mathbb{J}$) with state u ($u \in \mathbb{U}$) due to the interaction of individuals of the k–th population ($k \in \mathbb{J}$) with state v ($v \in \mathbb{U}$) with individuals of the l–th population ($l \in \mathbb{J}$) with state w ($w \in \mathbb{U}$) is described by the function

$$A = A(j, u; k, v; l, w), \qquad A : \left(\mathbb{J} \times \mathbb{U}\right)^3 \to \mathbb{R}_+ . \tag{9}$$

Consider the stochastic system of N–individuals with infinitesimal gener-ator given by

$$\Lambda_N \phi\left(j_1, u_1, j_2, u_2, \dots, j_N, u_N\right) = \frac{1}{N} \sum_{\substack{1 \leq n, m \leq N \\ n \neq m}} a(j_n, u_n, j_m, u_m) \tag{10}$$

$$\times \left(\sum_{k \in \mathbb{J}} \int_{\mathbb{U}} A\left(k, v; j_n, u_n, u_m\right) \right.$$

$$\times \phi\left(j_1, u_1, \dots, j_{n-1}, u_{n-1}, k, v, j_{n+1}, u_{n+1}, \dots, j_N, u_N\right) dv$$

$$\left. - \phi\left(j_1, u_1, \dots, j_N, u_N\right) \right),$$

where ϕ is an appropriate test function.

Under the assumptions that A is a transition function and a is a measur-able bounded function, the operator Λ_N is the generator for a Markov jump process — see [40], Ch. 4.

We assume

$$0 \leq A(j, u; k, v, l, w), \qquad 0 \leq a(j, u, k, v) \leq c_a < \infty, \tag{11}$$

for all j, k, l in $\mathbb{J}$ and a.a. u, v, w in $\mathbb{U}$, where c_a is a positive constant, and

$$\sum_{j \in \mathbb{J}} \int_{\mathbb{U}} A(j, u; k, v, l, w) \, du = 1, \tag{12}$$

for all $k, l \in \mathbb{J}$ and a.a. $v, w \in \mathbb{U}$ such that $a(k, v, l, w) > 0$.

Assume that the system is initially distributed according to the probability density $F^N \in L_1^{(N)}$, where $L_1^{(N)}$ is the space equipped with the norm

$$\|f\|_{L_1^{(N)}} = \sum_{j_1 \in \mathbb{J}} \int_{\mathbb{U}} \cdots \sum_{j_N \in \mathbb{J}} \int_{\mathbb{U}} |f(j_1, u_1, \ldots, j_N, u_N)| \, du_1 \ldots du_N. \tag{13}$$

The time evolution is described by the probability density

$$f^N(t) = \exp\left(t \Lambda_N^*\right) F^N. \tag{14}$$

It satisfies (in $L_1^{(N)}$)

$$\partial_t f^N = \Lambda_N^* f^N; \qquad f^N\big|_{t=0} = F^N, \tag{15}$$

where

$$\Lambda_N^* f(j_1, u_1, \ldots, j_N, u_N) = \tag{16}$$

$$= \frac{1}{N} \sum_{\substack{1 \leq n, m \leq N \\ n \neq m}} \left(\sum_{k \in \mathbb{J}} \int_{\mathbb{U}} A(j_n, u_n; k, v, j_m, u_m) a(k, v, j_m, u_m) \right.$$

$$\times f(j_1, u_1, \ldots, j_{n-1}, u_{n-1}, k, v, j_{n+1}, u_{n+1}, \ldots, j_N, u_N) \, dv$$

$$\left. - a(j_n, u_n, j_m, u_m) f(j_1, u_1, \ldots, j_N, u_N) \right).$$

Under Assumptions (11) the operator Λ_N^* is a bounded linear operator in the space $L_1^{(N)}$. Therefore the Cauchy Problem (15) has the unique solution (14) in $L_1^{(N)}$ for all $t \geq 0$. Moreover, by standard argument, we see that the solution is nonnegative for nonnegative initial data and the $L_1^{(N)}$–norm is preserved

$$\|f^N(t)\|_{L_1^{(N)}} = \|F^N\|_{L_1^{(N)}} = 1, \qquad \text{for } t > 0. \tag{17}$$

Therefore $\exp\left(t \Lambda_N^*\right)$ defines a continuous (linear) semigroup of Markov operators — continuous stochastic semigroup — cf. [13, 26, 79].

An alternative approach at the microscopic level (Mi) can base on the stochastic differential equations — cf. [62] and [26].

We assume that all functions are symmetric

$$f^N(j_1, u_1, \ldots, j_N, u_N) = f^N(j_{r_1}, u_{r_1}, \ldots, j_{r_N}, u_{r_N}), \tag{18}$$

for all $j_1, \ldots, j_N$ in $\mathbb{J}$, and a.a. $u_1, \ldots, u_N$ in $\mathbb{U}$ and for any permutation $\{r_1, \ldots, r_N\}$ of the set $\{1, \ldots, N\}$.

We introduce the s–individual marginal density ($1 \le s < N$)

$$f^{N,s}(j_1, u_1, \ldots, j_s, u_s) = \tag{19}$$

$$= \sum_{j_{s+1} \in \mathbb{J}} \int_{\mathbb{U}} \cdots \sum_{j_N \in \mathbb{J}} \int_{\mathbb{U}} f^N(j_1, u_1, \ldots, j_N, u_N) \, du_{s+1} \ldots du_N,$$

and $f^{N,N} = f^N$.

The function f^N satisfies Eq. (15) iff $f^{N,s}$ satisfy the following finite hierarchy of equations

$$\partial_t f^{N,s} = \frac{s}{N} \Lambda_s^* f^{N,s} + \frac{N-s}{N} \Theta_{s+1} f^{N,s+1}, \tag{20}$$

for $s = 1, 2, \ldots, N$, where

$$(\Theta_{s+1} f)(j_1, u_1, \ldots, j_s, u_s) =$$

$$= \sum_{n=1}^{s} \left(\sum_{k \in \mathbb{J}} \int_{\mathbb{U}} \sum_{l \in \mathbb{J}} \int_{\mathbb{U}} A(j_n, u_n; k, v, l, w) a(k, v, l, w) \right.$$

$$\times f(j_1, u_1, \ldots, j_{n-1}, u_{n-1}, k, v, j_{n+1}, u_{n+1}, \ldots, j_s, u_s, l, w) \, dv \, dw$$

$$\left. - \sum_{k \in \mathbb{J}} \int_{\mathbb{U}} a(j_n, u_n, k, v) f(j_1, u_1, \ldots, j_s, u_s, k, v) \, dv \right),$$

Taking N sufficiently large we may expect that the solution of the finite hierarchy (20) approximates solution of the following infinite hierarchy of equations

$$\partial_t f^s = \Theta_{s+1} f^{s+1}, \qquad s = 1, 2, \ldots . \tag{21}$$

The integral versions of hierarchies (20) and (21) read

$$f^{N,s}(t) = F^{N,s} + \frac{s}{N} \int_0^t \Lambda_s^* f^{N,s}(t_1) \, dt_1 + \tag{22}$$

$$+ \frac{N-s}{N} \int_0^t \Theta_{s+1} f^{N,s+1}(t_1) \, dt_1, \qquad s = 1, \ldots, N,$$

and

$$f^s(t) = F^s + \int_0^t \Theta_{s+1} f^{s+1}(t_1) \, dt_1, \qquad s = 1, 2, \ldots, \tag{23}$$

respectively.

Definition 1. *A admissible hierarchy* $\{f^s\}_{s=1,2,3,\ldots}$ *is a sequence of functions* f^s *satisfying (for $s = 1, 2, \ldots$)*

(i) f^s *is a probability density on* $(\mathbb{J} \times \mathbb{U})^s$;

(ii) $f^s(j_1, u_1, \ldots, j_s, u_s) = f^s(j_{r_1}, u_{r_1}, \ldots, j_{r_s}, u_{r_s})$ *for $j_1, \ldots, j_s$ in $\mathbb{J}$ and a.a.* $u_1, \ldots, u_s$ *in $\mathbb{U}$ and for any permutation $\{r_1, \ldots, r_s\}$ of the set $\{1, \ldots, s\}$;*

(iii) $f^s(j_1, u_1, \ldots, j_s, u_s) = \sum\limits_{j_{s+1} \in \mathbb{J}} \int\limits_{\mathbb{U}} f^{s+1}(j_1, u_1, \ldots, j_{s+1}, u_{s+1}) \, \mathrm{d}u_{s+1}$ *for j_1,* $\ldots, j_s$ *in $\mathbb{J}$ and a.a. $u_1, \ldots, u_s$ in $\mathbb{U}$.*

We have ([69])

Theorem 1. *Let $\{F^s\}_{s=1,2,\ldots}$ be an admissible hierarchy. Then, for all $t > 0$, there exists a unique hierarchy $\{f^s(t)\}_{s=1,2,\ldots}$, $f^s(t) \in L_1^{(s)}$ $(s = 1, 2, \ldots)$, that is a solution of* Eq. (23) *with initial data $f^s(0) = F^s$ $(s = 1, 2, \ldots)$. Moreover $\{f^s(t)\}_{s=1,2,\ldots}$, for all $t > 0$, is an admissible hierarchy.*

Proof. By (11) we have

$$\|\Theta_{s+1} f\|_{L_1^{(s)}} \leq c_1 s \|f\|_{L_1^{(s+1)}}, \tag{24}$$

and

$$\sum_{j_1 \in \mathbb{J}} \int_{\mathbb{U}} \cdots \sum_{j_1 \in \mathbb{J}} \int_{\mathbb{U}} (\Theta_{s+1} f)(j_1, u_1, \ldots, j_s, u_s) \, \mathrm{d}u_1 \ldots \mathrm{d}u_s = 0, \tag{25}$$

for all $f \in L_1^{(s+1)}$ and $s = 1, 2, \ldots$, where $c_1 = 2c_a$ is a constant. Moreover,

$$\sum_{j_s \in \mathbb{J}} \int_{\mathbb{U}} (\Theta_{s+1} f)(j_1, u_1, \ldots, j_s, u_s) \, \mathrm{d}u_s = (\Theta_s \bar{f})(j_1, u_1, \ldots, j_{s-1}, u_{s-1}), \tag{26}$$

for all $f \in L_1^{(s+1)}$ and $s = 1, 2, \ldots$, where

$$\bar{f}(j_1, u_1, \ldots, j_s, u_s) = \tag{27}$$
$$\sum_{j_{s+1} \in \mathbb{J}} \int_{\mathbb{U}} f(j_1, u_1, \ldots, j_{s-1}, u_{s-1}, j_{s+1}, u_{s+1}, j_s, u_s) \, \mathrm{d}u_{s+1}.$$

Iterating Eq. (23) we obtain the following *"perturbation series"*

$$f^s(t) = F^s + \sum_{m=1}^{\infty} \int_0^t \int_0^{t_1} \int_0^{t_2} \cdots \int_0^{t_{m-1}} \Theta_{s+1} \Theta_{s+2} \cdots \Theta_{s+m} F^{s+m} \, \mathrm{d}t_m \ldots \mathrm{d}t_1, \tag{28}$$

where it is understood that $t_0 = t$.

Existence of a solution, its uniqueness and representation in the form of (28) are directly obtained on the time interval $[0, t_*]$ on which the series, defined by the r.h.s. of (28), is convergent.

The $L_1^{(s)}$–norm of the mth–term of the series in the r.h.s. of (28) can be estimated by

$$(c_1 t)^m \frac{(s+m-1)!}{m!(s-1)!} .$$

Consequently, by

$$\frac{s(s+1)\ldots(s+m-1)}{m!} \leq 2^{s+m-1}, \qquad (29)$$

if $t_* = \frac{1}{4c_1}$, then the series converges uniformly on $[0, t_*]$.

The hierarchy $\{f^s\}_{s=1,2,\ldots}$ satisfies, for all $t \in]0, t_*]$ the property (ii), and by (26), the property (iii) of Definition 1.

The nonnegativity of the solution can be proved on a time interval $[0, t^*]$, where $0 < t^* \leq t_*$, by standard arguments.

By (25) we have

$$\|f^s(t)\|_{L_1^{(s)}} = \|F^s\|_{L_1^{(s)}}, \qquad \text{for } 0 \leq t \leq t^*, \qquad s = 1, 2, \ldots . \qquad (30)$$

It follows that $\{f^s(t)\}_{s=1,2,\ldots}$ is an admissible hierarchy for all $t \in [0, t^*]$.

Assuming $\{f^s(t^*)\}_{s=1,2,\ldots}$ as initial datum, we can repeat the same arguments for $t \in [t^*, 2t^*]$ and so on. This completes the proof of the theorem.
□

We assume now that the stochastic system starts with chaotic (i.e. factorized) probability density and we consider the hierarchy (23) with initial data

$$F^s = F \otimes \ldots \otimes F = (F)^{s \otimes}, \qquad s = 1, 2, \ldots \qquad (31)$$

i.e. s–fold outer product of a probability density F defined on $\mathbb{J} \times \mathbb{U}$. We may see that the propagation of chaos is held and the solution $f^s(t)$ to Eq. (23) is the s–product of solution $f(t)$ of the following bilinear equation (the generalized kinetic model GKM) — [76]

$$\partial_t f(t, j, u) = G[f](t, j, u) - f(t, j, u) Lf(t, j, u), \qquad (j, u) \in \mathbb{J} \times \mathbb{U}, \qquad (32)$$

where G is the gain term,

$$G[f](t, j, u) = \qquad (33)$$
$$= \sum_{k \in \mathbb{J}} \int_{\mathbb{U}} \sum_{l \in \mathbb{J}} \int_{\mathbb{U}} A(j, u; k, v, l, w) a(k, v, l, w) f(t, k, v) f(t, l, w) \, dv \, dw,$$

and fLf is the loss term,

$$Lf(t, j, u) = \sum_{k \in \mathbb{J}} \int_{\mathbb{U}} a(j, u, k, v) f(t, k, v) \, dv. \qquad (34)$$

Therefore we have ([69])

Corollary 1. *Let F be a probability density on $\mathbb{J} \times \mathbb{U}$. Then, for each $t_0 > 0$, there exists an admissible hierarchy $\{f^s\}_{s=1,2,\ldots}$ such that*

(i) *it is a unique solution of Eq. (23) with chaotic initial data (31),*
(ii) *$f^s(t)$ is chaotic*

$$f^s(t) = \left(f(t) \right)^{s \otimes}, \tag{35}$$

for all $0 < t \leq t_0$ and $s = 1, 2, \ldots$, where $f(t)$ is the unique solution in $L_1^{(1)}$ of Eq. (32) with the initial datum F.

We may now formulate the main result, namely the theorem stating that the solution of Eq. (32) is approximated by the solutions of Eq. (15) as $N \to \infty$ ([69] — the proof follows the line of [74]).

Theorem 2. *Let F be a probability density on $\mathbb{J} \times \mathbb{U}$. Then, for each $t_0 > 0$, there exists N_0 such that for $N \geq N_0$*

$$\sup_{[0,t_0]} \| f^{N,1} - f \|_{L_1^{(1)}} \leq \frac{c}{N^\eta}, \tag{36}$$

where the nonnegative functions $f^{N,s} \in L_1^{(s)}$ $(s = 1, \ldots, N)$ form the unique solution of Eq. (22) corresponding to the initial datum

$$f^{N,s}(0) = (F)^{s \otimes}, \qquad s = 1, \ldots, N; \tag{37}$$

$f \in L_1^{(1)}$ is the unique, nonnegative solution of Eq. (32) corresponding to the initial datum F; η and c are positive constants that depend on t_0.

Proof. Let $\{f^{N,s}\}_{s=1,\ldots,N}$ and $\{f^s\}_{s=1,2,\ldots}$ be the solutions of Eq. (22) and Eq. (23) corresponding to the initial data (37) and (31), respectively.
    If $N < s$ we assume $f^{N,s} \equiv 0$.
    We consider the difference

$$\Delta^{N,s}(t) = f^{N,s}(t) - f^s(t), \qquad t > 0. \tag{38}$$

It satisfies, for $s = 1, \ldots, N$ and $0 \leq t_1 < t \leq t_0$, the following equation

$$\Delta^{N,s}(t) = \mathcal{G}^{N,s}(t, t_1) + \int_{t_1}^{t} \Theta_{s+1} \Delta^{N,s+1}(t_2) \, dt_2, \tag{39}$$

where

$$\mathcal{G}^{N,s}(t, t_1) = \Delta^{N,s}(t_1) + \frac{s}{N} \int_{t_1}^{t} \left(\Lambda_s^* f^{N,s}(t_2) - \Theta_{s+1} f^{N,s+1}(t_2) \right) dt_2. \tag{40}$$

Let $\tau_k = t^* k$, where $t^* = \frac{1}{8c_1}$, c_1 is defined in (24), and $k = 1, 2, \ldots, \left[\frac{t_0}{t^*} \right] + 1$.

We want to prove

$$\sup_{[\tau_{k-1},\tau_k]} \left\| \Delta^{N,s} \right\|_{L_1^{(s)}} \leq \frac{2^{5k}}{N^{\eta_k}}, \tag{41}$$

for s such that

$$2^s \leq N^{\frac{\eta_k}{2}}, \tag{42}$$

where $\eta_k = \frac{1}{2^{2k+2}}$. We have

$$\left\| \mathcal{G}^{N,s}(t,\tau_{k-1}) \right\|_{L_1^{(s)}} \leq \left\| \Delta^{N,s}(\tau_{k-1}) \right\|_{L_1^{(s)}} + 2c_1 t^* \frac{s^2}{N}, \tag{43}$$

and

$$\left\| \Delta^{N,s}(t) \right\|_{L_1^{(s)}} \leq \left\| \mathcal{G}^{N,s}(t,\tau_{k-1}) \right\|_{L_1^{(s)}} + c_1 s \int_{\tau_{k-1}}^{t} \left\| \Delta^{N,s+1}(t_2) \right\|_{L_1^{(s+1)}} dt_2, \tag{44}$$

for $t \in [\tau_{k-1}, \tau_k]$. If (41) is satisfied for $k-1$ then by (43) we obtain

$$\left\| \mathcal{G}^{N,s}(t,\tau_{k-1}) \right\|_{L_1^{(s)}} \leq \frac{2^{5(k-1)}}{N^{\eta_{k-1}}} + \frac{s^2}{N}. \tag{45}$$

If moreover N is sufficiently large then by (42)

$$\left\| \mathcal{G}^{N,s}(t,\tau_{k-1}) \right\|_{L_1^{(s)}} \leq \frac{2^{1+5(k-1)}}{N^{\eta_{k-1}}}. \tag{46}$$

Using (46) we may iterate (44) up to the largest m for which

$$2^{s+m} \leq N^{\frac{\eta_{k-1}}{2}}, \tag{47}$$

and assume that

$$2^m \geq \frac{N^{\frac{\eta_{k-1}-\eta_k}{2}}}{2^{5k-1}}. \tag{48}$$

We have

$$\left\| \Delta^{N,s}(t) \right\|_{L_1^{(s)}} \leq \frac{2^{1+5(k-1)}}{N^{\eta_{k-1}}} \left(1 + \sum_{l=1}^{m-1} \frac{s(s+1)\dots(s+l-1)}{4^l \, l!} \right) \tag{49}$$

$$+ 2 \frac{s(s+1)\dots(s+m-1)}{4^m \, m!},$$

for $t \in [\tau_{k-1}, \tau_k]$, where the last term on the r.h.s. of (49) is obtained by using the obvious estimate

$$\left\| \Delta^{N,s} \right\|_{L_1^{(s)}} \leq 2. \tag{50}$$

By (49) and (45) it follows that

$$\sup_{[\tau_{k-1},\tau_k]} \|\Delta^{N,s}\|_{L_1^{(s)}} \leq \frac{2^{s+1+5(k-1)}}{N^{\eta_{k-1}}} + 2^{s-m}. \tag{51}$$

Now the proof of (41) follows by induction on k. We first show that (41) is satisfied for $k = 1$ and then assume that it holds for $k - 1 \geq 1$. Using (42), (48) and (51) we obtain

$$\sup_{[\tau_{k-1},\tau_k]} \|\Delta^{N,s}\|_{L_1^{(s)}} \leq 2^{5k-4} N^{\frac{\eta_k}{2}-\eta_{k-1}} + 2^{5k-1} N^{\eta_k - \frac{\eta_{k-1}}{2}} \tag{52}$$

and therefore (41) for every k follows. □

Theorem 2 shows that the solution of the Boltzmann–like bilinear integro–differential system of equations can be approximated by the solutions of linear equations describing the stochastic system of individuals — provided that the parameters of the stochastic system are suitably chosen.

The estimates are not optimized. One can hope that some of them can be improved to make them uniform with respect to t_0.

The above approach is an example of the methods of kinetic theory (statistical physics) which may successfully be applied to population theory. Completely different approaches were given in papers [46] and [89].

3 Generalized Kinetic Models

The general class of bilinear systems of Boltzmann–like integro–differential equations (32) — the so–called Generalized Kinetic Models (GKM) — describing the dynamics of entities undergoing kinetic (stochastic) interactions was proposed and analyzed in [76]. This type of equations can model interactions between pairs of entities of various populations at the mesoscopic scale (**Me**) — cf. Section 1.

The class of equations (32) can be regarded as a generalization of the Jäger and Segel kinetic model [51] as well as those of Arlotti and Bellomo [4, 5], Arlotti, Bellomo and Lachowicz [7], Lachowicz and Wrzosek [75], Othmer, Dunbar and Alt [88], Geigant, Ladizhansky and Mogilner [43]. In some applications Assumption (12) is relaxed to a non–probabilistic case

$$\sum_{j \in \mathbb{J}} \int_\mathbb{U} A(j, u; k, v, l, w)\,\mathrm{d}u < \infty, \tag{53}$$

for all $k, l \in \mathbb{J}$ and a.a. $v, w \in \mathbb{U}$ such that $a(k, v, l, w) > 0$.

Two particular cases can be distinguish (cf. [7]):

- Destructive case

$$\sum_{j \in \mathbb{J}} \int_\mathbb{U} A(j, u; k, v, l, w)\,\mathrm{d}u < 1, \tag{54}$$

for all $k, l \in \mathbb{J}$ and a.a. $v, w \in \mathbb{U}$ such that $a(k, v, l, w) > 0$;

- Proliferative case

$$\sum_{j \in \mathbb{J}} \int_{\mathbb{U}} A(j, u; k, v, l, w) \, \mathrm{d}u > 1 \,, \tag{55}$$

for all $k, l \in \mathbb{J}$ and a.a. $v, w \in \mathbb{U}$ such that $a(k, v, l, w) > 0$.

The non–probabilistic interpretation of the function $f = f(t, j, u)$, $f : \mathbb{R}_+ \times \mathbb{J} \times \mathbb{U} \to \mathbb{R}_+$, a solution to Eq. (32), is such that it defines the density of entities of the j–th population with state u at time $t \geq 0$. Then the total number of individuals at time $t \geq 0$ is given by

$$\sum_{j \in \mathbb{J}} \int_{\mathbb{U}} f(t, j, u) \, \mathrm{d}u \,. \tag{56}$$

Various models known in literature can be covered by the general model (32).

Example 6. Let $|\mathbb{J}| = 1$ (e.g. $\mathbb{J} = \{1\}$) and $\mathbb{U} = [0, 1]$. Then Eq. (32) is the Jäger–Segel model [51], where the state u ($u \in [0, 1]$), called dominance, is related to the social behavior of a certain population of interacting insects. The Jäger–Segel model corresponds to the conservative case (12).

Example 7. Let $\mathbb{J} = \{1\}$, $\mathbb{U} = [-\pi, \pi]$,

$$A(1, u; 1, v; 1, w) = \tilde{A}(v - u, v - w) \,, \qquad a(1, u; 1, v) = \tilde{a}(u - v) \,, \quad (57)$$

and $\tilde{A}$, $\tilde{a}$ be given 2π–periodic (with respect to all variables) functions. Geigant, Ladizhansky and Mogilner [43] proposed this kind of model in order to describe angular self–organization of the actin cytoskeleton as a process of changing of filament orientation in course of specific actin–actin interaction. The model corresponds to the conservative case (12). Geigant *et al* assumed that the frequency of interaction a is angle–independent $a = \mathrm{const}$ and that $A = A_\sigma$ converges to the appropriate δ distribution as $\sigma \downarrow 0$, where $\sigma > 0$ is the uncertainty of turning parameter, corresponding to the case of purely attracting interactions. They showed (on the formal way) that in the limiting case $\sigma = 0$ (complete alignment in one direction) a single δ–peak is a stable equilibrium solution.

Example 8. The models for which $|\mathbb{J}| < \infty$, $\mathbb{U} = [0, 1]$, and

$$A(j, u; k, v, l, w) = \delta_{j,k} \, \tilde{A}_{k,l}(u; v, w) \,, \tag{58}$$

where $\delta_{j,j} = 1$ and $\delta_{j,k} = 0$ for $j \neq k$, and $\tilde{A}_{k,l}$ are given functions, were proposed and analyzed by Arlotti and Bellomo [4, 5]. These models refer to a more restrictive case than (12), in which the number of entities of each subpopulation is conserved, i.e.

$$\int_{\mathbb{U}} \tilde{A}_{j,k}(u; v, w) \, \mathrm{d}u = 1 \,, \tag{59}$$

for all j, k in $\mathbb{J}$ and a.a. v, w in $\mathbb{U}$ such that $a(j, v, k, w) > 0$.

Example 9. Condition (58) was removed in [7]. Such a framework was generalized in order to include space diffusion processes and interactions with time delay. It was observed that for some choices of functions A and a (constant with respect to u, v and w) Eq. (32) leads to various important models known in the literature, such as e.g. the Kermack–McKendrick (*SIR*) model in theory of epidemics.

The Kermack–McKendrick model reads

$$\dot{\rho}_1 = -\gamma_1 \rho_1 \rho_2\,, \tag{60}$$
$$\dot{\rho}_2 = \gamma_1 \rho_1 \rho_2 - \gamma_2 \rho_2\,,$$
$$\dot{\rho}_3 = \gamma_2 \rho_2\,,$$

where ρ_1, ρ_2, ρ_3 are densities of susceptibles, infectives and removed individuals, respectively, and γ_1 and γ_2 are given positive constants (parameters of the model), $\dot{\rho} = \frac{d}{dt}\rho$.

Let $\mathbb{J} = \{0, 1, 2, 3\}$, $\mathbb{U} = [0, 1]$ and $A(j, u; k, v, l, w)$, $a(j, u, k, v)$ be independent of u, v, w. Assume

$$A(2; 1, 2) = A(2; 2, 1) = 1\,, \quad A(i; 1, 2) = A(i; 2, 1) = 0\,, \quad i = 0, 1, 3, \tag{61}$$
$$A(3; 2, 0) = A(3; 0, 2) = A(0; 2, 0) = A(0; 0, 2) = \frac{1}{2}\,,$$
$$A(i; 2, 0) = A(i; 0, 2) = 0\,, \quad i = 1, 2\,;$$

$$a(1, 2) = a(2, 1) = \gamma_1\,, \quad a(2, 0) = a(0, 2) = \gamma_2\,, \tag{62}$$

while all other $a(j, k)$ are assumed to be equal 0.

Equation (32) with A and a defined above is the GKM corresponding to Eq. (60) in the sense that if $f = (f_0, f_1, f_2, f_3)(t)$ is a solution to Eq. (32) then $\bar{f}_j(t) = \int\limits_0^1 f(t, j, u)\, du$, $j = 1, 2, 3$, is the solution to Eq. (60); the constant density $\bar{f}_0$ plays an auxiliary rôle. Note that A satisfies (12).

Example 10. If $\mathbb{J} = \mathbb{N} = \{1, 2, 3, \ldots\}$ and

$$A(j, u; k, v, l, w) = \frac{1}{2}\delta_{j, k+l}\,, \qquad a(j, u, k, v) = a_{j, k}\,, \tag{63}$$

where $a_{k, l}$ are given and independent of u, v, w, for $k, l \in \mathbb{J}$, then Eq. (32) is the Smoluchowski infinite system of equations ([12, 98]) describing the binary coagulation of colloids

$$\partial_t f_j = \frac{1}{2}\sum_{k=1}^{j-1} a_{j-k, k} f_{j-k} f_k - f_j \sum_{k=1}^{\infty} a_{j, k} f_k\,. \tag{64}$$

The size of entities (clusters) is characterized by an integer $j \in \mathbb{J}$ identified with the number of identical elementary entities.

The coagulation model (64) refers to the conservative case corresponding to a different setting than (12), namely,

$$\sum_{j \in \mathbb{J}} j \Big(G[f] - fLf \Big)(j) = 0 \,, \tag{65}$$

for all functions f, is formally satisfied, which corresponds to the total mass conservation law in the coagulation process. In the standard setting it corresponds to the destructive case (54).

Example 11. If $\mathbb{J} = \mathbb{N}$ and

$$A(j, u; k, v; l, w) = \frac{1}{2} \delta_{j,k+l} \, \tilde{A}_{k,l}(u; v, w) \,, \qquad a(j, u; k, v) = a_{j,k}(u, v) \,, \tag{66}$$

where $\tilde{A}_{k,l}$ and $a_{k,l}$ are given (measurable) functions of u, v, w and v, w, respectively, for $k, l \in \mathbb{N}$,

$$\int_{\mathbb{U}} \tilde{A}_{k,l}(u; v, w) \, du = 1 \qquad\qquad \text{for } k, l \in \mathbb{J} \text{ and a.a. } v, w \text{ in } \mathbb{U} \,, \tag{67}$$

then Eq. (32) is the nonlocal coagulation model proposed in [75]. Actually, in [75], the more general nonlocal coagulation-fragmentation model including diffusion according to Fick's law was proposed and analyzed. The variables v and w were interpreted as the positions (before the interaction) of interacting clusters in the physical space $\mathbb{U}$.

Example 12. The continuous coagulation equation may be defined analogously like the discrete one in Example 10. The size of entities is now characterized by a real nonnegative number $s \in [0, \infty[$. The discrete and the continuous Smoluchowski coagulation equations may unified in the following notation

$$\partial_t f = Q_1[f] \,, \qquad t > 0 \,, \quad s \in \mathbb{S} \,, \tag{68}$$

where $\mathbb{S}$ is either $\mathbb{N}$ or $[0, \infty[$, $f = f(t, s)$ is the density of clusters of size s at time $t \geq 0$,

$$Q_1[f](s) = \tfrac{1}{2} \int_{\mathbb{S}} \chi(s' < s) a(s - s', s') f(s - s') f(s') \, d\lambda(s') \tag{69}$$
$$- f(s) \int_{\mathbb{S}} a(s, s') f(s') \, d\lambda(s') \,,$$

λ is the counting measure in the case $\mathbb{S} = \mathbb{N}$ or the Lebesgue measure in the case $\mathbb{S} = [0, \infty[$, $a(s, s')$ is the coagulation rate.

Another coagulation model was proposed by Oort and van de Hulst (and then by Safronov) — cf. references in [73] to describe the process of aggregation of protoplanetary bodies in astrophysics. The Oort–Hulst–Safronov coagulation equation reads

$$\partial_t f = Q_0[f], \qquad s \in [0, \infty[\tag{70}$$

$$Q_0[f](s) = -\partial_s \left(f(s) \int_0^s s' \alpha(s, s') f(s') \, ds' \right) - f(s) \int_s^\infty \alpha(s, s') f(s') \, ds' \tag{71}$$

In [73] the following class of generalized coagulation equations was introduced

$$\partial_t f = Q_{GC}[f], \qquad t > 0, \quad s \in [0, \infty[, \tag{72}$$

where

$$Q_{GC}[f](r) = \tfrac{1}{2} \int_0^\infty \int_0^\infty A(s; s', s'') a(s', s'') f(s') f(s'') \, ds' \, ds'' \tag{73}$$

$$- f(s) \int_0^\infty a(s, s') f(s') \, ds',$$

A is the weighted probability that the interaction of a cluster of size s' and another cluster of size s'' generates a cluster of size s and is a nonnegative function satisfying

$$A(s; s', s'') = A(s; s'', s'), \qquad s, s', s'' \in [0, \infty[, \tag{74}$$

$$\int_0^\infty s\, A(s; s', s'') \, ds = s' + s'', \qquad s', s'' \in [0, \infty[. \tag{75}$$

Condition (75) ensures that the total volume is preserved during the coagulation reaction. In fact we have

$$\int_0^\infty Q_{GC}[f] \phi \, ds = \tag{76}$$

$$= \int_0^\infty \int_0^{s'} \left(\left(\int_0^\infty A(s; s', s'') \phi(s) \, ds \right) - \phi(s') - \phi(s'') \right) a(s', s'') f(s') f(s'') \, ds'' \, ds',$$

for any test function ϕ.

In [73] a family of generalized coagulation equations connecting the continuous ($\mathbb{S} = [0, \infty[$) Smoluchowski (68) and the Oort–Hulst–Safronov (70) coagulation equations was introduced. For $\varepsilon \in \,]0, 1]$ and $s', s'' \in [0, \infty[$ it was defined

$$A_\varepsilon(s; s', s'') = \delta\left(s - s' \vee s'' - \varepsilon s' \wedge s'' \right) + (1 - \varepsilon) \delta\left(s - s' \wedge s'' \right), \tag{77}$$

where

$$s' \vee s'' = \max\left\{ s', s'' \right\}, \qquad s' \wedge s'' = \min\left\{ s', s'' \right\},$$

δ is the Dirac distribution,

$$a_\varepsilon(s', s'') = \frac{a(s', s'')}{\varepsilon}, \tag{78}$$

Putting A_ε instead of A and a_ε instead of a in Eq. (72) we set $Q_{GC} = Q_\varepsilon$ and consider

$$\partial_t f = Q_\varepsilon[f], \qquad t > 0, \quad s \in [0, \infty[, \tag{79}$$

It is straightforward to see that the choice $\varepsilon = 1$ yields Eq. (68) in the case $\mathbb{S} = [0, \infty[$. On the other hand in [73] the convergence of the weak solution f_ε with the initial datum $f^{(0)}$ towards a weak solution to the Oort–Hulst–Safronov equation with the same initial datum was proved.

Example 13. Various forms of Eq. (32) have been developed to describe the interaction between tumor cells and the immune system — see [1,8,9,20–23,59] and the bibliography therein. The parameter $u \in \mathbb{U}$ has been related to the *activation state* of the entities involved in the process. These models were not characterized by any of the assumptions (12), (54), (55).

Example 14. Mathematical structures similar to Eq. (32) were used in [94] (see references therein) in the context of unequal crossover of genetic sequences containing sections with repeated units. The discrete variable $j \in \{0, 1, \ldots\}$ describes the number of repeated units (the model does not contain a continuous variable u).

Example 15. Let $\mathbb{J} = \mathbb{N}_{-1}$ and

(i) $A_{j,k,l} = A(j, u; k, v; l, w)$ be independent of u, v, w, for $j, k, l \in \mathbb{J}$;
(ii) $a_{j,k} = a(j, u; k, v)$ be independent of u, v, for $j, k \in \mathbb{J}$;
(iii) $A_{-1,k,l} = 0$ for $k, l \in \mathbb{J}$, $a_{-1,k} = 0$ for $k \in \mathbb{J}$;
(iv) $a_{j,-1} = a_j$ for $j \in \mathbb{N}_0 = \{0, 1, 2, \ldots\}$, where $\{a_j\}_{j \in \mathbb{N}_0}$ is a given sequence such that $a_j \geq 0$ for $j \in \mathbb{N}_0$, $a_{j,k} = 0$ for $(j, k) \in \mathbb{N}_0^2$;
(v) $A_{0,k,-1} a_{k,-1} = \delta_{1,k} d_k$ for $k \in \mathbb{J}$, $A_{j,k,-1} a_{k,-1} = \delta_{j-1,k} b_k + \delta_{j+1,k} d_k$ for $j \in \mathbb{N}$, $k \in \mathbb{J}$, where $\{b_j\}_{j \in \mathbb{N}_0}$, $\{d_j\}_{j \in \mathbb{N}}$ are given sequences such that $b_j \geq 0$ for $j \in \mathbb{N}_0$, $d_j \geq 0$ for $j \in \mathbb{N}$.

Assuming that $f(0, -1) = 1$ one formally has $f(t, -1) = 1$ for $t > 0$. Therefore the nonlinear model Eq. (32) leads (formally) to the infinite system of linear ODEs

$$\begin{aligned}
\dot{f}_0 &= -a_0 f_0 + d_1 f_1, \\
\dot{f}_j &= -a_j f_j + b_{j-1} f_{j-1} + d_{j+1} f_{j+1}, \qquad j \in \mathbb{N},
\end{aligned} \tag{80}$$

where $f_j = f(t, j)$, $j \in \mathbb{N}_0$, is independent of u.

This system is a birth–and–death–type system of population dynamics that can model (cf. [13, 15–17] and references therein) a population of cancer cells characterized by different levels of drug resistance (the gene amplification–deamplification process with cell proliferation).

In [15–17] the chaotic behavior (in the sense of Devaney [37]) of the corresponding linear semigroups was studied under various assumptions on the coefficients a_n, b_n, d_n.

Example 16. Some interesting new interpretations and particularizations of the general model (32) are also possible and remain to be developed. The discrete variable can number the enzyme binding sites or the DNA complexes connecting AT or CG pairs, whereas u can describe the states of binding sites or complexes, respectively. In such a case the model (32) can be considered as a generalization of the models in [100] — Chapter 7.

These examples show that Eq. (32) is a very general structure which can be particularized in various important models. Paper [76] was a first step in the description of the mathematical properties of Eq. (32). In particular [76] provided some existence and uniqueness theorems, discussed its equilibrium solutions, and studied its diffusive limit. The existence of unstable equilibrium solutions which are inhomogeneous with respect to the (j, u)–variable was proved. The case when only homogeneous equilibrium solutions exist was specified. Under suitable scaling it was proved that the one–dimensional version of Eq. (32) is asymptotically equivalent to the (nonlinear) porous medium equation ([10, 19]) also used in mathematical biology as the model for density dependent population dispersal.

We assume the conservative case, i.e. (12) is satisfied and $|\mathbb{U}| < \infty$. Let $l_p(\mathbb{J})$ be the Banach space (with the norm $\| \cdot \|_{l_p(\mathbb{J})}$) of real–valued functions whose p–power is summable [bounded] on $\mathbb{J}$ for $1 \le p < \infty$ [$p = \infty$]; $L_q(\mathbb{U})$ be the Banach space (with the norm $\| \cdot \|_{L_q(\mathbb{U})}$) of measurable real–valued functions whose q–power is integrable [essentially bounded] on $\mathbb{U}$ for $1 \le q < \infty$ [$q = \infty$]. Let $L_{p,q}(\mathbb{J} \times \mathbb{U})$ $(1 \le p \le \infty, 1 \le q \le \infty)$ be the Banach space (of functions on $\mathbb{J} \times \mathbb{U}$) equipped with the norm

$$\|f\|_{p,q} = \left\| \left(\|f\|_{L_q(\mathbb{U})} \right) \right\|_{l_p(\mathbb{J})}.$$

It is easy to see that if (11) together with (12) are satisfied then the operator $f \to G[f] - fLf$ is locally Lipschitz continuous in $L_{1,1}(\mathbb{J} \times \mathbb{U})$ and hence there exists a unique solution to the Cauchy problem for Eq. (32) in $L_{1,1}(\mathbb{J} \times \mathbb{U})$ on some time interval $[0, t_0]$, where $t_0 > 0$ depends only on a_0 and on the $L_{1,1}$–norm of initial datum. Moreover, both operators G and L are monotone and therefore, by standard arguments, we can prove that the solution is nonnegative (in $L_{1,1}(\mathbb{J} \times \mathbb{U})$) provided that the initial datum is nonnegative.

By (32), the local (in time) solution can be prolonged onto $\mathbb{R}_+$ by the usual continuation argument. Thus, we have ([76])

Theorem 3. *Let* (11) *and* (12) *be satisfied. Then, for every nonnegative initial datum* $f_0 \in L_{1,1}(\mathbb{J} \times \mathbb{U})$ *and every* $t > 0$, *there exists a unique solution*

$$f \in C^0\big([0,t]; L_{1,1}(\mathbb{J} \times \mathbb{U})\big) \tag{81}$$

to the Cauchy problem for Eq. (32). *Moreover*

$$f \in C^1\big(]0,t[; L_{1,1}(\mathbb{J} \times \mathbb{U})\big), \tag{82}$$

$$f(t) \geq 0, \qquad \|f(t)\|_{1,1} = \|f_0\|_{1,1}, \qquad \forall\, t > 0. \tag{83}$$

$\square$

In the following we assume that $\mathbb{U} = \mathbb{T}^d$, where $\mathbb{T}^d$ is the d–dimensional torus, i.e. the rectangular parallelepiped $[0,1]^d$ with the identified opposite faces. If necessary, a function on $\mathbb{T}^d$ can be interpreted as a periodic function on $\mathbb{R}^d$. In kinetic theory (cf. [48]), the assumption that the particle positions are in a torus has a clear physical meaning. The domain for the wave vector variable in semiconductors theory (cf. [92]) is usually assumed to be $\mathbb{T}^3$. The periodic structures were also considered in Example 7.

The basic assumption is the following

Assumption 1 *Let*

$$0 \leq \; A(j,u;k,v,l,w) = \tfrac{1}{2}\big(A_{k,l}^{(j)}(u-v) + A_{l,k}^{(j)}(u-w)\big), \tag{84}$$
$$0 \leq \; a(j,u,k,v) = a_{j,k}(u-v),$$

for a.a. (j,u), (k,v), (l,w) *in* $\mathbb{J} \times \mathbb{U}$, *where* $A_{k,l}^{(j)}$ *and* $a_{j,k}$ *are given measurable functions defined on* $\mathbb{T}^d$.

We can then relax (11) to include unbounded a and obtain the following local existence result ([76]) in the space $L_{1,\infty}(\mathbb{J} \times \mathbb{T}^d)$

Lemma 1. *Let* (12) *and Assumption* 1 *be satisfied. If additionally there exists a constant* $c_2 < \infty$

$$\sup_{j,k \in \mathbb{J}} \int_{\mathbb{T}^d} a_{j,k}(u)\,\mathrm{d}u \leq c_2, \tag{85}$$

then, for every nonnegative initial datum $f_0 \in L_{1,\infty}(\mathbb{J} \times \mathbb{T}^d)$, *there exists* $t_1 > 0$ *and a unique, nonnegative, solution*

$$f \in C^0\big([0,t_1]; L_{1,\infty}(\mathbb{J} \times \mathbb{T}^d)\big) \tag{86}$$

of the Cauchy problem for Eq. (32). *Moreover*

$$f \in C^1\big(]0,t_1[; L_{1,\infty}(\mathbb{J} \times \mathbb{T}^d)\big), \tag{87}$$

and

$$\|f(t)\|_{1,1} = \|f_0\|_{1,1} \leq \|f_0\|_{1,\infty}, \qquad \forall\, t \in]0,t_1]. \tag{88}$$

Proof. From (12) and Assumption 1 we have

$$\sum_{j\in\mathbb{J}} \int_{\mathbb{T}^d} A_{k,l}^{(j)}(u-v)\,\mathrm{d}v = 1, \qquad \text{for a.a. } u\in\mathbb{T}^d, \quad \forall\, k,l\in\mathbb{J}. \tag{89}$$

From (85) and (89) it follows that the operator $f \to G[f] - fLf$ is locally Lipschitz continuous in $L_{1,\infty}(\mathbb{J}\times\mathbb{T}^d)$. This gives the existence result (where t_1 depends only on c_2 and on $\|f_0\|_{1,\infty}$). Moreover (87), (88) are satisfied. $\square$

Under more restrictive assumptions on A a global existence result in $L_{1,\infty}$ can be obtained ([76])

Theorem 4. *Let* (12), *Assumption 1 and* (85) *be satisfied. If*

$$\sup_{k,l\in\mathbb{J}} \sum_{j\in\mathbb{J}} \sup_{u\in\mathbb{T}^d} A_{k,l}^{(j)}(u) \leq A_1 < \infty, \tag{90}$$

then, for every nonnegative initial datum $f_0 \in L_{1,\infty}(\mathbb{J}\times\mathbb{T}^d)$ and every $t > 0$, there exists a unique, nonnegative solution

$$f \in C^0\big([0,t]; L_{1,\infty}(\mathbb{J}\times\mathbb{T}^d)\big)$$

to the Cauchy problem for Eq. (32). *Moreover*

$$f \in C^1\big(]0,t[; L_{1,\infty}(\mathbb{J}\times\mathbb{T}^d)\big), \tag{91}$$

and

$$\|f(t)\|_{1,1} = \|f_0\|_{1,1} \leq \|f_0\|_{1,\infty}, \qquad \forall\, t > 0. \tag{92}$$

Proof. The proof follows from Lemma 1 and the following estimate

$$\|G[f]\|_{1,\infty} \leq A_1 c_2 \|f\|_{1,1}\|f\|_{1,\infty}. \tag{93}$$

$\square$

Following Ref. [76] we consider the equilibrium solutions, i.e. the class of nonnegative functions h that satisfy

$$G[h] = hLh. \tag{94}$$

The existence of equilibrium solution corresponding to Eq. (32) with $|\mathbb{J}| < \infty$, $\mathbb{U} = [0,1]$ and under the assumptions (58), (59) was considered in Ref. [5] (actually in Ref. [5] the triple interactions were also included). Under the assumption that $a(j,u,k,v) = a_{j,k}(u,v)$ are independent of $u,v \in [0,1]$ and positive it was proved that there exists at least one equilibrium solution. However, Ref. [5] neither deliver information on the possible number of equilibrium solutions nor on their stability.

Lachowicz and Wrzosek [76] showed that the set of solutions to Eq. (94) may contain nonnegative functions from some finite dimensional linear space of functions. Assume that both A and a are periodic functions with respect to each variable on $\mathbb{Z} \times \mathbb{R}^d$ with period $(p, 1, \ldots, 1) \in \mathbb{Z} \times \mathbb{R}^d$, where $p > 0$ is an integer. This leads to assumption that $\mathbb{J} \times \mathbb{U} = \mathbb{Z}_p \times \mathbb{T}^d$, where $\mathbb{Z}_p$ is the group of integers $\mathbb{Z}$ modulo p. Moreover assume that the operators G and L in Eq. (32) are expressed by means of convolution. This leads to the following modification of Assumption 1.

Assumption 2 *Let*

$$0 \leq A(j, u; k, v, l, w) = \tfrac{1}{2} \Big(\alpha(j - k, u - v) + \alpha(j - l, u - w) \Big), \qquad (95)$$
$$0 \leq a(j, u, k, v) = \beta(j - k, u - v),$$

for all $j, k, l \in \mathbb{Z}_p$ and a.a. $u, v, w \in \mathbb{T}^d$, where α and $\beta \not\equiv 0$ are given measurable functions defined on $\mathbb{Z}_p \times \mathbb{T}^d$.

Then Eq. (94) reads

$$\alpha * H_h = H_h \,, \qquad (96)$$

where

$$H_h(j, u) = h(j, u) \, (\beta * h)(j, u) =$$
$$h(j, u) \sum_{k \in \mathbb{Z}_p} \int_{\mathbb{T}^d} \beta(j - k, u - v) h(k, v) \, dv \,, \qquad j \in \mathbb{Z}_p \,, \ u \in \mathbb{T}^d \,.$$

In the sequel some elements of Fourier analysis will be used. For any function $f \in L_{2,2}(\mathbb{Z}_p \times \mathbb{T}^d)$ the Fourier transform $\mathfrak{F} f$ is defined as follows

$$(\mathfrak{F} f)_{\mathbf{n}} = \sum_{j \in \mathbb{Z}_p} \int_{\mathbb{T}^d} \exp\Big(- 2\pi i \, (\frac{n_0}{p} j + n \cdot u) \Big) f(j, u) \, du \,, \qquad i = \sqrt{-1}, \quad (97)$$

for $\mathbf{n} = (n_0, n) \in \mathbb{Z}_p \times \mathbb{Z}^d$. Let

$$\mathcal{S}(f) = \big\{ \mathbf{n} \in \mathbb{Z}_p \times \mathbb{Z}^d \ : \ (\mathfrak{F} f)_{\mathbf{n}} \neq 0 \big\} \,. \qquad (98)$$

For $f_1, f_2 \in L_{2,2}(\mathbb{Z}_p \times \mathbb{T}^d)$, such that $\mathcal{S}(f_1) \cap \mathcal{S}(f_2) \neq \emptyset$, set

$$\mathcal{H}(f_1, f_2) =$$
$$= \big\{ \mathbf{n} \in \mathbb{Z}_p \times \mathbb{Z}^d \ : \quad \mathbf{n} = \mathbf{m}_1 + \mathbf{m}_2 \,, \ \mathbf{m}_1 \in \mathcal{S}(f_1) \,, \ \mathbf{m}_2 \in \mathcal{S}(f_1) \cap \mathcal{S}(f_2) \big\} \,.$$

The following theorem holds ([76])

Theorem 5. *Let Assumption 2 and (12) be satisfied.*

(i) *If $\alpha, \beta \in L_{1,1}(\mathbb{Z}_p \times \mathbb{T}^d)$ then any nonnegative constant function is an equilibrium solution.*

(ii) *Suppose that*

$$\alpha, \beta \in L_{2,2}(\mathbb{Z}_p \times \mathbb{T}^d), \tag{99}$$

let

$$\mathcal{K} = \left\{ \mathbf{n} \in \mathbb{Z}_p \times \mathbb{Z}^d \ : \ (\mathfrak{F}\alpha)_{\mathbf{n}} = 1 \right\}. \tag{100}$$

Then $|\mathcal{K}| < \infty$ and the set of equilibrium solutions contains the set

$$\mathcal{V} = \left\{ h \in L_{2,2}(\mathbb{Z}_p \times \mathbb{T}^d) \ : \ h \geq 0, \ \mathcal{H}(h, \beta) \subset \mathcal{K} \right\}. \tag{101}$$

Proof. Let $h \geq 0$ be a constant. Then $H_h = h^2 \|\beta\|_{1,1}$ and (i) results from (12).

To prove (ii) note that — by Young's inequality — (99) ensures that $H_h \in L_{2,2}(\mathbb{Z}_p \times \mathbb{T}^d)$ provided $h \in L_{2,2}(\mathbb{Z}_p \times \mathbb{T}^d)$. The function h satisfies Eq. (96) if and only if

$$(\mathfrak{F}\alpha)_{\mathbf{n}} = 1 \qquad \text{or} \qquad (\mathfrak{F}H_h)_{\mathbf{n}} = 0, \tag{102}$$

for any $\mathbf{n} \in \mathbb{Z}_p \times \mathbb{Z}^d$.

Note that $\mathbf{0} \in \mathcal{K}$ and $\mathbf{0} \in \mathcal{S}(h) \cap \mathcal{S}(\beta)$, for each nonnegative nonzero function h. Moreover, $|\mathcal{K}| < \infty$ since the Fourier series corresponding to α is convergent. By (102) it follows that h satisfies (96) if and only if

$$\mathcal{S}(H_h) \subset \mathcal{K}. \tag{103}$$

On the other hand one has

$$H_h(j, u) = \sum_{\mathbf{n}=(n_0, n) \in \mathcal{S}(h)} (\mathfrak{F}h)_{\mathbf{n}} \exp\left(-2\pi i(\tfrac{n_0}{p} j + n \cdot u)\right) \tag{104}$$

$$\times \sum_{\mathbf{m}=(m_0, m) \in \mathcal{S}(h) \cap \mathcal{S}(\beta)} (\mathfrak{F}\beta)_{\mathbf{m}} (\mathfrak{F}h)_{\mathbf{m}} \exp\left(-2\pi i(\tfrac{m_0}{p} j + m \cdot u)\right),$$

hence,

$$\mathcal{S}(H_h) \subset \mathcal{H}(h, \beta), \tag{105}$$

for each nonnegative nonzero function h. Therefore the condition

$$\mathcal{H}(h, \beta) \subset \mathcal{K} \tag{106}$$

implies Condition (103), which completes the proof.

$\square$

From Theorem 5 it follows ([76])

Corollary 2. *Let Assumption 2, (12), and (99) be satisfied. If β is such that*

$$\beta(j, u) = \sum_{\mathbf{n} \in \mathcal{K}'} (\mathfrak{F}\beta)_{\mathbf{n}} \exp\left(-2\pi i(\frac{n_0}{p} j + n \cdot u)\right), \tag{107}$$

where $\mathcal{K}' = \left((\mathbb{Z}_p \times \mathbb{Z}^d) \setminus \mathcal{K} \right) \cup \{\mathbf{0}\}$ and $\mathcal{K}$ is defined by (100), then the set of equilibrium solutions contains the set

$$W = \left\{ h \in C(\mathbb{Z}_p \times \mathbb{T}^d) : h \geq 0, \right. \tag{108}$$

$$h(j, u) = \sum_{\mathbf{n} \in \mathcal{K}} h_{\mathbf{n}} \exp \left(-2\pi i \left(\tfrac{n_0}{p} j + n \cdot u \right) \right), \, \bar{h}_{-\mathbf{n}} = h_{\mathbf{n}} \in \mathbb{C} \left. \right\}.$$

$\square$

Now a natural question arises: *Are there nonnegative functions such that* $(\mathfrak{F}\alpha)_0 = 1$ *and* $(\mathfrak{F}\alpha)_{\mathbf{n}} = 1$, *for some* $\mathbf{n} \neq \mathbf{0}$? In the case of positive answer the existence of nonconstant nonnegative solution to Eq. (96) follows from Theorem 5. For simplicity we consider the scalar case: $p = 1$, $d = 1$ and have ([76])

Proposition 1. *For fixed positive integer* m *there exist a periodic function* $\alpha \in C([0, 1])$ *such that*

$$\alpha \geq 0, \qquad \int_0^1 \alpha(u) \, du = 1, \tag{109}$$

and

$$\mathcal{K} = \left\{ n \in \mathbb{Z} : (\mathfrak{F}\alpha)_n = 1 \right\} = \{0, m, -m\}. \tag{110}$$

Proof. We consider an auxiliary problem — the following second order o.d.e. with periodic boundary conditions

$$-cg'' + \lambda g = F, \qquad \text{in }]0, 1[, \tag{111}$$
$$g(0) = g(1), \qquad g'(0) = g'(1),$$

where $c, \lambda > 0$. Given $F \in L_2(0, 1)$ there exists the unique solution g to Problem (111) and $g \in W^{2,2}(0, 1)$, where $W^{2,2}(0, 1)$ is the Sobolev space ([2]). The solution can be expressed as a Fourier series with Fourier coefficients depending on c, λ and F. Moreover by the maximum principle if $F \geq 0$, a.e. in $]0, 1[$, then $g \geq 0$ a.e. in $]0, 1[$. We will choose the constants c, λ and construct the function F in such a way that g satisfies (109) and (110). In other words we demand real Fourier coefficients to be

$$g_0 = 2, \quad g_m^c = 2, \quad g_m^s = 0, \tag{112}$$

where

$$g(u) = \frac{g_0}{2} + \sum_{n=1}^{\infty} g_n^c \cos(2\pi n u) + \sum_{n=1}^{\infty} g_n^s \sin(2\pi n u), \tag{113}$$

and

$$g_n^c = 2 \int_0^1 g(u) \cos(2\pi n u) \, du, \quad g_n^s = 2 \int_0^1 g(u) \sin(2\pi n u) \, du, \quad n > 0. \tag{114}$$

Let

$$F(u) = \frac{F_0}{2} + \sum_{n=1}^{\infty} F_n^c \cos(2\pi nu) + \sum_{n=1}^{\infty} F_n^s \sin(2\pi nu) \qquad (115)$$

and F_n^c, F_n^s, $n \geq 1$ are defined as in (114). The following relations link the Fourier coefficients of F and g

$$\lambda g_0 = F_0, \qquad 4c\pi^2 n^2 g_n^c + \lambda = F_n^c, \qquad 4c\pi^2 n^2 g_n^s + \lambda = F_n^s, \qquad . \qquad (116)$$

Keeping in mind (112) we set

$$\lambda = \frac{F_0}{2}, \qquad F_m^s = \frac{F_0}{2}, \qquad c = \frac{F_m^c - \frac{F_0}{2}}{8\pi^2 m^2}. \qquad (117)$$

Now it is sufficient to construct the function F such that

$$F_m^s = \int_0^1 F(u)\, du = \frac{F_0}{2}, \qquad (118)$$

$$F_m^c > \int_0^1 F(u)\, du = \frac{F_0}{2}. \qquad (119)$$

To this end we choose

$$F = \gamma \chi_{[u_m - \varepsilon, u_m + \delta]}, \qquad (120)$$

where $\chi_{\mathcal{A}}$ is the characteristic function of the set $\mathcal{A}$, $u_m = \frac{1}{12m}$ and $\gamma > 0$, $\varepsilon, \delta \in]0, u_m[$ will be specified later. Then we have

$$F_m^s = 2\gamma \left(\int_{u_m - \varepsilon}^{u_m} \sin(2\pi mu)\, du + \int_{u_m}^{u_m + \delta} \sin(2\pi mu)\, du \right) \qquad (121)$$

$$= 2\gamma \left(\frac{\varepsilon}{2} - P(\varepsilon) + \frac{\delta}{2} + Q(\delta) \right),$$

where

$$P(\varepsilon) = \frac{\varepsilon}{2} - \int_{u_m - \varepsilon}^{u_m} \sin(2\pi mu)\, du > 0, \qquad (122)$$

$$Q(\delta) = \frac{\delta}{2} + \int_{u_m}^{u_m + \delta} \sin(2\pi mu)\, du > 0.$$

Hence (118) is equivalent to the equality $Q(\delta) = P(\varepsilon)$. On the other hand

$$2 \int_0^{u_m + \delta} F(u) \cos(2\pi mu)\, du > \int_0^{u_m + \delta} F(u)\, du, \qquad (123)$$

for any $\delta, \varepsilon \in]0, u_m]$, since $2\cos(2\pi mu) > 1$, for $u \in [0, u_m + \delta[$. Therefore (119) follows.

Both functions $P = P(\varepsilon)$ and $Q = Q(\delta)$ are continuous and increasing. Moreover it is easy to check that $P(u_m) < Q(u_m)$. Therefore, for any $\varepsilon \in]0, u_m[$ there exists a unique $\delta \in]0, u_m[$ such that

$$P(\varepsilon) = Q(\delta), \quad \text{and} \quad 2 \int_{u_m - \varepsilon}^{u_m + \delta} F(u) \cos(2\pi mu) \, du > \gamma(\varepsilon + \delta). \qquad (124)$$

Due to the Sobolev embedding ([2]) one has $g \in C^{1,\nu}([0,1])$, $\nu \in]0, \frac{1}{2}]$. Testing Eq. (111) with smooth functions compactly supported in $[0, u_m - \varepsilon[\cup]u_m + \delta, 1]$ we conclude that

$$\text{supp } g \subset [u_m - \varepsilon, u_m + \delta], \qquad (125)$$

and hence that g is periodic with period 1. This completes the proof.

□

Example 17. Let $p = 1$, $d = 1$, and α be such that (109) together with

$$(\mathfrak{F}\alpha)_1 = 1 \qquad (126)$$

is satisfied (by Proposition 1 such a function exists). Moreover let

$$\beta(u) = \frac{\beta_0}{2} + \sum_{n=2}^{\infty} \beta_n^c \cos(2\pi nu) + \sum_{n=2}^{\infty} \beta_n^s \sin(2\pi nu). \qquad (127)$$

Then for all constants c_1, c_2, and c_3 such that $c_1 > \sqrt{2} \max\{|c_2|, |c_3|\}$ the function

$$h(u) = c_1 + c_2 \cos(2\pi u) + c_3 \sin(2\pi u) \qquad (128)$$

is an equilibrium solution.

Example 17 shows that for the case $\beta = \text{const}$ (studied in Ref. [5]) there are nonconstant equilibrium solutions. On the other hand we have ([76])

Corollary 3. *If $\beta = \text{const}$ then $\mathcal{S}(H_h) = \mathcal{S}(h)$ and h is a solution to Eq. (96) if and only if*

$$\mathcal{S}(h) \subset \mathcal{K}. \qquad (129)$$

Therefore, if $|\mathcal{K}| = 1$ (i.e. $\mathcal{K} = \{0\}$) then only the constant functions are equilibrium solutions.

An interesting question is whether the equilibrium solutions are stable. The following negative answer can be given ([76])

Corollary 4. *Let the conditions of Corollary 2 be satisfied, β be defined by (107) and $|\mathcal{K}| \geq 2$. Then none of the equilibrium solutions of the set $\mathcal{W}$ defined by (108) can be asymptotically stable with respect to the norms in $L_{p,q}$ and in C^0, for $1 \leq p, q \leq \infty$.*

Proof. The Fourier coefficients $h_{\mathbf{n}} = \bar{h}_{-\mathbf{n}} \in \mathbb{C}$ in (108) can be arbitrary. Therefore, for any of the above–mentioned norms and for each function $h \in \mathcal{W}$, it is possible to find another function $\tilde{h} \in \mathcal{W}$, $h \neq \tilde{h}$, (with the same $L_{1,1}$–norm), which is as close to h as we wish.   □

4 Diffusive Limit

Lachowicz and Wrzosek [76] studied the diffusive limit for Eq. (32). Consider the one–dimensional case of Eq. (32), i.e. with $|\mathbb{J}| = 1$, and let $\mathbb{U}$ be either $\mathbb{R}^d$ or d–dimensional torus $\mathbb{T}^d$. In Section 4 of [76] the asymptotic behavior of solutions when the range of interaction described by the support of the functions A and a shrinks was studied.

Let $\| \cdot \|^{(m,p)}$ be the norm in the Sobolev space $W^{m,p}(\mathbb{U})$ and C_B^m denote the Banach space of m–times differentiable functions equipped with the usual norm.

Assumption 3 *Let $0 < \sigma < \frac{1}{2}$ and $A_\sigma : \mathbb{U} \to \mathbb{R}_+$, $a_\sigma : \mathbb{U} \to \mathbb{R}_+$ be given by*

$$A_\sigma(\xi) = \frac{d}{\sigma^d \kappa_d} \chi(|\xi| < \sigma) , \qquad a_\sigma(\xi) = b \chi(|\xi| < \sigma) , \qquad \xi \in \mathbb{U}, \qquad (130)$$

where $\chi(\text{truth}) = 1$, $\chi(\text{false}) = 0$, κ_d is the total surface measure of unit ball in $\mathbb{U}$. Then we assume

$$A(u; v; w) = A_\sigma(u - v) , \qquad (131)$$

and

$$a(u; v) = a_\sigma(u - v) . \qquad (132)$$

The Cauchy problem for Eq. (32) with Assumption 3 reads

$$\partial_t f = A_\sigma \star \big((a_\sigma \star f) f \big) - (a_\sigma \star f) f , \qquad \text{in } L_1(\mathbb{U}) , \qquad (133)$$

$$f\big|_{t=0} = f_0 ,$$

where the convolution operator is considered for the functions defined on $\mathbb{U} = \mathbb{R}^d$ or $\mathbb{U} = \mathbb{T}^d$.

Let g be the mild solution ([19]) in $L_1(\mathbb{U})$ to the porous medium equation ([10])

$$\partial_t g = \frac{b \, \omega_d}{2 \, d \, (d+2)} \sigma^{d+2} \Delta(g)^2 , \qquad \text{in } L_1(\mathbb{U}) , \qquad (134)$$

$$g\big|_{t=0} = f_0 ,$$

where $\Delta = \sum_{i=1}^d \partial_{x_i}^2$.

The following theorem ([76]) shows that given $t_0 > 0$ solutions to (133) and (134) are asymptotically close to each other on $[0, t_0] \times \mathbb{U}$.

Theorem 6. *Let f and g be the solutions to the Cauchy problems for Eq. (133) and Eq. (134), respectively, defined on $[0, t_0] \times \mathbb{U}$, both with the same non–negative initial datum $f_0 \in W^{d+3,1}(\mathbb{U})$. Then there exists a constant c, depending on t_0, such that*

$$\sup_{t \in [0, t_0]} \|f(t) - g(t)\|_{L_1(\mathbb{U})} \leq c\sigma^{d+3} . \tag{135}$$

To prove Theorem 6 we need the following regularity result

Lemma 2. *Let $f_0 \in W^{d+4,1}(\mathbb{U})$, $f_0 \geq 0$ and $f^\varepsilon \in C^0([0, t_0]; L_1(\mathbb{U})$ for $t_0 > 0$ be the solution to Eq. (133). Then, for every $t \in [0, t_0]$, $f^\varepsilon(t, \cdot) \in C_B^3(\mathbb{U})$ and there exists a constant $c(t_0, b, \|f_0\|^{d+3,1})$ such that*

$$\sup_{t \in [0, t_0]} \|f^\varepsilon(t)\|_{C_B^3(\mathbb{U})} \leq c(t_0, b, \|f_0\|^{d+3,1}) . \tag{136}$$

Proof. (of Lemma 2). Let $\partial^\alpha f$ denote the generalized derivative of a function $f \in W^{|\alpha|,1}(\mathbb{U})$, where $\alpha = (\alpha_1, \dots, \alpha_d)$ is a multiindex. The space derivative $\partial^\alpha f^\varepsilon$ satisfies the following linear integro–differential equation

$$\partial_t(\partial^\alpha f^\varepsilon) = A^\varepsilon \star \left((a_\varepsilon \star f^\varepsilon)\partial^\alpha f^\varepsilon + (a_\varepsilon \star \partial^\alpha f^\varepsilon)f^\varepsilon \right) \tag{137}$$
$$- (a_\varepsilon \star f^\varepsilon)\partial^\alpha f^\varepsilon - (a_\varepsilon \star \partial^\alpha f^\varepsilon)f^\varepsilon + H_\varepsilon^\alpha , \qquad \text{in } L_1(\mathbb{U}) ,$$
$$\partial^\alpha f^\varepsilon \Big|_{t=0} = \partial^\alpha f_0 ,$$

where $H_\varepsilon^\alpha = 0$, for $|\alpha| = 1$, and

$$H_\varepsilon^\alpha = A_\varepsilon \star \sum_{\mathbb{I}} \tfrac{\alpha!}{\beta!\gamma!} (a_\varepsilon \star \partial^\gamma f_\varepsilon)\partial^\beta f^\varepsilon \tag{138}$$
$$- \sum_{\mathbb{I}} \tfrac{\alpha!}{\beta!\gamma!} (a_\varepsilon \star \partial^\gamma f_\varepsilon)\partial^\beta f^\varepsilon , \qquad \text{for } |\alpha| > 1 ,$$

and
$$\mathbb{I} = \left\{ (\beta, \gamma) : \quad \beta + \gamma = \alpha, \ (\beta, \gamma) \neq (0, \alpha), \ (\beta, \gamma) \neq (\alpha, 0) \right\} .$$

Hence $\partial^\alpha f^\varepsilon \in C^0([0, t_0]; L_1(\mathbb{U}))$, for $|\alpha| \leq d + 4$. By the Sobolev embedding theorem (Ref. [2]) we have

$$f^\varepsilon(t, \cdot) \in C_B^3(\mathbb{U}) , \qquad \text{for every } t \in [0, t_0] .$$

Estimate (136) is independent of ε. Indeed assuming α in (137) such that $|\alpha| = 1$ and using (12) we obtain

$$\sup_{t \in [0, t_0]} \|\partial^\alpha f^\varepsilon\|_{L_1(\mathbb{U})} \leq \|\partial^\alpha f_0\|_{L_1(\mathbb{U})} \exp\left(4b\|f_0\|_{L_1(\mathbb{U})}\right) . \tag{139}$$

Consequently, for $|\alpha| = 2$, $\sup\limits_{t\in[0,t_0]}\|H_\varepsilon\|_{L_1(\mathbb{U})}$ is independent of ε since it contains derivatives of order lower then $|\alpha|$. Thus for $|\alpha| = 2$ we obtain from (137)

$$\sup_{t\in[0,t_0]}\|\partial^\alpha f^\varepsilon\|_{L_1(\mathbb{U})} \leq \tag{140}$$

$$\left(\|\partial^\alpha f_0\|_{L_1(\mathbb{U})} + t_0 \sup_{t\in[0,t_0]}\|H_\varepsilon^\alpha\|_{L_1(\mathbb{U})}\right) \exp\left(4b\|f_0\|_{L_1(\mathbb{U})}t_0\right)$$

and similarly for higher derivatives.

$\square$

Proof. (of Theorem 6). We fix $(t,u) \in]0,t_0[\times\mathbb{U}$ and introduce generalized polar coordinates centered at u. The classical k–th derivative in direction η of function f is denoted by $\partial_\eta^k f$. The Taylor expansion of the function $\mathbb{U} \ni v \to f(t,v)$ at point u reads

$$f^\varepsilon(t,u+r\eta) = W_\varepsilon(t,u,r\eta) + R_\varepsilon(t,u,r\eta), \tag{141}$$

where

$$W_\varepsilon(t,u,r\eta) = f^\varepsilon(t,u) + r\partial_\eta^1 f^\varepsilon(t,u) + \tfrac{r^2}{2}\partial_\eta^2 f^\varepsilon(t,u), \tag{142}$$

$$R_\varepsilon(t,u,r\eta) = \tfrac{r^2}{6}\partial_\eta^3 f^\varepsilon(u+\theta\eta),$$

and $r \in \mathbb{R}_+$, $\eta \in \mathbb{S}^{d-1} = \{\eta \in \mathbb{R}^d : |\eta| = 1\}$, for some $\theta = \theta(u,r\eta)$, $\theta \in]0,r[$.

From now on, we skip in the notation both the index ε and the t–variable, for simplicity. Substituting (141) to the *lost term* in (133) we obtain

$$fLf(u) = b\,f(u) \int\limits_0^\varepsilon \int\limits_{\mathbb{S}^{d-1}} r^{d-1}\Big(f(u) + r\partial_\eta^1 f(u) \tag{143}$$

$$+\tfrac{r^2}{2}\partial_\eta^2 f(u) + R(u,r\eta)\Big)\,\mathrm{d}r\,\mathrm{d}\eta$$

$$= b\,\omega_d f^2(u)\tfrac{\varepsilon^d}{d} + bf(u)\Big(\tfrac{\varepsilon^{d+2}}{2d+2} \int\limits_{\mathbb{S}^{d-1}} \partial_\eta^2 f(u)\,\mathrm{d}\eta$$

$$+\int\limits_0^\varepsilon \int\limits_{\mathbb{S}^{d-1}} r^{d-1}R(u,r\eta)\Big)\,\mathrm{d}r\,\mathrm{d}\eta.$$

For the *gain term* in (133) we have

$$G[f](u) = \tfrac{b\,d}{\omega_d\varepsilon^d} \int\limits_0^\varepsilon \int\limits_{\mathbb{S}^{d-1}} \int\limits_0^\varepsilon \int\limits_{\mathbb{S}^{d-1}} r_1^{d-1}r_2^{d-1}f(u+r_1\eta_1) \tag{144}$$

$$\times f(u+r_1\eta_1+r_2\eta_2)\,\mathrm{d}r_1\,\mathrm{d}\eta_1\,\mathrm{d}r_2\,\mathrm{d}\eta_2$$

$$= \tfrac{b\,d}{\omega_d\varepsilon^d} \int\limits_0^\varepsilon \int\limits_{\mathbb{S}^{d-1}} \int\limits_0^\varepsilon \int\limits_{\mathbb{S}^{d-1}} r_1^{d-1}r_2^{d-1}\Big(W(u,r_1\eta_1)W(u,r_1\eta_1+r_2\eta_2)$$

$$+W(u,r_1\eta_1)R(u,r_1\eta_1+r_2\eta_2)$$

$$+R(u,r_1\eta_1)W(u,r_1\eta_1+r_2\eta_2)\Big)\,\mathrm{d}r_1\,\mathrm{d}\eta_1\,\mathrm{d}r_2\,\mathrm{d}\eta_2$$

After laborious computations we arrive at

$$G[f](u) = \frac{b\omega_d}{d}\varepsilon^d f^2(u) + \frac{3}{2}f(u)\frac{b\varepsilon^{d+2}}{d+2} \int\limits_{\mathbb{S}^{d-1}} \partial_\eta^2 f(u)\,\mathrm{d}\eta$$

$$+f(u)\frac{b\varepsilon^{d+2}}{d+2} \int\limits_{\mathbb{S}^{d-1}} \left(\partial_\eta^1 f(u)\right)^2 \mathrm{d}\eta + \mathcal{R}(u)\,,$$

where

$$\mathcal{R}(u) = \tag{145}$$

$$= \varepsilon^{d+4}\left(\frac{b}{4(d+4)} \int\limits_{\mathbb{S}^{d-1}} \left(\partial_\eta^2 f(u)\right)^2 \mathrm{d}\eta + \frac{bd}{(d+2)\,\omega_d}\left(\int\limits_{\mathbb{S}^{d-1}} \left(\partial_\eta^2 f(u)\right)^2 \mathrm{d}\eta\right)\right)$$

$$+\frac{bd}{\omega_d\varepsilon^d} \int\limits_0^\varepsilon \int\limits_{\mathbb{S}^{d-1}} \int\limits_0^\varepsilon \int\limits_{\mathbb{S}^{d-1}} r_1^{d-1}r_2^{d-1}\Big(W(u,r_1\eta_1)R(u,r_1\eta_1 + r_2\eta_2)$$

$$+R(u,r_1\eta_1)W(u,r_1\eta_1 + r_2\eta_2)\Big)\,\mathrm{d}r_1\,\mathrm{d}\eta_1\,\mathrm{d}r_2\,\mathrm{d}\eta_2\,.$$

Finally we obtain from (143) and (145)

$$(G[f] - fLf)(u) = \tag{146}$$

$$= \frac{b\varepsilon^{d+2}}{d+2} \int\limits_{\mathbb{S}^{d-1}} \left(f(u)\partial_\eta^2 f(u) + \left(\partial_\eta^1 f(u)\right)^2\right)\mathrm{d}\eta + \mathcal{R}(u)$$

$$= \frac{b}{2}\frac{\varepsilon^{d+2}}{d+2} \int\limits_{\mathbb{S}^{d-1}} \partial_\eta^2 f^2(u)\,\mathrm{d}\eta + \mathcal{R}(u) = \frac{b\,\omega_d\,\varepsilon^{d+2}}{2\,d\,d+2}\Delta f^2(u)\,\mathrm{d}\eta + \mathcal{R}(u)\,.$$

In what follows we denote by $\mathfrak{C}$ a constant which depends only on b, d, ω_d and t_0. From (145), (140) and Lemma 2 we conclude that

$$\sup\limits_{[0,t_0]\times \mathbb{U}} |\mathcal{R}| \le \mathfrak{C}\varepsilon^{d-1}\,, \tag{147}$$

for $\varepsilon \in\,]0,1[$.

In the case $\mathbb{U} = \mathbb{R}^d$ we have also

$$\sup\limits_{t\in[0,t_0]} \int\limits_{\mathbb{R}^d} |\mathcal{R}(t,u)|\,\mathrm{d}u \le \mathfrak{C}\varepsilon^{d-1}\,, \tag{148}$$

All components in $\mathcal{R}$ can be estimated in a similar way. For instance, using Lemma 2, we have

$$\frac{bd}{\omega_d\varepsilon^d} \int\limits_{\mathbb{R}^d} \left|\int\limits_0^\varepsilon \int\limits_{\mathbb{S}^{d-1}} \int\limits_0^\varepsilon \int\limits_{\mathbb{S}^{d-1}} r_1^{d-1}r_2^{d-1}f(u)\frac{r_1^3}{6}\right. \tag{149}$$

$$\times\partial_{\eta_1}^3 f(u + \theta\eta_1)\,\mathrm{d}r_1\,\mathrm{d}\eta_1\,\mathrm{d}r_2\,\mathrm{d}\eta_2\bigg|\,\mathrm{d}u$$

$$\le \frac{bd}{\omega_d\varepsilon^d} \sum\limits_{|\alpha|\le 3} \|\partial^\alpha f\|_{L_\infty(\mathbb{U})}\frac{\omega_d}{d(d+3)}\varepsilon^{2d+3}\|f\|_{L_1(\mathbb{U})} \le \mathfrak{C}\varepsilon^{d+3}\|f\|_{L_1(\mathbb{U})}\,.$$

To complete the proof we compare the solution to the equation

$$\partial_t f^\varepsilon = \frac{b\,\omega_d}{2\,d\,(d+2)}\varepsilon^{d+2}\Delta(f^\varepsilon)^2 + \mathcal{R}\,, \qquad\qquad \text{in } L_1(\mathbb{U})\,, \qquad (150)$$

with the solution g^ε to Eq. (134), both with the same initial datum

$$f^\varepsilon\Big|_{t=0} = g^\varepsilon\Big|_{t=0} = f_0\,. \qquad (151)$$

Using the comparison property of the mild solutions to the porous medium equation ([19, 24]) and (146) we obtain (135), which completes the proof.

□

5 Links in the Space–Homogeneous Case

Usually the description of biological populations is carried out at *macroscopic* level of interacting subpopulations of the system, e.g. in terms of the Lotka–Volterra–type equations. However in many cases the description at *microscopic* level of interacting individuals (e.g. cells) seems to be more adequate.

In this section, following [70], rigorous relationships between the 3 levels of description are stated

(**Mi**) microscopic level of stochastically interacting individuals, in terms of continuous linear semigroups of Markov operators, Eq. (14);

(**Me**) mesoscopic level of a distribution function related to a test individual, in terms of nonlinear semigroups related to bilinear GKM (32);

(**Ma**) macroscopic level of densities of interacting subpopulations, in terms of dynamical systems related to bilinear Lotka–Volterra–type equations.

It is worth to point out that some natural modifications of the equations at level (**Mi**) can lead to ℓ–linear equations at level (**Ma**) for given $\ell \in \{3, 4, \ldots\}$. On the other hand the further approximation to the solutions of macroscopic system, that are not necessarily ℓ–linear (for some $\ell \in \mathbb{N}$), is possible by using Tikhonov's Theorem ([101]) at level (**Ma**) — cf. the theory of Michaelis and Menten in [87], Chapter 2, A.

The present approach may offer a theoretical basis for the modeling of biological processes — e.g. such as the competition between tumor and immune system — at the level of interacting individuals (cells) — see Example 19 as well as those defined in the context of game theory — see [83].

Various relationships between particle systems and ODEs or PDEs equations have been discussed by various authors. The interested reader is addressed to [36, 39, 86] and references therein.

The novelty of the approach in [70] lies in the relating 3 different levels of description and in controlling the rates of approximations. The approach can be applied to a wide variety of important examples and allows for various generalizations.

Following [70] we consider the following general system of equations

$$\dot{\rho}_j = \sum_{k=1}^{r} \alpha_{j,k}\rho_k + \rho_j \sum_{k=1}^{r} \beta_{j,k}\rho_k, \qquad j = 1, 2, \ldots, r, \qquad (152)$$

where $\alpha_{j,j}$, $\beta_{j,k}$ ($j, k \in \{1, 2, \ldots, r\}$) are real constants (they can be positive, negative or zero), $\alpha_{j,k}$ ($j \neq k$) are non–negative constants; $\rho_j = \rho_j(t)$; $t \geq 0$ is the time variable.

The parameters $\alpha_{j,j}$ are intrinsic growth or decay rates of the j-subpopulation, and $\beta_{j,k}$ are the interaction rates (positive, negative or zero) between the j–th and k–th subpopulations.

System (152) such that

$$\alpha_{j,k} = 0, \quad \beta_{j,k}\beta_{k,j} < 0 \quad \text{for all } j \neq k \qquad (153)$$
$$\text{and} \quad \beta_{j,j} = 0 \quad \text{for all } j = 1, \ldots, r$$

is called a *Lotka–Volterra system*;
if

$$\alpha_{j,k} = 0, \quad \beta_{j,k}\beta_{k,j} < 0 \quad \text{for all } j \neq k \qquad (154)$$
$$\text{and} \quad \beta_{j,j} < 0 \quad \text{for all } j = 1, \ldots, r$$

it is called a *Verhulst–Volterra system*;
and if

$$\alpha_{j,k} = 0 \quad \text{for all } j \neq k, \quad \beta_{j,k} \leq 0 \quad \text{for all } j, k = 1, \ldots, r \qquad (155)$$
$$\text{and} \quad \text{for any } j = 1, \ldots, r \text{ there is } k = 1, \ldots, r \text{ such that } \beta_{j,k} < 0$$

it is called a *competitive system*.

If Eq. (152) is *conservative*, i.e.

$$\sum_{j,k=1}^{r} \alpha_{j,k}\rho_k + \sum_{j,k=1}^{r} \beta_{j,k}\rho_j\rho_k = 0, \qquad (156)$$
$$\text{for all } \rho_j \in [0, \infty[, \qquad j = 1, \ldots, r,$$

then one may apply the idea of Example 9. Then by Theorem 2 one may approximate the solutions of Eq. (152) by the system of N interacting entities given by (15). For example in the case of the *SIR* model (60) one immediately has

Theorem 7. *Let A and a be given by (61) and (62), respectively. Let F be a probability density on $\{0, 1, 2, 3\} \times [0, 1]$. Then, for each $t_0 > 0$, there exists N_0 such that for $N \geq N_0$*

$$\sup_{[0,t_0]} \sum_{j=1}^{3} |\bar{f}_j^{N,1} - \rho_j| \leq \frac{c}{N^\eta}, \qquad (157)$$

where the nonnegative function $f^N(t) \in L_1^{(N)}$ is the unique solution of Eq. (15) *corresponding to the initial datum* (37) *and* (ρ_1, ρ_2, ρ_3) *is the unique, nonnegative solution of* Eq. (60) *corresponding to the initial datum* $(\bar{F}_1, \bar{F}_2, \bar{F}_3)$; η, c *are positive constants that depend on* t_0; $\bar{f}_j = \int_0^1 f(j, u) \, du$.

Proof. The proof starts with the observation that Eq. (152), if (156) holds, has a global nonnegative solution for any nonnegative initial data. Let f be the global solution in $L_1^{(1)}$ of Eq. (32) with A and a defined by (61) and (62). Then we have

$$\sup_{[0, t_0]} \sum_{j=1}^{3} |\bar{f}_j^{N,1} - \rho_j| \leq \|f^{N,1} - f\|_{L_1^{(1)}}, \tag{158}$$

and by Theorem 2 we obtain (157).

$\square$

Now the problem appears how to relate the conservative (i.e. satisfying (12)) system (15) with a general system (152) not necessarily satisfying (156).

First we consider the general totally competitive system i.e. Eq. (152) such that

$$\alpha_{j,k} = 0 \text{ for all } j \neq k, \qquad \alpha_j = \alpha_{j,j} > 0 \text{ for all } j = 1, \ldots, r, \tag{159}$$
$$\beta_{j,k} < 0 \text{ for all } j, k = 1, 2, \ldots, r.$$

This system was considered in [67] and it was shown that the conservative (i.e. satisfying (12)) Eq. (15) results in the general totally competitive system (159) which is not conservative. In fact, let $\mathbb{J} = \{1, 2, \ldots, r\}$ and $\mathbb{U} = [0, R]$, with properly chosen $R > 0$. Let moreover $A(j, u; k, v, l, w)$ be independent of v and w and such that

$$A(j, u; k, v, l, w) = \begin{cases} A_{k,l}^{(j)}(u) & \text{for } j = l \\ 0 & \text{for } j \neq l, \end{cases} \tag{160}$$

where

$$\int_0^R A_{k,j}^{(j)}(u) \, du = 1, \quad \beta_{k,j} \int_0^R u \, A_{k,j}^{(j)}(u) \, du = \alpha_j, \qquad \forall \, k \in \{1, \ldots, r\}. \tag{161}$$

$R > 0$ is chosen here such that Conditions (161) may be satisfied.

Let $a(j, u, k, v)$ be independent of u and

$$a(j, u, k, v) = \beta_{j,k} v. \tag{162}$$

It is easy to see that (12) is satisfied and therefore the solution of the corresponding Eq. (32) *a priori* satisfies

$$\sum_{j=1}^{r} \int_0^R f(t, j, u) \, du = \sum_{j=1}^{r} \int_0^R f(0, j, u) \, du \tag{163}$$

provided that the initial data $f(0, j, u)$ are nonnegative. Let now

$$\hat{f}_j(t) = \int_0^R u\, f(t, j, u)\, du\,. \tag{164}$$

Using (160), (161), (162) and (163) we see that

$$\Big(\rho_1(t), \ldots, \rho_r(t)\Big) = \Big(\hat{f}_1(t), \ldots, \hat{f}_r(t)\Big)$$

is a solution of Eq. (152) with (159). Then by Theorem 2 we have ([67])

Theorem 8. *Let F be a probability density on $\{1, 2, \ldots, r\} \times [0, R]$. Then, for each $t_0 > 0$, there exists N_0 such that for $N \geq N_0$*

$$\sup_{[0, t_0]} \sum_{j=1}^r |\hat{f}_j^{N,1} - \rho_j| \leq \frac{c}{N^\eta}\,, \tag{165}$$

where the nonnegative functions $f^N \in L_1^{(N)}$ is the unique solution of Eq. (15) corresponding to the initial datum (37) and $(\rho_1, \ldots, \rho_r)$ is the unique, non-negative solution of Eq. (152) with (159) corresponding to the initial datum $(\hat{F}_1, \ldots, \hat{F}_r)$; $\hat{f}_j$ is given by (164); η and c are positive constants that depend on t_0.

Example 18. The simplest but important example of an equation (152) satisfying (159) is the logistic equation

$$\dot{\rho} = \alpha \rho - \beta \rho^2\,, \tag{166}$$

where α and β are positive constants. Let A and a be such that

$$\int_0^R A(u)\, du = 1\,, \qquad \int_0^R u\, A(u)\, du = \frac{\alpha}{\beta} \tag{167}$$

$$a(u, v) = \beta v\,, \tag{168}$$

where R is chosen such that $R > \frac{\alpha}{\beta}$. The corresponding (15) is defined by the linear operator

$$\Lambda_N^* f(u_1, \ldots, u_N) = \tag{169}$$

$$= \frac{\beta}{N} \sum_{\substack{1 \leq n, m \leq N \\ n \neq m}} u_m \left(A(u_n) \int_0^R f(u_1, \ldots, u_{n-1}, v, u_{n+1}, \ldots, u_N)\, dv \right.$$

$$\left. - f(u_1, \ldots, u_N) \right)$$

and the corresponding equation GKM (32) reads

$$\partial_t f(t, u) = \beta\, \hat{f}(t) \Big(A(u)\, \bar{f}(t) - f(t, u) \Big)\,, \tag{170}$$

where

$$\bar{f} = \int\limits_0^R f(u)\,\mathrm{d}u\,, \qquad \hat{f} = \int\limits_0^R u\,f(u)\,\mathrm{d}u\,. \tag{171}$$

It is easy to see that $\bar{f}(t) = \bar{F} = 1$, for any $t > 0$, where F is the initial datum being a density probability on $[0, R]$. Equation (170) is the GKM corresponding to the logistic equation (166). By Theorem 8 we may approximate the solution ρ to Eq. (166) by the solutions f^N of Eq. (15) with (169): for each $t_0 > 0$, there exists N_0 such that for $N \geq N_0$

$$\sup_{[0,t_0]} |\hat{f}^{N,1} - \rho| \leq \frac{c}{N^\eta}\,, \tag{172}$$

The situation is more complex for a general system (152). In the rest of this section, following [70] (see also [68]) it is shown that under suitable assumption the conservative (i.e. probabilistic (12)) model (15) results in System (152).

Let $\mathbb{J} = \{0, 1, \ldots, r\}$; $\mathbb{U} = \mathbb{R}_+$;

$$a_R(j, u, k, v) = a^*(j, u, k, v)\chi(u \leq R)\chi(v \leq R)\,, \tag{173}$$

$$a^*(j, u, k, v) = \begin{cases} b_{j,k}\,v & \text{for } j, k = 1, \ldots, r \\ b_{j,0} & \text{for } j = 1, \ldots, r,\ k = 0 \\ 0 & \text{for } j = 0,\ k = 0, \ldots, r\,, \end{cases} \tag{174}$$

$$b_{j,k} \geq 0\,, \qquad \forall\ j, k = 0, \ldots, r\,, \tag{175}$$

where $R \geq R_0 > 0$, $\chi(\text{true}) = 1$, $\chi(\text{false}) = 0$;

If $a^*(k, ., l, .) \equiv 0$ then $A_R(j, .; k, ., l, .) \equiv 0\ \forall\ j$; If $a^*(k, ., l, .) \not\equiv 0$, for some k, l, then

$$A_R(j, u; k, v, l, w) = \mathcal{A}_{j,k,l}^{(R)}(u, v)\chi(u \leq R)\chi(v \leq R)\chi(w \leq R)\,, \tag{176}$$

for $j, k, l = 0, \ldots, r$,

$$\mathcal{A}_{j,k,l}^{(R)}(u, v) = \frac{\mathcal{A}_{j,k,l}(u, v)}{\sum\limits_{j'=1}^{r} \int\limits_0^R \mathcal{A}_{j',k,l}(u', v)\,\mathrm{d}u'}\,, \tag{177}$$

and $\mathcal{A}_{j,k,l} \geq 0$ satisfies

$$\sum_{j'=1}^{r} \int\limits_0^{R_0} \mathcal{A}_{j',k,l}(u, v)\,\mathrm{d}u \geq c_3 > 0\,, \qquad \sum_{j'=1}^{r} \int\limits_0^{\infty} \mathcal{A}_{j',k,l}(u, v)\,\mathrm{d}u = 1\,, \tag{178}$$

$$\int\limits_0^{\infty} u\,\mathcal{A}_{j,k,l}(u, v)\,\mathrm{d}u = B_{j,k,l}\,v\,, \qquad \forall\ v > 0\,,$$

for all $j, k, l = 0, \ldots, r$, where c_3 is a constant. Moreover, we assume

$$\mathcal{A}_{0,k,l} \equiv 0, \qquad \forall \quad k, l = 0, \ldots, r, \tag{179}$$

$$\mathcal{A}_{j,k,l} \equiv 0, \text{ if } \quad j \neq k, \quad j, k, l = 1, \ldots, r. \tag{180}$$

Write

$$B_{j,l} = B_{j,j,l}, \qquad \text{for } j, l = 1, \ldots, r. \tag{181}$$

Note that the functions A_R are characterized by a singular behavior with respect to the variable v for $v = 0$. An example of such a function is

$$\mathcal{A}_j(u, v) = \frac{1}{v} \exp\left(-\frac{u}{v}\right), \qquad u > 0, \quad v > 0. \tag{182}$$

It is easy to see that (12) is satisfied and therefore the solution of Eq. (32) a priori satisfies

$$f(t, j, u) \geq 0, \qquad \sum_{j=0}^{r} \int_0^R f(t, j, u) \, du = \sum_{j=0}^{r} \int_0^R f(0, j, u) \, du, \tag{183}$$

$t > 0$, provided that the initial data $f(0)$ are non–negative and integrable. Moreover, we may assume that

$$f(t, 0, u) = \mathfrak{c}_0 \chi(u \leq R_0), \qquad \forall \, t \geq 0, \, \forall \, u \in [0, R], \tag{184}$$

where $\mathfrak{c}_0 > 0$, $R_0 > 0$ are constants.

Given parameters $\alpha_{j,k}$, $\beta_{j,k}$ $(j, k \in \{1, \ldots, r\})$ we assume that $B^{j,k}$, $b_{j,k}$ $(j, k \in \{0, 1, \ldots, r\})$, $\mathfrak{c}_0$, and R_0 are chosen such that

$$\left(B_{j,k} - 1\right) b_{j,k} = \beta_{j,k}, \qquad j, k \in \{1, 2, \ldots, r\}, \tag{185}$$

$$\left(B_{j,k,0} - \delta_{j,k}\right) b_{k,0} \, \mathfrak{c}_0 \, R_0 = \alpha_{j,k}, \qquad j, k \in \{1, 2, \ldots, r\},$$

If f is a solution of Eq. (32) with a_R and A_R given by (173)–(181) and (185), then

$$\left(\hat{f}_1(t), \ldots, \hat{f}_r(t)\right)$$

formally approximates a solution

$$\left(\rho_1(t), \ldots, \rho_r(t)\right)$$

of Eq. (152) in the limit $R \to \infty$ (for the simplicity of notation we do not indicate the R–dependence of f). The rigorous result will be discussed in what follows.

In order to cover the general case

$$\alpha_{j,j} \in \mathbb{R} \ \forall \ j = 1, \ldots, r, \qquad \alpha_{j,k} \geq 0 \quad \forall \ j \neq k, \ j, k = 1, \ldots, r, \quad (186)$$
$$\beta_{j,k} \in \mathbb{R} \qquad \forall \ j, k = 1, \ldots, r,$$

the limit $R \to \infty$ is considered.

Given initial data

$$(\rho_1, \ldots, \rho_r)\Big|_{t=0} = (\rho_1^{(0)}, \ldots, \rho_r^{(0)}) \in \mathbb{R}_+^r, \qquad (187)$$

Let now F be a probability density on $\mathbb{J} \times [0, R]$ such that

$$\hat{F}_j = \rho_j^{(0)}, \qquad \forall \ j = 1, \ldots, r. \qquad (188)$$

We introduce the following notation: L_1 is the space equipped with the norm

$$\|f\|_{L_1} = \sum_{j=0}^{r} \int_0^R |f(j, u)| \, du, \qquad R > 0,$$

and X_0 is the space equipped with the norm

$$\|f\|_0 = \sum_{j=0}^{r} \int_0^{\infty} |f(j, u)| \, du.$$

It is understood that any $f \in L_1$, $R > 0$, is in X_0 by taking the zero extension of f in $]R, \infty[$. We consider the following space $X_i \subset X_0$, $i = 1, 2$, equipped with the norm

$$\|f\|_i = \sum_{j=0}^{r} \int_0^{\infty} (1 + u^i)|f(j, u)| \, du.$$

Given a function $f \in L_1(0, \infty)$, let

$$\tilde{f} = \int_0^{\infty} u f(u) \, du. \qquad (189)$$

We need the following local existence result in X_2 ([70])

Lemma 3. *Let $a = a_\infty$ and $A = A_\infty$ be such that (12) with $\mathbb{J} \times \mathbb{R}_+$ is satisfied and*

$$0 \leq a_\infty(j, u, k, v) \leq c^a \left(1 + v^2\right), \qquad (190)$$

$$0 \leq A_\infty(j, u; k, v, l, w), \qquad (191)$$

$$\sum_{j'=1}^{r} \int_0^{\infty} u'^2 A_\infty(j', u'; k, v, l, w) \, du' \leq c^A \left(1 + v^2\right), \qquad (192)$$

for $j, k, l = 0, \ldots r$ and for a.a. u, v, w in $\mathbb{R}_+$, where c^a, c^A are constants. Then for every non–negative initial datum $F \in X_2$, there exists $t_* > 0$ and a unique, non–negative solution

$$f \in C^0([0, t_*]); X_2) \cap C^1(]0, t_*[\,; X_2)$$

of the Cauchy problem for Eq. (32).

Proof. The proof is a consequence of the fact that the operator given by the r.h.s. of Eq. (32) is locally Lipschitz continuous in X_2. The non–negativity follows by the standard method.

$\square$

Note that the function A of the form (182) satisfies Conditions (191), (192).

It is obvious that given any choice of parameters $\alpha_{j,k}$, $\beta_{j,k}$ as in (186) there is $t_1 > 0$ such that the unique solution $(\rho_1, \ldots, \rho_r)$ to System (152) with given initial data (187) exists in $[0, t_1]$. This is the basis for the following

Assumption 4 *Given parameters $\alpha_{j,k}$, $\beta_{j,k}$ as in (186) and $(\rho_1^{(0)}, \ldots, \rho_r^{(0)}) \in \mathbb{R}_+^r$, we assume that $t_1 > 0$ is such that the unique solution $(\rho_1, \ldots, \rho_r)$ to System (152) with given initial data (187) exists in $[0, t_1]$.*

The essence of Assumption 4 is the "size" of t_1 but not its existence. In general case (186) of Eq. (152) a blow-up can appear in finite time. However, for large class of Systems (152) the solutions exist globally in time — cf. e.g. Chapter 21 in [55] and references therein. The existence of solutions on any finite interval of time is, for example, guaranteed under the following biologically reasonable assumptions (additional to (186)) including a class of Lotka–Volterra, Verhulst–Volterra or competitive systems

$$\alpha_{j,k} = 0, \qquad \forall\, j \neq k, \tag{193}$$

and either

$$\beta_{j,k} \leq 0 \qquad \forall\, j, k = 1, \ldots, r, \tag{194}$$

or

$$\beta_{j,k}\beta_{k,j} \leq 0 \quad \forall\, j \neq k, \qquad \beta_{j,k}\beta_{j,l} \geq 0 \quad \forall\, j, k, l, \qquad \beta_{j,j} \leq 0, \quad \forall\, j. \tag{195}$$

Let now F be a probability density on $\mathbb{J} \times \mathbb{R}_+$ such that

$$\tilde{F}(j) = \rho_j^{(0)}, \qquad \forall\, j = 1, \ldots, r. \tag{196}$$

Consider Eq. (32) with the functions $a = a_\infty$, $A = A_\infty$ obtained from (173)–(181) and (185) by sending formally $R \to \infty$. We have ([70])

Lemma 4. *Let* Assumption 4 *be satisfied and* $a = a_\infty$, $A = A_\infty$ *in Eq. (32);* *Let moreover* A_∞ *satisfy* (192) *and* $F \in X_2$ *be a probability density on* $\mathbb{J} \times \mathbb{R}_+$ *such that* (196) *is satisfied. Then there exists a unique non–negative solution*

$$f_\infty \in C^0([0, t_1]); X_2) \cap C^1(]0, t_1[; X_2)$$

of the Cauchy problem for Eq. (32).

Proof. It is easy to see that the assumptions of Lemma 3 are satisfied and, therefore, there is a unique non–negative solution f_∞ of Eq. times (32) with the initial datum $F \in X_2$ at some time interval $[0, t_*]$, $0 < t_* \leq t_1$. The solution f_∞ can be extended to the whole interval $[0, t_1]$. In fact, considering that the solution f_∞ is *a priori* non–negative and introducing $|f|_i = \int_0^\infty u^i |f(u)| \, du$, we see that $|f_\infty(t, j)|_1 = \tilde{f}^{(\infty)}(t, j)$, $j = 0, \ldots, r$, satisfies Eq. (152) and

$$|f_\infty(0, j)|_1 = \tilde{f}_\infty(0, j) = \rho_j^{(0)}. \tag{197}$$

By Assumption 4 the following *a priori* estimate holds

$$\sup_{t \in [0, t_1]} \sum_{j=0}^r |f_\infty(t, j)|_1 \leq c, \tag{198}$$

hence by Gronwall's Lemma

$$\sup_{t \in [0, t_1]} \sum_{j=0}^r |f_\infty(t, j)|_0 \leq c, \tag{199}$$

and finally by (192)

$$\sup_{t \in [0, t_1]} \sum_{j=0}^r |f_\infty(t, j)|_2 \leq c, \tag{200}$$

where various constants are denoted by "c". Thus we conclude that the following *a priori* estimate is satisfied

$$\sup_{t \in [0, t_1]} \| f_\infty(t) \|_2 \leq c. \tag{201}$$

□

Given the initial datum (187) let F be a probability density on $\mathbb{J} \times \mathbb{R}_+$ such that

$$F(j, u) = 0, \qquad \text{for all } j = 1, \ldots, r, \quad \text{and a.a.} \quad u > R_0, \tag{202}$$

and (196) is satisfied, for some $R_0 > 0$. From now on we assume that R_0 is fixed and $R > R_0$.

Obviously, if $a = a_R$, $A = A_R$ are given by (173)–(181) together with (185), then the unique solution f to Eq. (32) satisfies

$$\sup_{0 \leq t \leq \infty} \| f(t) \|_1 \leq c_R , \tag{203}$$

for some R–dependent constant c_R. In order to get asymptotic results, it is however required that

$$\sup_{0 \leq t \leq t_2} \| f(t) \|_1 \leq c , \tag{204}$$

for some $t_2 > 0$, where the constant denoted by "c" is R–independent (but it may depend on t_2). Clearly, given initial datum F, such that $\| F \|_1$ is independent of R, it is possible to find $t_2 > 0$ such that (204) is satisfied. Note that under Assumption (193)–(195) the condition (204) is satisfied for any $t_2 > 0$. In the general case we need however

Assumption 5 *Given parameters $\alpha_{j,k}$, $\beta_{j,k}$ as in (186) and $(\rho_1^{(0)}, \ldots, \rho_r^{(0)}) \in \mathbb{R}_+^r$, $a = a_R$ and $A = A_R$ are given by (173–181) together with (185), we assume that $t_2 > 0$ is such that the unique solution f to Eq. (32) with given initial data F is such that (204) holds.*

The main result of this section reads as follows ([70])

Theorem 9. *Let Assumptions 4, 5 be satisfied, $a = a_R$, $A = A_R$ in Eq. (15) be given by (173)–(181) and such that (185) is satisfied; Let the corresponding A_∞ satisfy (192); Let F be a probability density on $\mathbb{J} \times \mathbb{R}_+$ such that (196) and (202) are satisfied. Then there exists $N_0 > 0$ such that for $N \geq N_0$ and $R > R_0$*

$$\sup_{t \in [0, t_3]} \sum_{j=1}^r | \hat{f}^{N,1}(t, j) - \rho_j(t) | \leq \frac{c_R}{N^{\eta_R}} + \frac{c}{R} , \tag{205}$$

where the non–negative function $f^N \in L_1^{(N)}$ is the unique solution of Eq. (15) corresponding to the initial datum $F^{N \otimes}$; and $(\rho_1, \ldots, \rho_r)$ is the unique non–negative solution of Eq. (152) corresponding to the initial datum (187); $t_3 = \min \{ t_1, t_2 \}$; η_R and c_R are positive constants that depend on R; c is a constant.

Proof. Given parameters $\alpha_{j,k}$, $\beta_{j,k}$, let a_R and A_R be given by (173)–(181) and such that (185) is satisfied. Then the linear operator Λ_N^* in (15) is bounded in $L_1^{(N)}$ with $\mathbb{J} \times [0, R]$. Consequently, there exists a unique solution f^N of Problem (15) in $L_1^{(N)}$ and is given by (14), with the initial datum F^N such that (37) is satisfied. Clearly the solution f^N depends on R but, for the simplicity of notation, we do not indicate the R–dependence.

The assumption of Theorem 2 are satisfied and therefore there exists N_0 such that

$$\sup_{[0, t_1]} \| f^{N,1} - f \|_{L_1^{(1)}} \leq \frac{c_R'}{N^{\eta_R}} , \qquad \forall N \geq N_0 , \tag{206}$$

where $f^{N,1}$ is the 1–individual marginal density given by (19); f is the unique non–negative solution of Eq. (32) corresponding to the initial datum F; c'_R and η_R are positive constants depending on R. The R–dependence of f is not indicated. By (206) we have

$$\sup_{[0,t_1]} \sum_{j=1}^{r} \left| \hat{f}^{N,1}(t,j) - \hat{f}(t,j) \right| \leq \frac{c''_R}{N^{\eta_R}}, \qquad \forall\, N \geq N_0, \tag{207}$$

with some R–dependent constant c''_R, where $\bar{f}$ is given by (164).

The last step of the proof is to compare the solution f to the R–truncated equation and the solution f_∞. To this end we note that the following estimates are satisfied

$$\sum_{j=0}^{r} \int_0^R u \sum_{k=1}^{r} b_{k,0} \int_R^\infty \mathcal{A}_{j,k,0}(u,v) f_\infty(t,k,v)\, dv\, du \leq \frac{c}{R} \| f_\infty(t) \|_2,$$

$$\sum_{j=0}^{r} \int_0^R u \sum_{k=1}^{r} b_{k,0} \int_0^R \left| \mathcal{A}_{j,k,0}(u,v) - \mathcal{A}^{(R)}_{j,k,0}(u,v) \right| f_\infty(t,k,v)\, dv\, du$$
$$\leq \frac{c}{R} \| f_\infty(t) \|_2,$$

$$\sum_{j=0}^{r} \int_0^R u \sum_{k=1}^{r} b_{k,0} \int_0^R \mathcal{A}^{(R)}_{j,k,0}(u,v) \left| f_\infty(t,k,v) - f(t,k,v) \right| dv\, du$$
$$\leq \frac{c}{c_3} \sum_{j=0}^{r} \int_0^R u \left| f_\infty(t,k,u) - f(t,k,u) \right| du,$$

$$\sum_{j=0}^{r} \int_0^R u \sum_{l=1}^{r} b_{j,l} \tilde{f}^\infty(t,l) \int_R^\infty \mathcal{A}_{j,j,l}(u,v) f_\infty(t,j,v)\, dv\, du \leq \frac{c}{R} \| f_\infty(t) \|_1^2,$$

$$\sum_{j=0}^{r} \int_0^R u \sum_{l=1}^{r} b_{j,l} \left| \tilde{f}^\infty(t,l) - \hat{f}(t,l) \right| \int_0^R \mathcal{A}_{j,j,l}(u,v) f_\infty(t,j,v)\, dv\, du$$
$$\leq c \| f_\infty(t) \|_1 \lfloor f_\infty(t) - f(t) \rfloor_1 + \frac{c}{R} \| f_\infty(t) \|_2^2,$$

$$\sum_{j=0}^{r} \int_0^R u \sum_{l=1}^{r} b_{j,l} \hat{f}(t,l) \int_0^R \mathcal{A}^{(R)}_{j,j,l}(u,v) \left| \tilde{f}^\infty(t,j,v) - \hat{f}(t,j,v) \right| dv\, du$$
$$\leq \frac{c}{c_3} \| f(t) \|_1 \lfloor f_\infty(t) - f(t) \rfloor_1,$$

where

$$\rfloor f \lfloor_1 = \sum_{j=0}^{r} \int_0^R u|f(j,u)| \, du \, ;$$

the terms $b_{j,0}(f_\infty - f)$, $(f_\infty - f)\sum_{k=1}^{r} b_{j,k} \tilde{f}^\infty(k)$, $f \sum_{k=1}^{r} b_{j,k}(\tilde{f}^\infty(k) - \hat{f}(k))$ can be estimated in the same way. Thus by Gronwall's Lemma the assertion follows.

□

Example 19. Consider the following model ([11]) describing the receptor-mediated regulation of tumor growth

$$\dot{\rho}_1 = \lambda_1 (\rho_4 + \rho_5) - \mu_1 \rho_1 \tag{208}$$
$$\dot{\rho}_2 = \lambda_2 \rho_1 - \mu_2 \rho_2$$
$$\dot{\rho}_3 = \lambda_3 \rho_1 - \mu_3 \rho_3$$
$$\dot{\rho}_4 = \lambda_4 \rho_1 - \mu_4 \rho_4 - \sigma_4 \rho_2 \rho_4$$
$$\dot{\rho}_5 = \lambda_5 \rho_1 - \mu_5 \rho_5 - \sigma_5 \rho_3 \rho_5 \, ,$$

where λ_j, μ_j, σ_j are non–negative parameters of the model; ρ_j, $j = 1, \ldots, 5$, are concentrations of cell population (j=1), of a growth inhibiting factor (j=2), of a growth stimulatory factor (j=3), of unoccupied receptors (j=4) and of the receptor–stimulator complex (j=5). Let $\mathbb{J} = \{0, 1, \ldots, 5\}$, mat $\mathbb{U} = [0, R]$, and a, A be given by (173)–(181) together with

$$B_{1,k,0} b_{k,0} \mathfrak{c}_0 R_0 = \lambda_1 \, , \quad k = 4, 5 \, , \qquad \left(B_{1,1,0} - 1\right) b_{1,0} \mathfrak{c}_0 R_0 = -\mu_1 \, , \tag{209}$$
$$\mathcal{A}_{1,k,0} = 0 \, , \quad k = 2, 3 \, , \qquad \qquad b_{1,k} = 0 \, , \quad k = 1, \ldots, 5 \, ;$$

if $j = 2, 3$, then

$$B_{j,1,0} b_{1,0} \mathfrak{c}_0 R_0 = \lambda_j \, , \qquad \qquad \left(B_{j,j,0} - 1\right) b_{j,0} \mathfrak{c}_0 R_0 = -\mu_j \, , \tag{210}$$
$$\mathcal{A}_{j,k,0} = 0 \, , \quad k \neq 1, \, k \neq j \, , \qquad b_{j,k} = 0 \, , \quad k = 1, \ldots, 5 \, ;$$

$$B_{4,1,0} b_{1,0} \mathfrak{c}_0 R_0 = \lambda_4 \, , \qquad \qquad \left(B_{4,4,0} - 1\right) b_{4,0} \mathfrak{c}_0 R_0 = -\mu_4 \, , \tag{211}$$
$$\mathcal{A}_{4,k,0} = 0 \, , \quad k \neq 1, \, k \neq 4 \, , \qquad \left(B_{4,2} - 1\right) b_{4,2} = -\sigma_4 \, ,$$
$$b_{4,k} = 0 \, , \quad k \neq 0, \, k \neq 2 \, ;$$

and finally

$$\left(B_{5,5,0} - 1\right) b_{5,0} \mathfrak{c}_0 R_0 = -\mu_5 \, , \qquad \mathcal{A}_{5,k,0} = 0 \, , \quad k = 2, 3, 4 \, , \tag{212}$$
$$B_{5,1,0} b_{1,0} \mathfrak{c}_0 R_0 = \lambda_5 \, , \qquad \qquad \left(B_{5,3} - 1\right) b_{5,3} = -\sigma_5 \, ,$$
$$b_{5,k} = 0 \, , \quad k \neq 0, \, k \neq 3 \, .$$

If F is a probability density on $\{0, 1, \ldots, 5\} \times \mathbb{R}_+$, such that (196) and (202) are satisfied, then, for each $t_0 > 0$, there exists N_0 such that for $N \geq N_0$

$$\sup_{t \in [0, t_0]} \sum_{j=1}^{5} |\hat{f}^{N,1}(t, j) - \rho_j(t)| \leq \frac{c_R}{N^{\eta_R}} + \frac{c}{R} \tag{213}$$

where the non–negative function $f^N \in L_1^{(N)}$ is the unique solution of Eq. (15) corresponding to the initial datum (37), and ρ is the unique non–negative solution of Eq. (152) corresponding to the initial datum (187); η and c_R are positive constants that depend on t_0 and R; c is a constant that depends on t_0.

6 Coagulation–Fragmentation Equations

The models presented in Example 12 are at the level of statistical description of test–particle (i.e. mesoscopic **(Me)** description). Using the analogous idea as in the case of (152) with (159) we show that the solutions of these mesoscopic models can be approximated by the solutions of systems describing the coagulation process of particles undergoing stochastic interactions (a "microscopic" **(Mi)** description) in terms of stochastic semigroups.

The mathematical relationships between the particle systems and various Smoluchowski coagulation equations were studied in a number of papers — e.g. see [35,38,50,78], and references therein. The present approach however is simpler and makes use of a general theory developed in the previous sections. We follow [71] and show that the solutions of Eq. (79) can be approximated by solutions of (linear) equations describing the dynamics of a suitable system of interacting particles.

For simplicity of notation we consider here only the continuous case $\mathbb{S} = [0, \infty[$ but the discrete case $\mathbb{S} = \mathbb{N}$ (as in Example 10) can be treated in the same way.

Consider a system composed of N interacting particles. Every particle $n \in \{1, 2, \ldots, N\}$ is characterized by $\mathbf{u}_n = (s_n, u_n)$, where $s_n \in \mathbb{S}$ characterizes the size of the n–particle and $u_n \in [0, 1]$ — its inner state. Here $\mathbb{S} = [0, \infty[$ and $\mathbb{U} = [0, 1]$. Actually u_n plays an auxiliary rôle but it may be related to the measure of "*coagulation intensity*" of the particle. The n–particle interacts with the m–particle and the interaction take place at random times. After the interaction both particles may merge or/and change their inner state.

Consider the stochastic system of N–particles given by Eq. (15) with

$$\Lambda_N^* f(s_1, u_1, \ldots, s_N, u_N) = \tag{214}$$

$$\frac{1}{N\varepsilon} \sum_{\substack{1 \leq n, m \leq N \\ n \neq m}} u_m \left(\int_0^1 \left(\chi((1+\varepsilon)s_m < s_n) \right. \right.$$

$$\times \alpha(s_n - \varepsilon s_m, s_m) B_1(u_n, v)$$

$$\times f(s_1, u_1, \ldots, s_{n-1}, u_{n-1}, s_n - \varepsilon s_m, v, s_{n+1}, u_{n+1}, \ldots, s_N, u_N)$$

$$+\chi(s_n < s_m)\alpha(s_n, s_m)B_2(u_n, v)$$

$$\times f\big(s_1, u_1, \ldots, s_{n-1}, u_{n-1}, s_n, v, s_{n+1}, u_{n+1}, \ldots, s_N, u_N\big)\bigg)\,\mathrm{d}v$$

$$-\alpha(s_n, s_m)f\big(s_1, u_1, \ldots, s_N, u_N\big)\bigg)$$

$$s_j \in [0, \infty[\,,\ u_j \in [0, 1]\,,\ j = 1, \ldots, N,$$

$\varepsilon \in\,]0, 1[$, B_1 and B_2 are measurable functions such that

$$\int\limits_0^1 B_i(u, u')\,\mathrm{d}u = 1\,, \qquad \int\limits_0^1 u B_i(u, u')\,\mathrm{d}u = u'\kappa_i\,, \tag{215}$$

$$\text{for a.a. } u' \in\ [0, 1]\,, \quad i = 1, 2\,, \qquad \kappa_1 = 1\,, \quad \kappa_2 = 1 - \varepsilon\,,$$

$$\chi(\text{true}) = 1\,, \qquad \chi(\text{false}) = 0\,.$$

We assume rather restrictive case

$$0 \le \alpha(s, s') \le c_\alpha\,, \qquad \text{for a.a. } s, s' \in\ [0, \infty[\,\times[0, 1]\,, \tag{216}$$

where c_α is a positive constant; However the general case of unbounded α can be treated by the usual approximation methods (cf. [73]).

We assume now that the stochastic system starts with chaotic (i.e. factorized) probability density (37). The corresponding GKM equation (32) reads

$$\partial_t f(t, s, u) = \Gamma_\varepsilon[f](t, s, u)\,, \qquad (s, u) \in [0, \infty[\,\times[0, 1]\,, \tag{217}$$

$$\Gamma_\varepsilon[f](s, u) = \frac{1}{\varepsilon}\int\limits_0^{\frac{s}{1+\varepsilon}}\int\limits_0^1 \alpha(s - \varepsilon s', s')B_1(u, v)f(s - \varepsilon s', v)\,\hat{f}(s')\,\mathrm{d}v\,\mathrm{d}s'$$

$$+\frac{1}{\varepsilon}\int\limits_s^\infty\int\limits_0^1 \alpha(s, s')B_2(u, v)f(s, v)\hat{f}(s')\,\mathrm{d}v\,\mathrm{d}s' - \frac{1}{\varepsilon}\int\limits_0^\infty \alpha(s, s')f(s, u)\hat{f}(s')\,\mathrm{d}s'\,,$$

where

$$\hat{f}(s) = \int\limits_0^1 u f(s, u)\,\mathrm{d}u\,. \tag{218}$$

From Theorem 2 we immediately have ([71])

Theorem 10. *Let F be a probability density on $[0, \infty[\,\times[0, 1]$. Then, for each $t_0 > 0$, there exists N_0 such that for $N \ge N_0$*

$$\sup_{[0, t_0]} \|f^{N,1} - f\|_{L_1^{(1)}} \le \frac{c}{N^\eta}\,, \tag{219}$$

where the nonnegative functions $f^N \in L_1^{(N)}$ is the unique solution of Eq. (15) *corresponding to the initial datum*

$$f^N(0) = (F)^{N \otimes} , \qquad (220)$$

$f \in L_1^{(1)}$ *is the unique, nonnegative solution of* Eq. (217) *corresponding to the initial datum* F; η *and* c *are positive constants that depend on* t_0.

Corollary 5. . *Under the assumptions of* Theorem 10

$$\sup_{t \in [0,t_0]} \int_0^\infty \left| \hat{f}^{N,1}(t,s) - \hat{f}(t,s) \right| ds \leq \frac{c}{N^\eta} , \qquad (221)$$

where $\hat{f}(t)$, *given by* (218), *is a unique solution of* Eq. (79) *corresponding to the initial datum* $\hat{F}$.

Corollary 218 shows that the solution of the coagulation bilinear integro–differential equations can be approximated by the solutions of linear equations describing the stochastic system of individuals — provided that the parameters of the stochastic system are suitably chosen.

There are some possible generalizations of the above result. With slight modifications one can consider k interacting clusters (another approach can be find in [60, 61]).

7 The Space–Inhomogeneous Case: Reaction–Diffusion Equations

We consider now the following general (spatially inhomogeneous) population system with diffusion

$$\partial_t \varrho_j - \sigma_j \Delta \varrho_j = \sum_{k=1}^r \alpha_{j,k} \varrho_k + \varrho_j \sum_{k=1}^r \beta_{j,k} \varrho_k , \qquad (222)$$

$j = 1, 2, \ldots, r$,
where $\alpha_{j,j}$, $\beta_{j,k}$ are real constants (positive, negative or zero);
$\alpha_{j,k}$ $(j \neq k)$, σ_j $(j \in \{1, 2, \ldots, r\})$ are non–negative constants;
$\varrho_j = \varrho_j(t,x)$; $t \geq 0$ is the time variable, $x \in \mathbb{T}^d$ is the space variable, $\mathbb{T}^d$ is the d–dimensional torus, $d \geq 1$.
The torus $\mathbb{T}^d$ is naturally isomorphic to the Cartesian product of d copies $\mathbb{S}^1 \times \ldots \times \mathbb{S}^1$ of the circle. Assuming that the space variable is in the torus is a kind of mathematical simplification, but in some cases can have a clear meaning (cf. Example 7).

In the case of System (222) more general parameters $\alpha_{j,k}$, $\beta_{j,k}$, σ_j that are (regular) functions of the space variable x, more general diffusion operators and systems with more general bilinear terms can be treated as well. The details are left to the reader.

Let $\mathbb{J} = \{0, 1, \ldots, r+1\}$; $\mathbb{U} = \mathbb{R}_+ \times \mathbb{T}^d$;

$$\varepsilon > 0, \qquad \kappa_\varepsilon^d = \frac{\varepsilon^d}{d}|\mathbb{S}^{d-1}|,$$

$$\mathbb{S}^{d-1} = \left\{ x \in \mathbb{R}^d : |x| = 1 \right\}, \qquad |\mathbb{S}^{d-1}| = \int\limits_{\mathbb{S}^{d-1}} \mathrm{d}x;$$

$$a_{R,\varepsilon}(j, u, x, k, v, y) = a^*(j, u, x, k, v, y)\chi(u \le R)\chi(v \le R), \qquad (223)$$

$$a^*(j, u, x, k, v, y) = \begin{cases} \frac{1}{\kappa_{\varepsilon^3}^d}\chi(|y - x| < \varepsilon^3)b_{j,k}v & \text{for } j, k = 1, \ldots, r \\ b_{j,k} & \text{for } j = 1, \ldots, r, \ k = 0, r+1 \\ 0 & \text{for } j = 0, r+1, \ k = 0, \ldots, r+1; \end{cases} \qquad (224)$$

If $a^*(k, ., ., l, ., .) \equiv 0$ then
$A_{R,\varepsilon}(j, ., .; k, ., ., l, ., .) \equiv 0 \ \forall j$;
If $a^*(k, ., ., l, ., .) \not\equiv 0$, for some k, l, then

$$A_{R,\varepsilon}(j, u, x; k, v, y, l, w, z) = \qquad (225)$$
$$= \frac{1}{\kappa_{\varepsilon^3}^d}\chi(|y - x| < \varepsilon^3) \, \mathcal{A}_{j,k,l}^{(R)}(u, v) \, \chi(u \le R)\chi(v \le R)\chi(w \le R),$$

$j, k = 0, \ldots, r+1$, $l = 0, \ldots, r$, where $\mathcal{A}_{j,k,l}^{(R)}(u, v)$ is given by (176)–(179) for $j, k, l = 0, \ldots, r$,

$$\mathcal{A}_{r+1,k,l}^{(R)} \equiv 0, \qquad \forall \, k, l = 0, \ldots, r+1; \qquad (226)$$

$$A_{R,\varepsilon}(j, u, x; k, v, y, r+1, w, z) = \qquad (227)$$
$$\frac{1}{\kappa_\varepsilon^d} \, \delta_{j,k} \, \chi(|y - x| < \varepsilon) \, \mathcal{A}_j^{(R)}(u, v) \, \chi(u \le R)\chi(v \le R)\chi(w \le R),$$

for $j, k = 1, \ldots, r$, $\delta_{j,j} = 1$, $\delta_{j,k} = 0$ $(j \ne k)$,

$$\mathcal{A}_j^{(R)}(u, v) = \frac{\mathcal{A}_j(u, v)}{\sum\limits_{j'=1}^{r} \int\limits_0^R \mathcal{A}_{j'}(u', v) \, \mathrm{d}u'}, \qquad (228)$$

and $\mathcal{A}_j \ge 0$ satisfies

$$\sum\limits_{j'=1}^{r} \int\limits_0^{R_0} \mathcal{A}_{j'}(u, v) \, \mathrm{d}u \ge c_4 > 0, \qquad \sum\limits_{j'=1}^{r} \int\limits_0^\infty \mathcal{A}_{j'}(u, v) \, \mathrm{d}u = 1, \qquad (229)$$

$$\int\limits_0^\infty u \, \mathcal{A}_j(u, v) \, \mathrm{d}u = v, \qquad \forall \, v > 0,$$

for all $j = 1, \ldots, r$, where c_4 is a constant.

The functions $A_{R,\varepsilon}$ in (225)–(229), by (178) and (228), are characterized by a singular behavior with respect to the variable v for $v = 0$. An example could be (182).

It is easy to see that (12) is satisfied and therefore the solution of Eq. (32) a priori satisfies

$$f(t, j, u, x) \geq 0 \,, \tag{230}$$

$$\sum_{j=0}^{r+1} \int_0^\infty \int_{\mathbb{T}^d} f(t, j, u, x) \, dx \, du = \sum_{j=0}^{r+1} \int_0^\infty \int_{\mathbb{T}^d} f(0, j, u, x) \, dx \, du \,,$$

$t > 0$, provided that the initial data $f(0)$ are non–negative and integrable. Moreover, we assume

$$f(t, 0, u, x) = \mathfrak{c}_0 \, \chi(u \leq R_0) \,, \qquad f(t, r+1, u, x) = \mathfrak{c}_{r+1} \chi(u \leq R_0) \,, \tag{231}$$
$$\forall \, t \geq 0, \ \forall \, u \in \mathbb{R}_+, \ \forall \, x \in \mathbb{T}^d \,,$$

$\mathfrak{c}_0 > 0$, $\mathfrak{c}_{r+1} > 0$, $R_0 > 0$ are constants.

Let

$$\sigma_j = \varepsilon^2 \sigma_j^* \,. \tag{232}$$

Given parameters $\sigma_j^* \geq 0$, $\alpha_{j,k}$, $\beta_{j,k}$ $(j, k \in \{1, \ldots, r\})$ independent of ε (note that (1.17) is related to a "weak diffusion case") we assume that $B_{j,k}$, $b_{j,k}$ $(j, k \in \{0, 1, \ldots, r+1\})$, $\mathfrak{c}_0$, $\mathfrak{c}_{r+1}$ and R_0 are chosen such that

$$\left(B_{j,k} - 1\right) b_{j,k} = \beta_{j,k} \,, \qquad j, k \in \{1, 2, \ldots, r\} \,, \tag{233}$$

$$(2\,\pi)^d \left(B_{j,k,0} - \delta_{j,k}\right) b_{k,0} \, \mathfrak{c}_0 \, R_0 = \alpha_{j,k} \,, \qquad j, k \in \{1, 2, \ldots, r\} \,, \tag{234}$$

$$\frac{(2\,\pi)^d}{2\,(d+2)} \, b_{j,r+1} \, \mathfrak{c}_{r+1} \, R_0 = \sigma_j^* \,, \qquad j \in \{1, 2, \ldots, r\} \,. \tag{235}$$

If

$$\hat{f}(t, j, x) = \int_0^R u \, f(t, j, u, x) \, du \,, \tag{236}$$

where f is a solution of Eq. (32) with $a_{R,\varepsilon}$ and $A_{R,\varepsilon}$ given by (223)–(229) and (233)–(235) then $\left(\hat{f}(t, 1, x), \ldots, \hat{f}(t, r, x)\right)$ is formally asymptotically $\mathcal{O}(\varepsilon^3)$– close to the solution $\left(\varrho_1(t, x), \ldots, \varrho_r(t, x)\right)$ of Eq. (222) with $\sigma_j = \mathcal{O}(\varepsilon^2)$ (for all $j = 1, \ldots, r$) in the limit $R \uparrow \infty$ and $\varepsilon \downarrow 0$ (for the simplicity of notation we do not indicate the (R, ε)–dependence of the function f and the ε–dependence of the functions ϱ_j).

On the other hand, assuming that

$$a(j, u, x, r+1, v, y) = \frac{1}{\varepsilon^2} b_{j,r+1} \,, \qquad j = 1, \ldots, r \,, \tag{237}$$

where

$$\frac{(2\pi)^d}{2(d+2)}\, b_{j,r+1}\, \mathfrak{c}_{r+1}\, R_0 = \sigma_j\,, \qquad (238)$$

and repeating the remaining assumptions (223)–(229), we see that

$$\left(\hat{f}(t,1,x),\dots,\hat{f}(t,r,x)\right)$$

is formally asymptotically $\mathcal{O}(\varepsilon)$—close to the solution $\left(\varrho_1(t,x),\dots,\varrho_r(t,x)\right)$ of Eq. (222) with the parameters σ_j, r_j, $\beta_{j,k}$ that are independent of ε (a *strong diffusion case*) in the limit $R\uparrow\infty$ and $\varepsilon\downarrow 0$.

The former (weak diffusion) case is discussed in this section. The latter (strong diffusion) case is a more delicate problem because of an uncontrolled singular behavior of $a_{R,\varepsilon}$, given by (237), in the limit $\varepsilon\downarrow 0$.

The strategy developed in the previous section may now be applied to the space inhomogeneous case. We need

Assumption 6 *Given parameters $\alpha_{j,k}$, $\beta_{j,k}$ as in (186) and*

$$\left(\varrho_1^{(0)},\dots,\varrho_r^{(0)}\right)\in C^3\left(\mathbb{T}^d;\mathbb{R}_+^r\right),$$

we assume that $t_1 > 0$ is such that the unique classical solution $(\varrho_1,\dots,\varrho_r)$ to System (222) with initial data

$$\left(\varrho_1,\dots,\varrho_r\right)\Big|_{t=0} = (\varrho_1^{(0)},\dots,\varrho_r^{(0)}) \qquad (239)$$

exists in $[0,t_1]$.

Let $a = a_{R,\varepsilon}$ and $A = A_{R,\varepsilon}$ be given by (223)–(229) and such that (233)–(235) is satisfied. We denote by $a_{\infty,\varepsilon}$ and $A_{\infty,\varepsilon}$ the functions obtained from $a_{R,\varepsilon}$ and $A_{R,\varepsilon}$, respectively, by sending formally $R\to\infty$. Similarly to Section 5 we assume

Assumption 7 *Let $A_{\infty,\varepsilon}$ be such that*

$$\sum_{j=1}^{r}\int_0^\infty u^2 A_{\infty,\varepsilon}(j,u,x;k,v,y,l,w,z)\,\mathrm{d}u \le c_\varepsilon\left(1+v^2\right), \qquad (240)$$

for $k,l = 0,\dots,r+1$, a.a. u, v, w in $\mathbb{R}_+$, a.a. x, y, z in $\mathbb{T}^d$ and $\varepsilon\in\,]0,\varepsilon_0[$, for some $\varepsilon_0 > 0$; the constant denoted by "c_ε" depends only on ε.

Let F be now a probability density on $\mathbb{J}\times\mathbb{R}_+\times\mathbb{T}^d$ such that

$$\tilde{F}(j,\,.\,) = \rho_j^{(0)} \qquad \text{for all } j = 1,\dots,r\,, \qquad (241)$$

where $\tilde{F}$ is given by (189),

$$F(j, u, x) = 0, \qquad \text{for all } j = 1, \dots, r, \quad \text{a.a. } u > R_0, \quad \text{a.a. } x \in \mathbb{T}^d, \quad (242)$$

for some $R_0 > 0$, R_0 is fixed and $R > R_0$.

We need the following spaces: $X_i^{m,1}$ is the space equipped with the norm

$$\| f \|_i^{m,1} = \| \left(\| f \|^{m,1} \right) \|_i, \qquad i, m = 0, 1, \dots,$$

where $\| . \|_i$ is given in Section 4 and $\| . \|^{m,1}$ is the norm in the Sobolev space $W^{m,1}(\mathbb{T}^d)$ ([2]). By the Sobolev Imbedding Theorem ([2]) we have ([70])

Lemma 5. *Let*

$$\Theta_+[f_1, f_2](x) = \left(\frac{1}{\kappa_\varepsilon^d} \right)^2 \int\limits_{\mathbb{T}^d} \int\limits_{\mathbb{T}^d} \chi(|y - x| < \varepsilon)\chi(|z - y| < \varepsilon) f_1(y) f_2(z) \, dz \, dy,$$

and

$$\Theta_-[f_1, f_2](x) = \frac{1}{\kappa_\varepsilon^d} f_1(x) \int\limits_{\mathbb{T}^d} \chi(|y - x| < \varepsilon) f_2(y) \, dy.$$

If $m \geq 2d$, then

$$\| \Theta_\pm[f_1, f_2] \|^{m,1} \leq c \| f_1 \|^{m,1} \| f_2 \|^{m,1}, \qquad (243)$$

where the constant denoted by "c" is independent of $\varepsilon > 0$.

$\square$

The analogue of Theorem 9 in the space inhomogeneous case reads as follows ([70])

Theorem 11. *Let* Assumption 6 *be satisfied, $a = a_{R,\varepsilon}$, $A = A_{R,\varepsilon}$ in Eq. (15) be given by (223)–(229) and such that (233)–(235) is satisfied; Let the corresponding $A_{\infty,\varepsilon}$ satisfy (240); Let F be a probability density on $\mathbb{J} \times \mathbb{U}$, such that $\tilde{F}(j, .) \in W^{m,1}(\mathbb{T}^d)$ with $m \geq \max\{d + 3, 2d\}$, $j \in \mathbb{J}$, and (241), (242) are satisfied. Then there exist $t_2 \in]0, t_1]$ and $N_0 > 0$ such that for $N \geq N_0$ and $R > R_0$*

$$\sup_{t \in [0,t_2]} \sum_{j=1}^r \int\limits_{\mathbb{T}^d} |\hat{f}^{N,1}(t, j, x) - \varrho_j(t, x)| \, dx \leq \frac{c''}{N^\eta} + \frac{c'}{R} + c\varepsilon^3, \qquad (244)$$

where the non–negative function $f^N \in L_1^{(N)}$ is the unique solution of Eq. (15) corresponding to the initial datum $F^{N \otimes}$; $\hat{f}$ is given by (164); $(\varrho_1, \dots, \varrho_r)$ is the unique non–negative solution of Eq. (222) corresponding to the initial datum (239); η and c'' are positive constants that depend on R and ε; c' is a positive constant that depends on ε; c is a constant.

Proof. The proof is similar to that of Theorem 9. First note that for each $t_0 > 0$

$$\sup_{t \in [0,t_0]} \sum_{j=1}^{r} \int_{\mathbb{T}^d} \left| \hat{f}^{N,1}(t,j,x) - \hat{f}(t,j,x) \right| dx \le \frac{c''}{N^\eta} \tag{245}$$

for all $N \ge N_0$, $R > R_0$, $\varepsilon \in]0,1]$, with some $N_0 > 1$, where the constant c'' may depend on R, ε and t_0, f is the unique solution of Eq. (32) with $a = a_{R,\varepsilon}$ and $A = A_{R,\varepsilon}$ in $L_1^{(1)} = X_0^{0,1}$. By Lemma 5, we conclude that there exist $t_3 > 0$ and a unique solution f to Eq. (32) with $a = a_{R,\varepsilon}$ and $A = A_{R,\varepsilon}$ in $X_1^{m,1}$ such that

$$\sup_{0 \le t \le t_3} \| f(t) \|_1^{m,1} \le c, \tag{246}$$

where both t_3 and the constant c are independent of R, ε.

Consider now Eq. (32) with the functions $a = a_{\infty,\varepsilon}$, $A = A_{\infty,\varepsilon}$. As previously, there exist $t_4 > 0$, $t_4 \le t_3$, and a unique solution f_∞ with initial data F, in $X_2^{m,1}$, and such that

$$\sup_{0 \le t \le t_4} \| f_\infty(t) \|_2^{m,1} \le c, \tag{247}$$

$$\sup_{t \in [0,t_4]} \sum_{j=1}^{r} \int_{\mathbb{T}^d} \left| \hat{f}(t,j,x) - \tilde{f}^\infty(t,j,x) \right| dx \le \frac{c}{R} \tag{248}$$

for all $R > R_0$, $\varepsilon \in]0,1]$; $t_4 > 0$ and the constants denoted by "c" are independent of R, ε; $\tilde{f}$ is given by (189).

Using the Taylor expansion of the function $\tilde{f}^\infty = \tilde{f}^\infty(t,j,x)$ with respect to the x–variable, in much the same way as in the proof of Theorem 6, we obtain (244), with $t_2 = \min\{t_1, t_3, t_4\}$.

$\square$

We have shown that the general structure defined by Eq. (15) can result both in *diffusion* terms and in *reaction* terms in Eq. (222). However the alternative approach is possible with a diffusion term in the microscopic description — cf. [102].

Example 20. Consider the logistic equation with weak diffusion

$$\partial_t \varrho - \varepsilon^2 \sigma^* \Delta \varrho = \alpha \varrho - \beta \varrho^2 , \tag{249}$$

subject to regular initial data

$$\varrho \Big|_{t=0} = \varrho^{(0)} \ge 0 , \tag{250}$$

where $\sigma^* > 0$, $\alpha > 0$, $\beta > 0$ are given parameters, $\varepsilon > 0$ is a small parameter. In this case it is unnecessary to send $R \to \infty$ (cf. Example 18). Let $\mathbb{J} = \{1,2\}$, $\mathbb{U} = [0,R] \times \mathbb{T}^d$, $R = \frac{2\alpha}{\beta}$ and a, A be given by

$$a(1, u, x, 1, v, y) = \beta v, \qquad (u, x), (v, y) \in [0, R] \times \mathbb{T}^d, \qquad (251)$$
$$a(1, u, x, 2, v, y) = b, \qquad (u, x), (v, y) \in [0, R] \times \mathbb{T}^d,$$
$$a(2, ., ., 1., .) = a(2, ., ., 2., .) = 0,$$

where b is a constant,

$$A(1, u, x; 1, v, y, 1, w, z) = \tfrac{\beta}{2\alpha}, \qquad (252)$$
$$A(1, u, x; 1, v, y, 2, w, z) = \tfrac{1}{\kappa_\varepsilon^d}\chi(|y - x| < \varepsilon)\mathcal{A}(u, v),$$

$$(u, x), (v, y) \in]0, R[\times \mathbb{T}^d,$$

$$A(j, ., .; k., ., l., .) = 0, \qquad (j, k, l) \notin \left\{(1, 1, 1), (1, 1, 2)\right\},$$

$$\mathcal{A}(u, v) = \tfrac{1}{v}\chi\left(\tfrac{v}{2} < u < \tfrac{3v}{2}\right)\chi\left(0 < v \le \tfrac{R}{3}\right) \qquad (253)$$
$$+ \tfrac{3}{2R}\chi\left(v - \tfrac{R}{3} < u < v + \tfrac{R}{3}\right)\chi\left(\tfrac{R}{3} < v \le \tfrac{2R}{3}\right)$$
$$+ \tfrac{1}{R-v}\chi\left(\tfrac{3v-R}{2} < u < \tfrac{v+R}{2}\right)\chi\left(\tfrac{2R}{3} < v < R\right).$$

Let F be a probability density on $\mathbb{J} \times \mathbb{U}$, such that

$$\hat{F}(1, .) \in W^{m,1}(\mathbb{T}^d) \qquad \text{with } m \ge \max\{d + 3, 2d\},$$

and

$$\hat{F}(1, .) = \varrho^{(0)}, \qquad F(2, ., .) \equiv \mathfrak{c}_2, \qquad (254)$$

where $\mathfrak{c}_2$ is a positive constant and $\hat{F}$ is defined by (164). We assume that b is such that

$$\frac{(2\pi)^d}{2(d+2)} b \mathfrak{c}_2 R = \sigma^*. \qquad (255)$$

Then there exist $t_1 > 0$ and $N_0 > 0$ such that for $N \ge N_0$ and $\varepsilon \in [0, \tfrac{1}{2}]$

$$\sup_{t \in [0, t_1]} \int_{\mathbb{T}^d} |\hat{f}^{N,1}(t, 1, x) - \varrho(t, x)| \, dx \le \frac{c'}{N^\eta} + c\varepsilon^3, \qquad (256)$$

where the non–negative function $f^N \in L_1^{(N)}$ is the unique solution of Eq. (15) corresponding to the initial datum (37), and ϱ is the unique non–negative solution of Eq. (249) corresponding to the initial datum (250); η, c' are positive constants that depend on t_1, ε; and c is a constant.

On the other hand one may note that the convergence result can be attained on any time interval $[0, t_0]$, $t_0 > 0$.

8 Reaction–Diffusion–Chemotaxis Equations

In this section following [72] we show that the theory developed in previous sections can be generalized to take into account reaction–diffusion–chemotaxis systems (i.e. reaction–diffusion equations with a chemotaxis–type term). Although we are interested in and motivated by a particular model of tissue invasion by a solid tumor, the method is quite general and can be applied (at least at the formal level) to a large class of systems at the macroscopic level including the Keller–Segel–type systems cf. [34, 56, 57, 90, 96].

There is a huge literature related to the rigorous derivation of chemotaxis equations from microscopic models. The interested reader is referred to [34, 56, 57, 90, 96] and references therein. We mention here Stevens's paper [99] in which she proved that for sufficiently large numbers of particles the dynamics of an interacting particle system can be approximated by the solution of chemotaxis systems and explicit error estimates can be given.

In Ref. [41] the idea of hydrodynamic limit was used to derive hyperbolic models for chemosensitive movements as a hydrodynamic limit of a velocity–jump process. In Ref. [84] (cf. [26, 27]) the stochastic modeling of a spatially structured biological population was investigated. Based on law of large numbers, the convergence of a system of stochastic differential equations describing the evolution of the mean–field spatial density of the population was proved.

Here we are dealing with a different approach. It is quite general in the sense that it can be applied to a large class of models at the level **(Ma)**. Moreover it relates the three scales of description: **(Mi)**, **(Me)** and **(Ma)**, and at every stage explicit error estimates are given. The methods may lead to new and more accurate modeling of complex processes.

A variety of models have been developed for various aspects of solid tumor growth. For example the reader is referred to Ref. [1, 6, 8, 23, 32, 33, 42, 81, 85] and references therein.

Following Chaplain and Anderson [32] (see also [31]) we consider the system of deterministic reaction-diffusion–chemotaxis equations that is able to model the invasive spatial spread of solid tumors. The model is able to capture some aspects of solid tumor growth and invasion at the tissue level (the macroscopic scale), but it fails at the cellular level and subsequently the subcellular level. The model is based on generic solid tumor growth at the avascular stage, and it describes the interactions between the tumor and the surrounding tissue. The key variables of the model are: tumor cell density, denoted by ϱ_1, ECM (the surrounding tissue or *extracellular matrix*) density, denoted by ϱ_2, and MDE (certain factors produced by the tumor cells and known as *matrix degrading* or *degradative enzymes*) concentration, denoted by ϱ_3. The model describes one key aspect of tissue invasion, namely the ability of tumor cells to produce and secrete MDEs and their migratory response. Chaplain and Anderson made assumptions that the tumor cells produce MDEs which degrade the ECM locally; the ECM degradation aids in tumor cells motility; the movement of tumor cells up to a gradient of ECM is referred to as *haptotaxis*;

tumor cell motion is driven only by random motility and haptotaxis; the proliferation of tumor cells is not taken into account.

With these assumptions the model (in dimensionless form) of Chaplain and Anderson reads

$$\partial_t \varrho_1 = \overbrace{d_1 \Delta \varrho_1}^{\text{random motility}} - \overbrace{\gamma \partial_x \cdot \left(\varrho_1 \partial_x \varrho_2 \right)}^{\text{haptotaxis}} \tag{257}$$

$$\partial_t \varrho_2 = - \overbrace{\nu \, \varrho_2 \, \varrho_3}^{\text{degradation}}$$

$$\partial_t \varrho_3 = \overbrace{d_3 \Delta \varrho_3}^{\text{diffusion}} + \overbrace{\alpha \, \varrho_1}^{\text{production}} - \overbrace{\beta \, \varrho_3}^{\text{decay}} \,,$$

where d_1, γ, ν, d_3, α, β are given positive constants (the macroscopic parameters).

In [32], in addition to the above continuum model, the authors consider a discrete, stochastic model based on a random walk, in which they include the proliferation and migration of individual tumor cells.

Following [31–33, 81] one may include proliferative terms in the system (257). For such a new, more general system, the methods developed in the present paper can be repeated with some technical modifications. Details are left to the reader.

The microscopic model that corresponds to the macroscopic model (257) is defined by Eq. (15). We are going to introduce the linear generator Λ_N^* that describes the evolution of the density probability at the microscopic scale that approximates the solution of Eq. (257).

In what follows, a (large) number N of entities (cells or factors) of 3 subpopulations is considered. In contrast to the general approach in Section 7, here we do not take into account auxiliary subpopulations related to the constant environment. We start from the general framework.

Every entity n ($n \in \{1, \ldots, N\}$) is characterized by $(j_n, u_n, x_n) \in \mathbb{J} \times \mathbb{U}$, where $j_n \in \mathbb{J} = \{1, 2, 3, \}$ is its subpopulation, $\mathbb{U} = [0, R] \times \Omega$, $u_n \in [0, R]$ — its (inner) state (its "*activity*" or "*fitness*"), $R > 0$, and $x_n \in \Omega$ — its position (of the center of mass), Ω is a domain in $\mathbb{R}^3$. Here, in order to avoid some additional technical difficulties, it is assumed that the domain Ω is the 3–dimensional torus $\mathbb{T}^3$ (analogously the case of $\Omega = \mathbb{T}^d$ or $\Omega = \mathbb{R}^d$, $d \geq 1$, may be considered).

In addition to the macroscopic parameters d_1, γ, ν, d_3, α, β, the microscopic model will contain the microscopic parameters N, $\varepsilon > 0$ and $R > 0$. The microscopic model will have a nonlocal (in space) character: in fact the two entities may interact up to some distance. The parameter ε describes a distance of possible interaction. Actually we will consider two characteristic distances of the order of ε and ε^3. The limit $\varepsilon \to 0$ will lead to local (in space) interactions (as in (257)). The parameter R has a more technical meaning: it is introduced in order to make the operator Λ_N^* bounded. However, it may be related to the possible range of the activity variable.

Let N, $\varepsilon > 0$ and $R \in]0, \infty[$ be given;

$$\kappa_i = \frac{\varepsilon^{3i}}{3}|\mathbb{S}^2|, \qquad \text{for } i = 1, 3; \quad \mathbb{S}^2 = \left\{ x \in \mathbb{R}^3 \ : \ |x| = 1 \right\}.$$

Consider $j = 1$ and

$$a_{1,1}(u, x, v, y) = a_{1,3}(u, x, v, y) = b_1\,, \tag{258}$$

$$a_{1,2}(u, x, v, y) = b_1 + \frac{b^{(1)}\,v}{\kappa_1}\chi(|y - x| < \varepsilon)\,,$$

where $\chi(\text{true}) = 1$, $\chi(\text{false}) = 0$ and b_1, $b^{(1)}$ are positive constants that are defined by the (macroscopic) parameters of model (257)

$$\frac{b_1}{10}\varepsilon^2 = d_1\,, \qquad \frac{b^{(1)}}{20}\varepsilon^2 = \gamma\,. \tag{259}$$

Moreover

$$A_{1,1}^{(1)}(u, x; v, y, w, z) = A_{1,3}^{(1)}(u, x; v, y, w, z) = \frac{A_1^{(1)}(u, v)}{\kappa_1}\chi(|y - x| < \varepsilon)\,, \tag{260}$$

$$A_{1,2}^{(1)}(u, x; v, y, w, z) = \tag{261}$$

$$\frac{1}{b_1 + \frac{b^{(1)}\,w}{\kappa_1}\chi(|z-y|<\varepsilon)}\left(\frac{b_1\,A_2^{(1)}(u,v)}{\kappa_1}\chi(|y - x| < \varepsilon) \right.$$

$$\left. + \frac{b^{(1)}\,A_2^{(1)}(u,v)\,w}{\kappa_3\,\kappa_1}\chi(|\tfrac{y+z}{2} - x| < \varepsilon^3)\chi(|z - y| < \varepsilon) \right)\,,$$

where $A_k^{(1)}$ are functions such that

$$A_k^{(1)} \geq 0\,, \quad k = 1, 2\,, \tag{262}$$

$$\int_0^\infty A_1^{(1)}(u, v)\,du = \tfrac{1}{2}\,, \qquad \int_0^\infty A_2^{(1)}(u, v)\,du = 1\,,$$

$$\int_0^\infty u\,A_k^{(1)}(u, v)\,du = v\,, \quad k = 1, 2\,, \quad v > 0$$

and

$$A_{k,l}^{(1)} \equiv 0 \qquad k = 2, 3\,,\ l = 1, 2, 3\,. \tag{263}$$

Consider $j = 2$ and

$$a_{2,3}(u, x, v, y) = \frac{b_2\,v}{\kappa_3}\chi(|y - x| < \varepsilon^3)\,, \tag{264}$$

where b_2 is a positive constant (defined by (269));

$$a_{2,k} \equiv 0 \qquad k = 1, 2\,. \tag{265}$$

Moreover
$$A_{k,l}^{(2)} \equiv 0 \qquad \forall\, (k, l) \neq (2, 3)\,, \tag{266}$$

and
$$A_{2,3}^{(2)}(u, x; v, y, w, z) = \frac{\mathcal{A}^{(2)}(u, v)}{\kappa_3} \chi(|y - x| < \varepsilon^3)\,, \tag{267}$$

where $\mathcal{A}^{(2)}$ is a function such that
$$\mathcal{A}^{(2)} \geq 0\,; \qquad \int\limits_0^\infty \mathcal{A}^{(2)}(u, v)\,\mathrm{d}u = 1\,, \tag{268}$$

$$\int\limits_0^\infty u\, \mathcal{A}^{(2)}(u, v)\,\mathrm{d}u = B^{(2)}\, v\,, \qquad v > 0\,,$$

$B^{(2)}$ and b_2 are chosen such that
$$\left(1 - B^{(2)}\right) b_2 = \nu\,, \tag{269}$$

where ν is the parameter of the macroscopic model (257).
Finally consider $j = 3$ and
$$a_{3,k}(u, x, v, y) = b_3 \qquad k = 1, 2, 3\,, \tag{270}$$

$$A_{3,k}^{(3)}(u, x; v, y, w, z) = \frac{\mathcal{A}_3^{(3)}(u, v)}{\kappa_1} \chi(|y - x| < \varepsilon)\,, \qquad k = 1, 2, 3\,, \tag{271}$$

$$A_{1,k}^{(3)}(u, x; v, y, w, z) = \frac{\mathcal{A}_1^{(3)}(u, v)}{\kappa_3} \chi(|y - x| < \varepsilon^3)\,, \qquad k = 1, 3\,, \tag{272}$$

and
$$A_{1,2}^{(3)}(u, x; v, y, w, z) = \frac{b_1\, \mathcal{A}_1^{(3)}(u, v)}{b_1 + \frac{b^{(1)} w}{\kappa_1} \chi(|z - y| < \varepsilon)} \frac{1}{\kappa_3} \chi(|y - x| < \varepsilon^3) \tag{273}$$

where $\mathcal{A}_k^{(3)}$ are functions such that
$$\mathcal{A}_k^{(3)} \geq 0\,, \qquad k = 1, 3\,; \tag{274}$$

$$\int\limits_0^\infty \mathcal{A}_1^{(3)}(u, v)\,\mathrm{d}u = \tfrac{1}{2}\,, \qquad \int\limits_0^\infty \mathcal{A}_3^{(3)}(u, v)\,\mathrm{d}u = 1\,,$$

$$\int\limits_0^\infty u\, \mathcal{A}_k^{(3)}(u, v)\,\mathrm{d}u = B_k^{(3)}\, v\,, \qquad v > 0\,, \quad k = 1, 3\,,$$

$$A_{2,k}^{(3)} \equiv 0 \qquad k = 1, 2, 3\,, \tag{275}$$

b_3, $B_k^{(3)}$, $k = 1, 3$, are positive constants such that
$$\frac{b_3\, B_3^{(3)}}{10}\, \varepsilon^2 = d_3\,, \qquad (1 - B_3^{(3)}) b_3 = \beta\,, \qquad b_1 B_1^{(3)} = \alpha\,, \tag{276}$$

d_3, α, β are parameters of the macroscopic model (257).

We next define

$$a(j, u, x, k, v, y) = a_{j,k}(u, x, v, y)\chi(u \le R)\chi(v \le R) , \tag{277}$$

where $a_{j,k}$, $j, k = 1, 2, 3$, were defined before.

Note that for fixed $R > 0$ the function a is bounded.

Let $\zeta > 0$ and $R_0 > 0$ be given numbers (independent of ε) such that if $a_{k,l} \not\equiv 0$, for some k, l, then

$$\sum_{j=1}^{3} \int_0^{R_0} \int_\Omega A_{k,l}^{(j)}(u, x; v, y, w, z)\, dx\, du \ge \zeta . \tag{278}$$

It is easy to see that such numbers exist. In what follows we assume that $R > R_0$. If $a_{k,l} \not\equiv 0$ then we define

$$A(j, u, x; k, v, y, l, w, z) = \tag{279}$$

$$\frac{A_{k,l}^{(j)}(u, x; v, y, w, z)}{\sum\limits_{j'=1}^{3} \int_0^{R} \int_\Omega A_{k,l}^{(j)}(u', x'; v, y, w, z)\, dx'\, du'} \chi(u \le R)\chi(v \le R)\chi(w \le R) .$$

The functions a and A defined by (277) and (279) on $\left(\{1,2,3\} \times [0, R] \times \Omega\right)^2$ and $\left(\{1,2,3\} \times [0, R] \times \Omega\right)^3$, respectively, satisfy (8)–(12). The operator Λ_N^* defined by (16) is a bounded operator in the space $L_1^{(N)}$ and the Cauchy Problem (15) has the unique solution $f^N(t) \in L_1^{(N)}$ for all $t \ge 0$. The solution preserves the non–negativity and $L_1^{(N)}$-norm of initial data. Therefore $\exp\left(t\Lambda_N^*\right)$ defines a continuous stochastic semigroup.

We assume that the process starts with chaotic (i.e. factorized) probability density (37).

In the limit $N \to \infty$ the linear equation (15) results in a bilinear system of Boltzmann–like integro–differential equations (32)

$$\partial_t f(t, j, u, x) = \Gamma[f](t, j, u, x) , \tag{280}$$

$$t > 0 , \quad (j, u, x) \in \{1, 2, 3\} \times [0, R] \times \Omega ,$$

where

$$\Gamma[f](j, u, x) = \sum_{k=1}^{3} \int_0^{R}\!\!\int_\Omega \sum_{l=1}^{3} \int_0^{R}\!\!\int_\Omega A(j, u, x; k, v, y, l, w, z)$$

$$\times a(k, v, y, l, w, z) f(k, v, y) f(l, w, z)\, dz\, dw\, dy\, dv$$

$$- f(t, j, u, x) \sum_{k=1}^{3} \int_0^{R}\!\!\int_\Omega a(j, u, x, k, v, y) f(t, k, v, y)\, dy\, dv ,$$

where a and A are given by (277) and (279).

Formally letting $R \to \infty$ one gets from Eq. (281) the following system of Boltzmann-like equations

$$\partial_t f(t, j, u, x) = \Gamma^\infty[f](t, j, u, x), \tag{281}$$
$$t > 0, \quad (j, u, x) \in \{1, 2, 3\} \times [0, \infty[\times \Omega,$$

where

$$\Gamma^\infty[f](1, u, x) = \tag{282}$$

$$= \frac{b_1}{\kappa_1} \mathcal{I}[f] \int_0^\infty \int_\Omega \mathcal{A}_1^{(1)}(u, v) \chi(|y - x| < \varepsilon) f(1, v, y) \, dy \, dv$$

$$+ \frac{b^{(1)}}{\kappa_3 \kappa_1} \int_0^\infty \int_\Omega \int_0^\infty \int_\Omega \mathcal{A}_2^{(1)}(u, v) \chi\left(\left|\frac{y+z}{2} - x\right| < \varepsilon^3\right) \chi\left(|z - y| < \varepsilon\right)$$

$$\times f(1, v, y) \, w \, f(2, w, z) \, dy \, dv \, dz \, dw$$

$$- b_1 f(1, u, x) \mathcal{I}[f] - \frac{b^{(1)}}{\kappa_1} f(1, u, x) \int_0^\infty \int_\Omega \chi(|y - x| < \varepsilon) \, v \, f(2, v, y) \, dy \, dv,$$

$$\Gamma^\infty[f](2, u, x) = \tag{283}$$

$$= \frac{b_2}{\kappa_3^2} \int_0^\infty \int_\Omega \int_0^\infty \int_\Omega \mathcal{A}^{(2)}(u, v) \chi(|y - x| < \varepsilon^3) \chi(|z - y| < \varepsilon^3)$$

$$\times f(2, v, y) \, w \, f(3, w, z) \, dy \, dv \, dz \, dw$$

$$- \frac{b_2}{\kappa_3} f(2, u, x) \int_0^\infty \int_\Omega \chi(|y - x| < \varepsilon^3) \, v \, f(3, v, y) \, dy \, dv,$$

$$\Gamma^\infty[f](3, u, x) = \tag{284}$$

$$= \frac{b_3}{\kappa_1} \mathcal{I}[f] \int_0^\infty \int_\Omega \mathcal{A}_3^{(3)}(u, v) \chi(|y - x| < \varepsilon) f(3, v, y) \, dy \, dv$$

$$+ \frac{b_1}{\kappa_3} \mathcal{I}[f] \int_0^\infty \int_\Omega \mathcal{A}_1^{(3)}(u, v) \chi(|y - x| < \varepsilon^3) f(1, v, y) \, dy \, dv - b_3 f(3, u, x) \mathcal{I}[f],$$

and $\mathcal{I}[f] = \sum_{l=1}^3 \int_0^\infty \int_\Omega f(l, w, z) \, dz \, dw$.

We note that any solution f to Eq. (281) is (formally) a density probability and consequently $\mathcal{I}[f] \equiv 1$. Hence, for $\tilde{f}$ given by (189),

$$\partial_t \tilde{f}(t, 1, x) = \frac{b_1}{\kappa_1} \int_\Omega \chi(|y - x| < \varepsilon) \tilde{f}(t, 1, y) \, dy - b_1 \tilde{f}(t, 1, x) \tag{285}$$

$$+ \frac{b^{(1)}}{\kappa_3 \kappa_1} \int_\Omega \int_\Omega \chi\left(\left|\frac{y+z}{2} - x\right| < \varepsilon^3\right) \chi\left(|z - y| < \varepsilon\right) \tilde{f}(t, 1, y) \tilde{f}(t, 2, z) \, dz \, dy$$

$$- \frac{b^{(1)}}{\kappa_1} \tilde{f}(t, 1, x) \int_\Omega \chi\left(|y - x| < \varepsilon\right) \tilde{f}(t, 2, y) \, dy.$$

This gives (see Section 4)

$$\partial_t \tilde{f}(t,1,x) = \tfrac{b_1}{10}\varepsilon^2\,\Delta\,\tilde{f}(t,1,x) - b^{(1)}\tilde{f}(t,1,x)\tilde{f}(t,2,x) \qquad (286)$$
$$-\tfrac{b^{(1)}}{10}\varepsilon^2\tilde{f}(t,1,x)\,\Delta\,\tilde{f}(t,2,x) + \mathcal{M}[f](t,x) + \mathcal{O}(\varepsilon^3)\,,$$

where

$$\mathcal{M}[f](t,x) = \tfrac{b^{(1)}}{\kappa_3\kappa_1} \int\limits_{\Omega}\int\limits_0^\varepsilon r^2 \int\limits_{\mathbb{S}^2} \chi\Big(|y-x| < \varepsilon^3\Big)$$
$$\times \tilde{f}(t,1,y - \tfrac{r}{2}n)\tilde{f}(t,2,y + \tfrac{r}{2})\,\mathrm{dn}\,\mathrm{dr}\,\mathrm{dy}\,.$$

Expanding (formally) the fourth term $\mathcal{M}$ in Taylor's series one obtains

$$\mathcal{M}[f](t,x) = b^{(1)}\,\tilde{f}(t,1,x)\tilde{f}(t,2,x) \qquad (287)$$
$$+\tfrac{b^{(1)}}{40\,\kappa_3}\varepsilon^2 \int\limits_{\Omega} \chi\Big(|y-x| < \varepsilon^3\Big)\tilde{f}(t,1,y)\,\Delta\,\tilde{f}(t,2,y)\,\mathrm{dy}$$
$$+\tfrac{b^{(1)}}{40\,\kappa_3}\varepsilon^2 \int\limits_{\Omega} \chi\Big(|y-x| < \varepsilon^3\Big)\,\Delta\,\tilde{f}(t,1,y)\tilde{f}(t,2,y)\,\mathrm{dy}$$
$$-\tfrac{b^{(1)}}{20\,\kappa_3}\varepsilon^2 \int\limits_{\Omega} \chi\Big(|y-x| < \varepsilon^3\Big)\partial_y\tilde{f}(t,1,y)\cdot\partial_y\tilde{f}(t,2,y)\,\mathrm{dy} + \mathcal{O}(\varepsilon^3)$$

Assuming that $f(t,j,\,.\,) \in C^2(\Omega)$, $j = 1, 2$, and applying Green's formula we obtain

$$\mathcal{M}[f](t,x) = b^{(1)}\,\tilde{f}(t,1,x)\tilde{f}(t,2,x) \qquad (288)$$
$$+\tfrac{b^{(1)}}{20\,\kappa_3}\varepsilon^2 \int\limits_{\Omega} \chi\Big(|y-x| < \varepsilon^3\Big)\tilde{f}(t,1,y)\,\Delta\,\tilde{f}(t,2,y)\,\mathrm{dy}$$
$$-\tfrac{b^{(1)}}{20\,\kappa_3}\varepsilon^2 \int\limits_{\Omega} \chi\Big(|y-x| < \varepsilon^3\Big)\partial_y\tilde{f}(t,1,y)\cdot\partial_y\tilde{f}(t,2,y)\,\mathrm{dy} + \mathcal{O}(\varepsilon^3)\,.$$

Expanding (288) in Taylor's series and substituting into (285) we obtain

$$\partial_t\tilde{f}(t,1,x) = \tfrac{b_1}{10}\varepsilon^2\,\Delta\,\tilde{f}(t,1,x) - \tfrac{b^{(1)}}{20}\varepsilon^2\partial_x\cdot\Big(\tilde{f}(t,1,x)\partial_x\tilde{f}(t,2,x)\Big) + \mathcal{O}(\varepsilon^3)\,. \qquad (289)$$

By similar arguments, we see that

$$\partial_t\tilde{f}(t,2,x) = \Big(B^{(2)} - 1\Big)b_2\tilde{f}(t,2,x)\tilde{f}(t,3,x) + \mathcal{O}(\varepsilon^3)\,. \qquad (290)$$

and

$$\partial_t\tilde{f}(t,3,x) = \Big(B^{(3)}_3 - 1\Big)b_3\tilde{f}(t,3,x) \qquad (291)$$
$$+\tfrac{b_3\,B^{(3)}}{10}\varepsilon^2\,\Delta\,\tilde{f}(t,3,x) + b_1 B^{(3)}_1\tilde{f}(t,1,x) + \mathcal{O}(\varepsilon^3)\,.$$

One may compare Eq. (289)–(291) (cf. (259), (269) and (276)) with Eq. (257). It follows that Eq. (285) is a nonlocal version of the local equation $(257)_1$, ect.

We assume that

$$\sigma_i = \varepsilon^2 \sigma_i^*, \quad i = 1, 3, \qquad \gamma = \varepsilon^2 \gamma^*, \tag{292}$$

where σ_i^* and γ^* are given ε–independent positive constants. The parameter $\varepsilon \in]0, 1[$ is assumed to be small and therefore Assumption (292) refers to weak diffusion and haptotaxis processes. In other words the diffusion and haptotaxis are small in comparison with other processes (the parameters α, β, ν are assumed to be ε–independent). This was actually the case in the situation considered in [31–33,81].

We may note that the parameters σ_i, γ that are ε–independent may be considered (at least at a formal level) — cf. strong diffusion case in Section 7.

Additionally to (262), (268) and (274) throughout this section we assume the following bounds are satisfied

$$\int_0^\infty u^2 \mathcal{A}_k^{(1)}(u, v)\, du < c, \quad k = 1, 2, \qquad \int_0^\infty u^2 \mathcal{A}^{(2)}(u, v)\, du < c, \tag{293}$$

$$\int_0^\infty u^2 \mathcal{A}_l^{(3)}(u, v)\, du < c, \quad l = 1, 3,$$

where positive constants are denoted by c.

It is easy to see that the functions $\mathcal{A}_1^{(1)}$, $\mathcal{A}_2^{(1)}$, $\mathcal{A}^{(2)}$, $\mathcal{A}_1^{(3)}$, $\mathcal{A}_3^{(3)}$, satisfying (262), (268), (274), (294), exist — cf. (182).

Let X_i, $i = 0, 1, 2, \ldots$, be the space equipped with the norm

$$\|f\|_i = \sum_{j=1}^3 \int_0^\infty \int_\Omega \left(1 + u^i\right) \left| f(j, u, x) \right| dx\, du. \tag{294}$$

Then we have ([72])

Theorem 12. *For every non–negative initial datum $F \in X_2$ and every $t_0 > 0$ there exists the unique solution*

$$f \in C^0\left([0, t_0]; X_2\right)$$

to the Cauchy problem for Eq. (281) and

$$\sup_{t \in [0, t_0]} \|f(t)\|_2 \le c, \tag{295}$$

$$f \in C^1\left(]0, t_0[\,; X_2\right), \qquad f(t) \ge 0, \quad t \in]0, t_0], \tag{296}$$

where c is a positive constant that depends on t_0 and ε.

Proof. The bilinear operator Γ^∞ is locally Lipschitz continuous in X_2 and hence there exists a unique solution f to the Cauchy problem for Eq. (281)

in X_2 on some time interval $[0, t_1]$. Moreover, by standard arguments, we can prove that the solution is non–negative and

$$\mathcal{I}[f](t) = \| f(t) \|_0 = \| F \|_0, \qquad t \in [0, t_1]. \tag{297}$$

By (281)–(284) we have the following bounds

$$\int_0^\infty \int_\Omega \left(1 + u^2\right) \left| f(t, 3, u, x) \right| \, dx \, du \leq c \exp(ct), \tag{298}$$

$$\int_0^\infty \int_\Omega \left(1 + u^2\right) \left| f(t, 2, u, x) \right| \, dx \, du \leq c \exp(c \exp(ct)), \tag{299}$$

$$\int_0^\infty \int_\Omega \left(1 + u^2\right) \left| f(t, 1, u, x) \right| \, dx \, du \leq c \exp(c \exp(c \exp(ct))), \tag{300}$$

where various positive constants, depending on $\| F \|_2$ and ε, are denoted by c. Therefore the statement follows.

$\square$

Let $X_i^{m,1}$, $i, m = 0, 1, 2, \ldots$, be the space equipped with the norm

$$\| \cdot \|_i^{m,1} = \| \left(\| \cdot \|^{m,1} \right) \|_i, \tag{301}$$

where $\| \cdot \|_i$ is given by (294) and $\| \cdot \|^{m,1}$ is the norm in the Sobolev space $W^{m,1}(\Omega)$.

By the Sobolev Embedding Theorem and Lemma 5, we obtain the existence of solutions with a ε–independent bound ([72])

Theorem 13. *Let $m \geq 6$. For every non–negative initial datum $F \in X_2^{m,1}$ and every $t_0 > 0$ there exists the unique solution*

$$f \in C^0 \left([0, t_0]; X_2^{m,1}\right)$$

to the Cauchy problem for Eq. (281) and

$$\sup_{t \in [0, t_0]} \| f(t) \|_2^{m,1} \leq c, \tag{302}$$

$$f \in C^1 \left(]0, t_0[\, ; X_2^{m,1}\right), \qquad f(t) \geq 0, \quad t \in]0, t_0], \tag{303}$$

where c is a positive constant that depends on t_0 but is independent of ε.

$\square$

We now are going to state the theorems, which define links between the solutions to various equations presented in this section.

We begin by showing that the distance between the solution of (280) and the solution of (281) can be controlled by R (R is fixed sufficiently large). The parameter $\varepsilon > 0$ is assumed to be fixed.

It is understood that any $f \in L_1^{(1)}$, $R > 0$, is in $X^{(0)}$ by taking the zero extension of f in $]R, \infty[$. We have

Theorem 14. *Let $F \in L_1^{(1)}$ for $R > 0$. Then, for each $t_0 > 0$,*

$$\sup_{t \in [0,t_0]} \sum_{j=1}^{3} \int_0^R \int_\Omega u \left| f_\infty(t, j, u, x) - f(t, j, u, x) \right| dx\, du \leq \frac{c}{R}, \qquad (304)$$

where $f_\infty \in X_2$ is the unique, non–negative solution of Eq. (281) corresponding to the initial datum F; $f \in L_1^{(1)}$ is the unique, non–negative solution of Eq. (280) corresponding to the initial datum F; and c is a positive constant that depends on t_0.

Proof. The solution $f_\infty \in X_2$ exists by Theorem 12. The rest of the proof follows by the same method as in the proof of Theorem 9.

□

Now we want to compare the solutions of Eq. (257) with those of Eq. (281) by using the expansions (289–291). Therefore we need an existence theorem for Eq. (257). This is actually an open problem (see however [25, 34] and references therein) and we restrict ourselves to a conditional result assuming

Assumption 8 *Given positive parameters d_1^*, γ^*, ν, d_3^*, α, β independent of ε and initial data*

$$\left(\varrho_1^{(0)}, \varrho_2^{(0)}, \varrho_3^{(0)} \right) \in C^m \left(\Omega; [0, \infty[^3 \right), \qquad \text{with some } m \geq 6, \qquad (305)$$

we assume that $t_1 > 0$ is such that the unique classical solution $(\varrho_1, \varrho_2, \varrho_3)$ to the Cauchy problem for System (281) with (292) exists in $[0, t_1]$.

Let F be a probability density on $\{1, 2, 3\} \times [0, \infty[\times \Omega$ such that

$$\tilde{F}(j, \, . \,) = \varrho_j^{(0)}, \qquad j = 1, 2, 3, \qquad (306)$$

$$F(j, u, x) = 0, \qquad \text{for} \quad j = 1, 2, 3, \quad u > R_0, \quad x \in \Omega, \qquad (307)$$

for some $R_0 > 0$, R_0 is fixed and $R > R_0$.

Using Taylor's expansion (289–291), in much the same way as in the proof of Theorem 6 (cf. Theorem 11), and using (304), we obtain the following result ([72])

Theorem 15. *Let* Assumption 8 *be satisfied and* $F \in X_2^{m,1}$ *be a probability density such that* (306) *and* (307) *are satisfied. Then for sufficiently large* N, R *and small* $\varepsilon > 0$

$$\sup_{t \in [0,t_1]} \sum_{j=1}^{3} \int_{\Omega} \left| \hat{f}^{N,1}(t,j,x) - \varrho_j(t,x) \right| \mathrm{d}x \le \frac{c''}{N^\eta} + \frac{c'}{R} + c\varepsilon^3 , \tag{308}$$

where the non–negative function $f^N \in L_1^{(N)}$ *is the unique solution of* Eq. (15) *and corresponding to the initial datum* $F^{N \otimes}$; η *and* c'' *are positive constants that depend on* R *and* ε; c' *is a positive constant that depends on* ε; c *is a constant.*

$\square$

Theorem 15 shows that if the solution to the nonlinear equation (257) exists then it may be approximated by the solutions of the linear equation (15). On the other hand, the linear equation (15) related to the microscopic description, and the nonlinear equation (281) at the mesoscopic scale, may play independently an important rôle in the mathematical description of the process in question, even if Eq. (257) does not possess smooth solutions.

Acknowledgment

The author acknowledges research support from the EU project "*Modeling, Mathematical Methods and Computer Simulations of Tumour Growth and Therapy*" — M3CSTuTh — Contract No. MRTN–CT–2004–503661 and the Polish SPUB–M Grant. The work bases on Author's Lectures on the Banach Centre and CIME course and workshop "FROM A MICROSCOPIC TO A MACROSCOPIC DESCRIPTION OF COMPLEX SYSTEMS", Będlewo, Poland, 4th – 9th September, 2006.

References

1. Adam, J.A., Bellomo, N., Eds.: A Survey of Models for Tumor–Immune System Dynamics. Birkhäuser, Boston (1997)
2. Adams, R.A.: Sobolev Spaces. Academic Press, New York (1975)
3. Agazzi, A., Faye, J., Eds.: The Problem of the Unity of Science. World Scientific, Singapore (2001)
4. Arlotti, L., Bellomo, N.: Population dynamics with stochastic interaction. Transport Theory Statist. Phys., **24**, 431–443 (1995)
5. Arlotti, L., Bellomo, N.: Solution of a new class of nonlinear kinetic models of population dynamics. Appl. Math. Lett., **9**, 65–70 (1996)
6. Arlotti, L., Bellomo, N., De Angelis, E., Lachowicz, M.: Generalized Kinetic Models in Applied Sciences, World Scientific, New Jersey (2003)
7. Arlotti, L., Bellomo, N., Lachowicz, M.: Kinetic equations modelling population dynamics. Transport Theory Statist. Phys., **29**, 125–139 (2000)

8. Arlotti, L., Gamba, A., Lachowicz, M.: A kinetic model of tumour/immune system cellular interactions. J. Theoret. Medicine, **4**, 39–50 (2002)
9. Arlotti, L., Lachowicz, M.: Qualitative analysis of a nonlinear integro-differential equation modelling tumor–host dynamics. Math. Comput. Modelling, **23**, 11–29 (1996)
10. Aronson, D.G.: The porous medium equation. In: Fasano, A., Primicerio, M. (eds) Nonlinear Diffusion Problems, Lecture Notes in Math. 1224, Springer, Berlin (1986), 1–46
11. Bajzer, Ž, Marušič, M., Vuk–Pavlovič, S.: Conceptual frameworks for mathematical modelling of tumor growth dynamics. Math. Comput. Model., **23**, 31–46 (1996)
12. Ball, J.M., Carr, J., The discrete coagulation-fragmentation equations: existence, uniqueness and density conservation. J. Statist. Phys., **61**, 203–234 (1990)
13. Banasiak, J.: Positivity in natural sciences. In this volume
14. Banasiak, J., Arlotti, L.: Positive Perturbation of Semigroups with Applications, Springer, New York (2005)
15. Banasiak, J., Lachowicz, M.: Chaos for a class of linear kinetic models. Compt. Rend. Acad. Sci. Paris, **329**, Série IIb, 439–444 (2001)
16. Banasiak, J., Lachowicz, M.: Topological chaos for birth–and–death–type models with proliferation. Math. Models Methods Appl. Sci., **12**, 755–775 (2002)
17. Banasiak, J., Lachowicz, M., Moszyński, M.: Chaotic behaviour of semigroups related to the process of gene amplification–deamplification with cell proliferation. Math. Biosci., to appear
18. Banasiak, J., Mika, J.: Singularly perturbed evolution equations with applications to kinetic theory, World Sci., River Edge (1995)
19. Barbu, V.: Nonlinear Semigroups and differential equations in Banach spaces, Noordhoff (1976)
20. Bellomo, N., Bellouquid, A., Delitala, M.: Mathematical topics on the modelling complex multicellular systems and tumor immune cells competition. Math. Models Methods Appl. Sci., **14**, 1683–1733 (2004)
21. Bellomo, N., Forni, G.: Dynamics of tumor interaction with the host immune system. Math. Comput. Modelling, **20**, 107–122 (1994)
22. Bellomo, N., Forni, G.: Looking for new paradigms towards a biological-mathematical theory of complex multicellular systems. Math. Models Methods Appl. Sci., **16**, 1001–1029 (2006)
23. Bellomo, N., Preziosi, L.: Modelling and mathematical problems related to tumour evolution and its interaction with immune system. Math. Comput. Model., **32**, 413–453 (2000)
24. Benilan, Ph., Brezis, H., Crandall, M.G.: A semilinear elliptic equation in $L_1(\mathbb{R}^N)$. Ann. Scuola Norm. Sup. Pisa, **2**, 523–555 (1975)
25. Biler, P., Existence and asymptotic of solutions for a parabolic–elliptic system with nonlinear no–flux boundary conditions. Nonlin. Anal., **12**, 1121–1136 (1992)
26. Capasso, V., Morale, D.: Rescaling stochastic processes: Asymptotics. In this volume
27. Capasso, V., Bakstein, D.: An Introduction to Continuous–Time Stochastic Processes, Birkhäuser, Boston (2005)
28. Cercignani, C.: Theory and Application of the Boltzmann Equation. Scottish Academic Press, Edinburgh (1975)

29. Cercignani, C.: The Grad limit for a system of soft spheres. Comm. Pure Appl. Math., **36**, 479–494 (1983)
30. Cercignani, C., Illner R., Pulvirenti M.: The Mathematical Theory of Dilute Gases, Springer, New York (1994)
31. Chaplain, M.A.J.: Modelling tumour growth. In this volume
32. Chaplain, M.A.J., Anderson, A.R.A.: Mathematical Modelling of Tissue Invasion. In: Cancer Modelling and Simulation. Ed. Preziosi L., Chapman & Hall/CRT, 269–297 (2003)
33. Chaplain, M.A.J., Lolas, G.: Mathematical modelling of cancer cell invasion of tissue: The role of the urokinase plasminogen activation system. Math. Models Methods Appl. Sci., **15**, 1685–1734 (2005)
34. Corrias, L., Perthame, B., Zaag, H.: Global solutions of some chemotaxis and angiogenesis systems in high space dimensions. Milan J. Math., **72**, 1–28 (2004)
35. Deaconu, M., Fournier, N.: Probabilistic approach of some discrete and continuous coagulation equation with diffusion. Stochastic Process. Appl., **101**, 83–111 (2002)
36. DeMasi, A., Ferrari, A., Lebowitz, J.L.: Reaction–diffusion equations for interacting particle systems. J. Statist. Phys., **44**, 589–644 (1986)
37. Devaney, R.L.: An Introduction to Chaotic Dynamical Systems. 2nd edn., Addison-Wesley, New York (1989)
38. Donnelly, P., Simons, S.: On the stochastic approach to cluster size distribution during particle coagulation. J. Phys. A: Math. Gen., **26**, 2755–2767 (1993)
39. Durrett, R., Neuhauser, C.: Particle systems and reaction–diffusion equations. Ann. Probab., **22**, 289–333 (1994)
40. Ethier, S.N., Kurtz, T.G.: Markov Processes, Characterization and Convergence. Wiley, New York (1986)
41. Filbert, F., Laurençot, P., Perthame, B.: Derivation of hyperbolic models for chemosensitive movement. J. Math. Biol., **50**, 189–207 (2005)
42. Friedman, A.: A hierarchy of cancer models and their mathematical challenges. Discr. Contin. Dyn. Systems, **4**, 147–159 (2004)
43. Geigant, E., Ladizhansky, K., Mogilner, A.: An intergrodifferential model for orientational distribution of F–Actin in cells. SIAM J. Appl. Math., **59**, 787–809 (1998)
44. Geymonat, L.: Filosofia e filosofia della scienza. Feltrinelli, Milano (1960)
45. Glass, L., Mackey, M.C.: From clocks to chaos. The rhythms of life. Princeton Univ. Press, Princeton (1988)
46. Goel, N.S., Maitra, S.C., Montroll, E.W.: On the Volterra and other nonlinear models of interacting populations. Rev. Modern Phys., **43**, 231–276 (1971)
47. Golse, F., Saint–Raymond, L.: The Navier–Stokes limit of the Boltzmann equation for bounded collision kernels. Invent. Math., **155**, 81–161 (2004)
48. Grad, H.: Asymptotic theory of the Boltzmann equation II. In: Laurmann, J. (ed) Rarefied Gas Dynamics vol. I, Academic Press, (1963), 26–59
49. Graham, C., Méléard, S.: Stochastic particle approximations for generalized Boltzmann models and convergence estimates. Ann. Probab., **25**, 115–132 (1997)
50. Guias, F.: Coagulation–fragmentation processes: relations between finite particle models and differential equations, Preprint 98–41 (SFB 359), Juli 1998, Heidelberg.
51. Jäger, E, Segel, L.: On the distribution of dominance in a population of interacting anonymous organisms. SIAM J. Appl. Math., **52**, 1442–1468 (1992)

52. Herberman, R.B.: NK Cells and Other Natural Effector Cells, Academic Press, New York (1982)
53. Heisenberg, W.: Der Teil und das Ganze. Piper, München (1969)
54. Heisenberg, W.: Physics and Beyond. Encounters and Conversations. Haper and Row, New York (1971)
55. Hofbauer, J., Sigmund, K.: The Theory of Evolution and Dynamical Systems. Mathematical Aspects of Selection. Cambridge Univ. Press, New York (1988)
56. Horstmann, D.: From 1970 until present: the Keller–Segel model in chemotaxis and its consequences. I. Jahresber. Deutsch Math.–Verein., **105**, 103–165 (2003)
57. Horstmann, D.: From 1970 until present: the Keller–Segel model in chemotaxis and its consequences. II. Jahresber. Deutsch Math.–Verein., **106**, 51–69 (2004)
58. Kac, M.: Probability and Related Topics in Physical Sciences, Wiley–Interscience, New York (1959)
59. Kolev, M., Kozłowska, E., Lachowicz, M.: A mathematical model for a single cell cancer — immune system dynamics. Math. Comput. Model., **41**, 1083–1095 (2005)
60. Kolokoltsov, V.N.: On extension of mollified Boltzmann and Smoluchowski equations to particle systems with a k–nary interaction. Russian J. Math. Phys., **10**, 268–295 (2003)
61. Kolokoltsov, V.N.: Hydrodynamic limit of coagulation-fragmentation type models of k–nary interacting particles. J. Statist. Phys., **115**, 1621–1653 (2004)
62. Lachowicz, M.: A system of stochastic differential equations modeling the Euler and the Navier-Stokes hydrodynamic equations. Japan J. Industr. Appl. Math., **10**, 109–131 (1993)
63. Lachowicz, M.: Asymptotic analysis of nonlinear kinetic equations: The hydrodynamic limit. In: Bellomo, N. (ed) Lecture Notes on the Mathematical Theory of the Boltzmann equation, World Scientific, Singapore (1995), 65–148
64. Lachowicz, M.: Hydrodynamic limits of some kinetic equations. In: Matsumura, A., Kawashima, S. (eds) Mathematical Analysis in Fluid and Gas Dynamics, RIMS Kokyuroku 1146, Kyoto (2000), 121–142
65. Lachowicz, M.: Competition tumor–immune system. In: Proceedings of the Sixth National Conference on Application of Mathematics in Biology and Medicine, Kraków (2000), 89–93
66. Lachowicz, M.: From microscopic to macroscopic description for generalized kinetic models. Math. Models Methods Appl. Sci., **12**, 985–1005 (2002)
67. Lachowicz, M.: Describing competitive systems at the level of interacting individuals. In: Proceedins of the Eight Nat. Confer. Appl. Math. Biol. Medicine, Kraków (2002), 95–100
68. Lachowicz, M.: From microscopic to macroscopic descriptions of complex systems. Comp. Rend. Mecanique (Paris), **331**, 733–738 (2003)
69. Lachowicz, M.: On bilinear kinetic equations. Between micro and macro descriptions of biological populations. In: Rudnicki, R. (ed) Banach Center Publ. 63, Warszawa (2004), 217–230
70. Lachowicz, M.: General population systems. Macroscopic limit of a class of stochastic semigroups. J. Math. Anal. Appl. **307**, 585–605 (2005)
71. Lachowicz, M.: Stochastic semigroups and coagulation equations. Ukrainian Math. J., **57**, 770–777 (2005)

72. Lachowicz, M.: Micro and meso scales of description corresponding to a model of tissue invasion by solid tumours. Math. Models Methods Appl. Sci., **15**, 1667–1683 (2005)
73. Lachowicz M., Laurençot Ph., Wrzosek D.: On the Oort–Hulst–Safronov coagulation equation and its relation to the Smoluchowski equation. SIAM J. Math. Anal., **34**, 1399–1421 (2003)
74. Lachowicz, M., Pulvirenti, M.: A stochastic particle system modeling the Euler equation. Arch. Rational Mech. Anal., **109**, 81–93 (1990)
75. Lachowicz, M., Wrzosek, D.: A nonlocal coagulation-fragmentation model. Appl. Math. (Warsaw), **27**, 45–66 (2000)
76. Lachowicz, M., Wrzosek, D.: Nonlocal bilinear equations. Equilibrium solutions and diffusive limit. Math. Models Methods Appl. Sci., **11**, 1393–1409 (2001)
77. Lanford III, O.: Time evolution of large classical systems. In: Moser, E.J. (ed) Lecture Notes in Physics 38, Springer, New York (1975), 1–111
78. Lang, R., Xanh, N.: Smoluchowski's theory of coagulation in colloids holds rigorously in the Boltzmann–Grad limit. Z. Wahrscheinlichkeits. verw. Geb., **54**, 227–280 (1980)
79. Lasota, A., Mackey, M.C.: Chaos, Fractals, and Noise, Springer, New York (1994)
80. Leontovich, M.A.: Fundamental equations of the kinetic theory of gases from the point of view of stochastic processes. Zhur. Exper. Teoret. Fiz. **5**, 211–231 (1935, in Russian)
81. Lolas, G.: Mathematical modelling of the urokinase plasminogen activation system and its role in cancer invasion of tissue. Ph.D. Thesis, Department of Mathematics, University of Dundee, June 2003.
82. McKean, H.P.: Speed of approach to equilibrium for Kac's caricature of maxwellian gas. Arch. Rational Mech. Anal., **21**, 347–367 (1966)
83. Miękisz, J.: Evolutionary game theory and population dynamics. In this volume
84. Morale, D., Capasso, V., Oelschläger, K.: An interacting particle system modelling aggregation behaviour: from individuals to populations. J. Math. Biol., **50** 49–66 (2005)
85. Murray, J.D.: Mathematical Biology, Springer, New York (2003)
86. Oelschläger, K.: On the derivation of reaction–diffusion equations as the limit dynamics of systems of moderately interacting processes. Probab. Theory Related Fields, **82**, 565–586 (1989)
87. O'Malley, R.E., Jr.: Singular Perturbation Methods for Ordinary Differential Equations, Springer, New York (1991)
88. Othmer, H.G., Dunbar, S.R., Alt, W.: Models of dispersal in biological systems. J. Math. Biol., **26**, 263–298 (1988)
89. Pavlotsky, I.P.: Vlasov equations for Volterra dynamical systems. Dokl. AN SSSR, **285**, 1985, 331–339 (1985)
90. Perthame, B.: PDE models for chemotactic movements: Parabolic, hyperbolic and kinetic. Appl. Math., **49**, 539–564 (2004)
91. Petrina, D.Ya., Gerasimenko, V.I., Malyshev, P.V.: Mathematical Foundations of Classical Statistical Mechanics. Gordon and Breach, New York (1989)
92. Poupaud, F.: On a system of nonlinear Boltzmann equation of semiconductor physics. SIAM J. Appl. Math., **50**, 1593–1606 (1990)
93. Pulvirenti, M., Wagner, W., Zavelani Rossi, M.B.: Convergence of particle schemes for the Boltzmann equation. European J. Mech. B Fluids, **13**, 339–351 (1994)

94. Redner, O., Baake, M.: Unequal crossover dynamics in discrete and continuous time. J. Math. Biol., **49**, 201–226 (2004)

95. Skorohod, A.V.: Stochastic Equation for Complex Systems, Nauka, Moscow (1983,in Russian) and Reidel Pub. Co., Dordrecht (1988)

96. Sleeman, B.D., Levine, H.A.: Partial differential equations of chemotaxis and angiogenesis. Math. Methods Appl. Sci., **24** 405–426 (2001)

97. Smoller, J.: Shock Waves and Reaction–Diffusion Equations, II ed., Springer, New York (1994)

98. Smoluchowski M.: Versuch einer mathematischen Theorie der kolloiden Lösungen. Z. Phys. Chem., **92**, 129–168 (1917)

99. Stevens, A.: The derivation of chemotakxis equations as limit dynamics of moderately interacting stochastic many–particle systems. SIAM J. Appl. Math. (electronic) **61**, 183–212 (2000)

100. Thompson, C.J.: Mathematical Statistical Mechanics. Princeton University Press, Princeton (1972)

101. Tikhonov, A.N., Vasileva, A.B., Sveshnikov, A.G.: Differential Equations. Nauka, Moscow (1985), in Russian

102. Wagner, W.: A stochastic particle system associated with the spatially inhomogeneous Boltzmann equation. Transport Theory Statist. Phys. **23**, 455–478 (1994)

103. Wagner, W.: A functional law of large numbers for Boltzmann type stochastic particle systems. Stochastic Anal. Appl., **14**, 591–636 (1996)

104. Waluś, W.: Computational methods for the Boltzmann equation, In Bellomo, N. (ed.), Lecture Notes on the Mathematical Theory of the Boltzmann equation, World Sci., Singapore (1995), 179–223

105. von Weizsacker, C.F.: Die Einheit der Natur. Hanser, Munchen (1971)

106. Wightman, A.S.: Hilbert's sixth problem: Mathematical treatment of the axioms of Physics. Proceedings of Symposia in Pure Math. Northern Illinois Univ., Amer. Math. Soc., Providence, **28**, 147–240 (1976)

Evolutionary Game Theory and Population Dynamics

Jacek Miękisz

Institute of Applied Mathematics and Mechanics, Faculty of Mathematics, Informatics and Mechanics, Warsaw University, Banacha 2, 02-097 Warsaw, Poland
miekisz@mimuw.edu.pl

Summary. Many socio-economic and biological processes can be modeled as systems of interacting individuals. The behaviour of such systems can be often described within game-theoretic models. We introduce fundamental concepts of evolutionary game theory and review basic properties of deterministic replicator dynamics and stochastic dynamics of finite populations. We discuss the problem of the selection of efficient equilibria and the dependence of the long-run behaviour of a population on various parameters such as the time delay, the noise level, and the size of the population.

1 Short Overview

We begin these lecture notes by a crash course in game theory. In particular, we introduce a fundamental notion of a Nash equilibrium. To address the problem of the equilibrium selection in games with multiple equilibria, we review basic properties of the deterministic replicator dynamics and stochastic adaptation dynamics.

We show the almost global asymptotic stability of an efficient equilibrium in the replicator dynamics with a migration between subpopulations. We also show that the stability of a mixed equilibrium depends on the time delay introduced in replicator equations. For large time delays, a population oscillates around its equilibrium.

We analyze the long-run behaviour of stochastic dynamics in well-mixed populations and in spatial games with local interactions. We review results concerning the effect of the number of players and the noise level on the stochastic stability of Nash equilibria. In particular, we present examples of games in which when the number of players increases or the noise level decreases, a population undergoes a transition between its equilibria. We discuss similarities and differences between systems of interacting players in spatial games maximizing their individual payoffs and particles in lattice-gas models minimizing their interaction energy.

In short, there are two main themes of our lecture notes: the selection of efficient equilibria (providing the highest payoffs to all players) in population dynamics and the dependence of the long-run behaviour of a population on various parameters such as the time delay, the noise level, and the size of the population.

2 Introduction

Many socio-economic and biological processes can be modeled as systems of interacting individuals; see for example econophysics bulletin [16] and statistical mechanics and quantitative biology archives [13]. One may then try to derive their global behaviour from individual interactions between their basic entities such as animals in ecological and evolutionary models, genes in population genetics and people in social processes. Such approach is fundamental in statistical physics which deals with systems of interacting particles. One can therefore try to apply methods of statistical physics to investigate the population dynamics of interacting individuals. There are however profound differences between these two systems. Physical systems tend in time to states which are characterized by the minimum of some global quantity, the total energy or free energy of the system. Population dynamics lacks such general principle. Agents in social models maximize their own payoffs, animals and genes maximize their individual darwinian fitness. The long-run behavior of such populations cannot in general be characterized by the global or even local maximum of the payoff or fitness of the whole population. We will explore similarities and differences between these systems.

The behaviour of systems of interacting individuals can be often described within game-theoretic models [14, 24, 25, 27, 36, 37, 48, 66, 67, 78, 97, 100, 103]. In such models, players have at their disposal certain strategies and their payoffs in a game depend on strategies chosen both by them and by their opponents. The central concept in game theory is that of a **Nash equilibrium**. It is an assignment of strategies to players such that no player, for fixed strategies of his opponents, has an incentive to deviate from his current strategy; no change can increase his payoff.

In Chapter 3, we present a crash course in game theory. One of the fundamental problems in game theory is the equilibrium selection in games with multiple Nash equilibria. In two-player symmetric games with two strategies, we may have two Nash equilibria: a payoff dominant (also called efficient) and a risk-dominant one. In the efficient equilibrium, players receive highest possible payoffs. The strategy is risk-dominant if it has a higher expected payoff against a player playing both strategies with equal probabilities. It is played by individuals averse to risk. One of the selection methods is to construct a dynamical system where in the long run only one equilibrium is played with a high frequency.

John Maynard Smith [46–48] has refined the concept of the Nash equilibrium to include the stability of equilibria against mutants. He introduced the fundamental notion of an **evolutionarily stable strategy**. If everybody plays such a strategy, then the small number of mutants playing a different strategy is eliminated from the population. The dynamical interpretation of the evolutionarily stable strategy was later provided by several authors [35, 93, 106]. They proposed a system of difference or differential replicator equations which describe the time-evolution of frequencies of strategies. Nash equilibria are stationary points of this dynamics. It appears that in games with a payoff dominant equilibrium and a risk-dominant one, both are asymptotically stable but the second one has a larger basin of attraction in the replicator dynamics.

In Chapter 4, we introduce **replicator dynamics** and review theorems concerning asymptotic stability of Nash equilibria [36, 37, 105]. Then in Chapter 5, we present our own model of the replicator dynamics [55] with a migration between two subpopulations for which an efficient equilibrium is almost globally asymptotically stable.

It is very natural, and in fact important, to introduce a **time delay** in the population dynamics; a time delay between acquiring information and acting upon this knowledge or a time delay between playing games and receiving payoffs. Recently Tao and Wang [91] investigated the effect of a time delay on the stability of interior stationary points of the replicator dynamics. They considered two-player games with two strategies and a unique asymptotically stable interior stationary point. They proposed a certain form of a time-delay differential replicator equation. They showed that the mixed equilibrium is asymtotically stable if a time delay is small. For sufficiently large delays it becomes unstable.

In Chapter 6, we construct two models of discrete-time replicator dynamics with a time delay [2]. In the social-type model, players imitate opponents taking into account average payoffs of games played some units of time ago. In the biological-type model, new players are born from parents who played in the past. We consider two-player games with two strategies and a unique mixed Nash equilibrium. We show that in the first type of dynamics, it is asymptotically stable for small time delays and becomes unstable for large ones when the population oscillates around its stationary state. In the second type of dynamics, however, the Nash equilibrium is asymptotically stable for any time delay. Our proofs are elementary, they do not rely on the general theory of delay differential and difference equations.

Replicator dynamics models population behaviour in the limit of the infinite number of individuals. However, real populations are finite. Stochastic effects connected with random matchings of players, mistakes of players and biological mutations can play a significant role in such systems. We will discuss various stochastic adaptation dynamics of populations with a fixed number of players interacting in discrete moments of time. In well-mixed populations, individuals are randomly matched to play a game [40, 54, 75].

The deterministic selection part of the dynamics ensures that if the mean payoff of a given strategy is bigger than the mean payoff of the other one, then the number of individuals playing the given strategy increases. However, players may mutate hence the population may move against a selection pressure. In spatial games, individuals are located on vertices of certain graphs and they interact only with their neighbours; see for example [5, 7, 17, 18, 30–34, 41, 44, 51, 52, 56, 62–64, 85, 86, 88, 103] and a recent review [89] and references therein. In discrete moments of times, players adapt to their opponents by choosing with a high probability the strategy which is the best response, i.e. the one which maximizes the sum of the payoffs obtained from individual games. With a small probability, representing the noise of the system, they make mistakes. The above described stochastic dynamics constitute ergodic Markov chains with states describing the number of individuals playing respective strategies or corresponding to complete profiles of strategies in the case of spatial games. Because of the presence of random mutations, our Markov chains are ergodic (irreducible and periodic) and therefore they possess unique stationary measures. To describe the long-run behavior of such stochastic dynamics, Foster and Young [22] introduced a concept of stochastic stability. A configuration of the system is **stochastically stable** if it has a positive probability in the stationary measure of the corresponding Markov chain in the zero-noise limit, that is the zero probability of mistakes. It means that in the long run we observe it with a positive frequency along almost any time trajectory.

In Chapter 7, we introduce the concept of stochastic stability and present a useful representation of stationary measures of ergodic Markov chains [23, 82, 104].

In Chapter 8, we discuss populations with random matching of players in **well-mixed populations**. We review recent results concerning the dependence of the long-run behavior of such systems on the number of players and the noise level. In the case of two-player games with two symmetric Nash equilibria, an efficient one and a risk-dominant one, when the number of players increases, the population undergoes twice a transition between its equilibria. In addition, for a sufficiently large number of individuals, the population undergoes another **equilibrium transition** when the noise decreases.

In Chapter 9, we discuss **spatial games**. We will see that in such models, the notion of a Nash equilibrium (called there a Nash configuration) is similar to the notion of a ground-state configuration in classical lattice-gas models of interacting particles. We discuss similarities and differences between systems of interacting players in spatial games maximizing their individual payoffs and particles in lattice-gas models minimizing their interaction energy.

The concept of stochastic stability is based on the zero-noise limit for a fixed number of players. However, for any arbitrarily low but fixed noise, if the number of players is large enough, the probability of any individual configuration is practically zero. It means that for a large number of players, to observe a stochastically stable configurations we must assume that players make mistakes with extremely small probabilities. On the other hand, it may happen that in the long run, for a low but fixed noise and sufficiently large number of players, the stationary configuration is highly concentrated on an ensemble consisting of one Nash configuration and its small perturbations, i.e. configurations where most players play the same strategy. We will call such configurations **ensemble stable.** It will be shown that these two stability concepts do not necessarily coincide. We will present examples of spatial games with three strategies where concepts of stochastic stability and ensemble stability do not coincide [51,53]. In particular, we may have the situation, where a stochastically stable strategy is played in the long run with an arbitrarily low frequency. In fact, when the noise level decreases, the population undergoes a sharp transition with the coexistence of two equilibria for some noise level. Finally, we discuss the influence of **dominated strategies** on the long-run behaviour of population dynamics.

In Chapter 10, we shortly review other results concerning stochastic dynamics of finite populations.

3 A Crash Course in Game Theory

To characterize a game-theoretic model one has to specify players, strategies they have at their disposal and payoffs they receive. Let us denote by $I = \{1, \ldots, n\}$ the set of players. Every player has at his disposal m different strategies. Let $S = \{1, \ldots, m\}$ be the set of strategies, then $\Omega = S^I$ is the set of strategy profiles, that is functions assigning strategies to players. The payoff of any player depends not only on his strategy but also on strategies of all other players. If $X \in \Omega$, then we write $X = (X_i, X_{-i})$, where $X_i \in S$ is a strategy of the i-th player and $X_{-i} \in S^{I-\{i\}}$ is a strategy profile of remaining players. The payoff of the i-th player is a function defined on the set of profiles,

$$U_i : \Omega \to R, \ \ i, \ldots, n$$

The central concept in game theory is that of a Nash equilibrium. An assignment of strategies to players is a **Nash equilibrium**, if for each player, for fixed strategies of his opponents, changing his current strategy cannot increase his payoff. The formal definition will be given later on when we enlarge the set of strategies by mixed ones.

Although in many models the number of players is very large (or even infinite as we will see later on in replicator dynamics models), their strategic interactions are usually decomposed into a sum of two-player games. Only recently, there have appeared some systematic studies of truly multi-player games [10, 11, 42, 74]. Here we will discuss only two-player games with two or three strategies. We begin with games with two strategies, A and B. Payoffs functions can be then represented by 2×2 payoff matrices. A general payoff matrix is given by

$$U = \begin{array}{c|cc} & A & B \\ \hline A & a & b \\ B & c & d, \end{array}$$

where U_{kl}, $k, l = A, B$, is a payoff of the first (row) player when he plays the strategy k and the second (column) player plays the strategy l.

We assume that both players are the same and hence payoffs of the column player are given by the matrix transposed to U; such games are called symmetric. In this classic set-up of static games (called matrix games or games in the normal form), players know payoff matrices, simultaneously announce (use) their strategies and receive payoffs according to their payoff matrices.

We will present now three main examples of symmetric two-player games with two strategies. We begin with an anecdote, then an appropriate game-theoretic model is build and its Nash equilibria are found.

Example 1 (Stag-hunt game)

Jean-Jacques Rousseau wrote, in his Discourse on the Origin and Basis of Equality among Men, about two hunters going either after a stag or a hare [24, 71]. In order to get a stag, both hunters must be loyal one to another and stay at their positions. A single hunter, deserting his companion, can get his own hare. In the game-theory language, we have two players and each of them has at his disposal two strategies: Stag (St) and Hare (H). In order to present this example as a matrix game we have to assign some values to animals. Let a stag (which is shared by two hunters) be worth 10 units and a hare 3 units. Then the payoff matrix of this symmetric game is as follows:

$$U = \begin{array}{c|cc} & St & H \\ \hline St & 5 & 0 \\ H & 3 & 3 \end{array}$$

It is easy to see that there are two Nash equilibria: (St, St) and (H, H).

In a general payoff matrix, if $a > c$ and $d > b$, then both (A, A) and (B, B) are Nash equilibria. If $a + b < c + d$, then the strategy B has a higher expected payoff against a player playing both strategies with the probability $1/2$. We say that B risk dominates the strategy A (the notion of the risk-dominance was introduced and thoroughly studied by Harsányi and Selten [29]). If at the same time $a > d$, then we have a selection problem of choosing between the payoff-dominant (Pareto-efficient) equilibrium (A, A) and the risk-dominant (B, B).

Example 2 (Hawk-Dove game)

Two animals are fighting for a certain territory of a value V. They can be either aggressive (hawk strategy - H) or peaceful (dove strategy - D). When two hawks meet, they accure the cost of fighting $C > V$ and then they split the territory. When two dove meets, they split the territory without a fight. A dove gives up the territory to a hawk. We obtain the following payoff matrix:

$$U = \begin{array}{c c c} & H & D \\ H & (V\text{-}C)/2 & V \\ D & 0 & V/2, \end{array}$$

The Hawk-Dove game was analyzed by John Maynard Smith [48]. It is also known as the Chicken game [76] or the Snowdrift game [31]. It has two non-symmetric Nash equilibria: (H, D) and (D, H).

Example 3 (Prisoner's Dilemma)

The following story was discussed by Melvin Dresher, Merill Flood, and Albert Tucker [4,72,83]. Two suspects of a bank robbery are caught and interrogated by the police. The police offers them separately the following deal. If a suspect testifies against his colleague (a strategy of defection - D), and the other does not (cooperation - C), his sentence will be reduced by five years. If both suspects testify, that is defect, they will get the reduction of only one year. However, if they both cooperate and do not testify, their sentence, because of the lack of a hard evidence, will be reduced by three years. We obtain the following payoff matrix:

$$U = \begin{array}{c c c} & C & D \\ C & 3 & 0 \\ D & 5 & 1 \end{array}$$

The strategy C is a **dominated strategy** - it results in a lower payoff than the strategy D, regardless of a strategy used by the other player. Therefore, (D, D) is the unique Nash equilibrium but both players are much better off when they play C - this is the classic Prisoner's Dilemma.

A novel behaviour can appear in games with three strategies.

Example 4 (Rock-Scissors-Paper game)

In this game, each of two players simultaneously exhibits a sign of either a scissors (S), a rock (R), or a paper (P). The game has a cyclic behaviour:

rock crashes scissors, scissors cut paper, and finally paper wraps rock. The payoffs can be given by the following matrix:

$$
U = \begin{array}{c|ccc} & R & S & P \\ \hline R & 1 & 2 & 0 \\ S & 0 & 1 & 2 \\ P & 2 & 0 & 1 \end{array}
$$

It is easy to verify that this game, because of its cyclic behavior, does not have any Nash equilibria as defined so far. However, we intuitively feel that when we repeat it many times, the only way not to be exploited is to mix randomly strategies, i.e. to choose each strategy with the probability $1/3$.

This brings us to a concept of a mixed stategy, a probabilty mass function on the set of pure strategies S. Formally, a **mixed strategy** x is an element of a simplex

$$
\Delta = \{x \in R^m, 0 \leq x_k \leq 1, \sum_{k=1}^{m} x_k = 1\}.
$$

By the support of a mixed strategy x we mean the set of pure strategies with positive probabilities in x. Payoffs of mixed strategies are defined as appropriate expected values. In two-player games, a player who uses a mixed strategy x against a player with a mixed strategy y receives a payoff given by

$$
\sum_{k,l \in S} U_{kl} x_k y_l.
$$

In general n-player games, profiles of strategies are now elements of $\Theta = \Delta^I$. We are now ready to define formally a Nash equilibrium.

Definition 1 $X \in \Theta$ is a **Nash equilibrium** *if for every $i \in I$ and every $y \in \Delta$,*

$$
U_i(X_i, X_{-i}) \geq U_i(y, X_{-i})
$$

In the mixed Nash equilibrium, expected payoffs of all strategies in its support should be equal. Otherwise a player could increase his payoff by increasing the probability of playing a strategy with the higher expected payoff. In two-player games with two strategies, we identify a mixed strategy with its first component, $x = x_1$. Then the expected payoff of A is given by $ax+b(1-x)$ and that of B by $cx+d(1-x)$. $x^* = (d-b)/(d-b+a-c)$ for which the above two expected values are equal is a mixed Nash equilibrium or more formally, a profile (x, x) is a Nash equilibrium.

In Examples 1 and 2, in addition to Nash equilibria in pure strategies, we have mixed equilibria, $x^* = 3/5$ and $x^* = V/C$ respectively. It is obvious that the Prisoner's Dilemma game does not have any mixed Nash equilibria . On the other hand, the only Nash equilibrium of the Rock-Scissors-Paper game is a mixed one assigning the probability $1/3$ to each strategy.

We end this chapter by a fundamental theorem due to John Nash [59, 60].

Theorem 1 *Every game with a finite number of players and a finite number of strategies has at least one Nash equilibrium.*

In any Nash equilibrium, every player uses a strategy which is a best reply to the profile of strategies of remaining players. Therefore a Nash equilibrium can be seen as a best reply to itself - a fixed point of a certain best-reply correspondence. Then one can use the Kakutani fixed point theorem to prove the above theorem.

4 Replicator Dynamics

The concept of a Nash equilibrium is a static one. Here we will introduce the classical replicator dynamics and review its main properties [36, 37, 105]. Replicator dynamics provides a dynamical way of achieving Nash equilibria in populations. We will see that Nash equilibria are stationary points of such dynamics and some of them are asymptotically stable.

Imagine a finite but a very large population of individuals. Assume that they are paired randomly to play a symmetric two-player game with two strategies and the payoff matrix given in the beginning of the previous chapter. The complete information about such population is provided by its strategy profile, that is an assignment of pure strategies to players. Here we will be interested only in the proportion of individuals playing respective strategies. We assume that individuals receive average payoffs with respect to all possible opponents - they play against the average strategy.

Let $r_i(t)$, $i = A, B$, be the number of individuals playing the strategy A and B respectively at the time t. Then $r(t) = r_A(t) + r_B(t)$ is the total number of players and $x(t) = \frac{r_1(t)}{r(t)}$ is a fraction of the population playing A.

We assume that during the small time interval ϵ, only an ϵ fraction of the population takes part in pairwise competitions, that is plays games. We write

$$r_i(t + \epsilon) = (1 - \epsilon)r_i(t) + \epsilon r_i(t)U_i(t); \quad i = A, B, \tag{1}$$

where $U_A(t) = ax(t) + b(1 - x(t))$ and $U_B(t) = cx(t) + d(1 - x(t))$ are average payoffs of individuals playing A and B respectively. We assume that all payoffs are not smaller than 0 hence r_A and r_B are always non-negative and therefore $0 \leq x \leq 1$.

The equation for the total number of players reads

$$r(t + \epsilon) = (1 - \epsilon)r(t) + \epsilon r(t)\bar{U}(t), \tag{2}$$

where $\bar{U}(t) = x(t)U_A(t) + (1 - x(t))U_B(t)$ is the average payoff in the population at the time t. When we divide (1) by (2) we obtain an equation for the frequency of the strategy A,

$$x(t + \epsilon) - x(t) = \epsilon \frac{x(t)[U_A(t) - \bar{U}(t)]}{1 - \epsilon + \epsilon\bar{U}(t)}. \tag{3}$$

Now we divide both sides of (3) by ϵ, perform the limit $\epsilon \to 0$, and obtain the well known differential replicator equation:

$$\frac{dx(t)}{dt} = x(t)[U_A(t) - \bar{U}(t)]. \tag{4}$$

The above equation can also be written as

$$\frac{dx(t)}{dt} = x(t)(1-x(t))[U_A(t)-U_B(t)] = (a-c+d-b)x(t)(1-x(t))(x(t)-x^*). \tag{5}$$

For games with m strategies we obtain a system of m differential equations for $x_k(t)$, fractions of the population playing the k-th strategy at the time t, $k = 1, \ldots, m$,

$$\frac{dx_k(t)}{dt} = x_k(t)[\sum_{l=1}^{m} U_{kl}x_l(t) - \sum_{k,l=1}^{m} U_{kl}x_k(t)x_l(t)], \tag{6}$$

where on the right hand-size of (6) there is a difference of the average payoff of the k-th strategy and the average payoff of the population. The above system of differential equations or analogous difference equations, called replicator dynamics was proposed in [35,93,106]. For any initial condition $x^0 \in \Delta$, it has the unique global solution, $\xi(x^0, t)$, which stays in the simplex Δ.

Now we review some theorems relating replicator dynamics and Nash equilibria [36, 37, 105]. We consider symmetric two-player games. We denote the set of strategies corresponding to symmetric Nash equilibria by

$$\Delta^{NE} = \{x \in \Delta : (x,x) \text{ is a Nash equilibrium}\}.$$

It follows from the definition of the Nash equilibrium (see also discussion in the previous chapter concerning mixed strategies) that

$$\Delta^{NE} = \{x \in \Delta : u(i,x) = \max_{z \in \Delta} u(z,x) \text{ for every } i \text{ in the support of } x\}.$$

It is easy to see that

$$\Delta^0 = \{x \in \Delta : u(i,x) = u(x,x) \text{ for every } i \text{ in the support of } x\}$$

is the set of stationary points of the replicator dynamics.

It follows that symmetric Nash equilibria are stationary points of the replicator dynamics.

Theorem 2 $S \cup \Delta^{NE} \subset \Delta^0$

The following two theorems relate stability of stationary points to Nash equilibria [36, 105].

Theorem 3 *If $x \in \Delta$ is Lyapunov stable, then $x \in \Delta^{NE}$.*

Theorem 4 *If $x^0 \in interior(\Delta)$ and $\xi(x^0, t) \rightarrow_{t \to \infty} x$, then $x \in \Delta^{NE}$.*

Below we present the replicator dynamics in the examples of two-player games discussed in the previous chapter. We write replicator equations and show their phase diagrams.

Stag-hunt game

$$\frac{dx}{dt} = x(1-x)(5x-3)$$

$$\bullet < - - - - - - - - - - - - \bullet - - - - - - - > \bullet$$

0   3/5   1

Hawk-Dove game

$$\frac{dx}{dt} = -x(1-x)(x - C/V)$$

$$\bullet - - - - - - - - - - - - > \bullet < - - - - - - - \bullet$$

0   V/C 1

Prisoner's Dilemma

$$\frac{dx}{dt} = -x(1-x)(x+1)$$

$$\bullet < - \bullet$$

0   1

We see that in the Stag-hunt game, both pure Nash equilibria are asymptotically stable. The risk-dominant one has the larger basin of attraction which is true in general because $x^* = (d-b)/(d-b+a-c) > 1/2$ for games with an efficient equilibrium and the risk dominant one.

In the Hawk-Dove game, the unique symmetric mixed Nash equilibrium is asymptotically stable.

In the Prisoner's Dilemma, the strategy of defection is globally asymptotically stable.

In the Rock-Scissors-Paper game, a more detailed analysis has to be done. One can show, by straithforward computations, that the time derivative of $lnx_1x_2x_3$ is equal to zero. Therefore $lnx_1x_2x_3 = c$ is an equation of a closed orbit for any constant c. The stationary point $(1/3, 1/3, 1/3)$ of the replicator dynamics is Lyapunov stable and the population cycles on a closed trajectory (which depends on the initial condition) around its Nash equilibrium.

5 Replicator Dynamics with Migration

We discuss here a game-theoretic dynamics of a population of replicating
who can migrate between two subpopulations or habitats [55]. We consider
symmetric two-player games with two strategies: A and B. We assume that
$a > d > c$, $d > b$, and $a + b < c + d$ in a general payoff matrix given in the
beginning of Chapter 3. Such games have two Nash equilibria: the efficient one
(A, A) in which the population is in a state with a maximal fitness (payoff)
and the risk-dominant (B, B) where players are averse to risk. We show that
for a large range of parameters of our dynamics, even if the initial conditions
in both habitats are in the basin of attraction of the risk-dominant equilibrium
(with respect to the standard replication dynamics without migration), in the
long run most individuals play the efficient strategy.

We consider a large population of identical individuals who at each time
step can belong to one of two different non-overlapping subpopulations or
habitats which differ only by their replication rates. In both habitats, they
take part in the same two-player symmetric game. Our population dynamics
consists of two parts: the standard replicator one and a migration between
subpopulations. Individuals are allowed to change their habitats. They move
to a habitat in which the average payoff of their strategy is higher; they do
not change their strategies.

Migration helps the population to evolve towards an efficient equilibrium.
Below we briefly describe the mechanism responsible for it. If in a subpopulation, the fraction of individuals playing the efficient strategy A is above
its unique mixed Nash equilibrium fraction, then the expected payoff of A is
bigger than that of B in this subpopulation, and therefore the subpopulation
evolves to the efficient equilibrium by the replicator dynamics without any
migration. Let us assume therefore that such fraction is below the Nash equilibrium in both subpopulations. Without loss of generality we assume that
initial conditions are such that the fraction of individuals playing A is bigger
in the first subpopulation than in the second one. Hence the expected payoff
of A is bigger in the first subpopulation than in the second one, and the expected payoff of B is bigger in the second subpopulation than in the first one.
This implies that a fraction of A-players in the second population will switch
to the first one and at the same time a fraction of B-players from the first
population will switch to the second one - migration causes the increase of
the fraction of individual of the first population playing A. However, any B-
player will have more offspring than any A-player (we are below a mixed Nash
equilibrium) and this has the opposite effect on relative number of A-players
in the first population than the migration. The asymptotic composition of the
whole population depends on the competition between these two processes.

We derive sufficient conditions for migration and replication rates such
that the whole population will be in the long run in a state in which most
individuals occupy only one habitat (the first one for the above described
initial conditions) and play the efficient strategy.

Let ϵ be a time step. We allow two subpopulations to replicate with different speeds. We assume that during any time-step ϵ, a fraction ϵ of the first subpopulation and a fraction $\kappa\epsilon$ of the second subpopulation plays the game and receives payoffs which are interpreted as the number of their offspring. Moreover, we allow a fraction of individuals to migrate to a habitat in which their strategies have higher expected payoffs.

Let r_s^i denote the number of individuals which use the strategy $s \in \{A, B\}$ in the subpopulation $i \in \{1, 2\}$. By U_s^i we denote the expected payoff of the strategy s in the subpopulation i:

$$U_A^1 = ax + b(1 - x), \quad U_B^1 = cx + d(1 - x),$$
$$U_A^2 = ay + b(1 - y), \quad U_B^2 = cy + d(1 - y),$$

where

$$x = \frac{r_A^1}{r_1}, \quad y = \frac{r_A^2}{r_2}, \quad r_1 = r_A^1 + r_B^1, \quad r_2 = r_A^2 + r_B^2;$$

x and y denote fractions of A-players in the first and second population respectively. We denote by $\alpha = \frac{r_1}{r}$ the fraction of the whole population in the first subpopulation, where $r = r_1 + r_2$ is the total number of individuals.

The evolution of the number of individuals in each subpopulation is assumed to be a result of the replication and the migration flow. In our model, the direction and intensity of migration of individuals with a given strategy will be determined by the difference of the expected payoffs of that strategy in both habitats. Individuals will migrate to a habitat with a higher payoff. The evolution equations for the number of individuals playing the strategy s, $s \in \{A, B\}$, in the habitat i, $i \in \{1, 2\}$, have the following form:

$$r_A^1(t + \epsilon) = R_A^1 + \Phi_A, \tag{7}$$

$$r_B^1(t + \epsilon) = R_B^1 + \Phi_B, \tag{8}$$

$$r_A^2(t + \epsilon) = R_A^2 - \Phi_A, \tag{9}$$

$$r_B^2(t + \epsilon) = R_B^2 - \Phi_B, \tag{10}$$

where all functions on the right-hand sides are calculated at the time t.

Functions R_s^i describe an increase of the number of the individuals playing the strategy s in the subpopulation i due to the replication:

$$R_s^1 = (1 - \epsilon)r_s^1 + \delta U_s^1 r_s^1, \tag{11}$$

$$R_s^2 = (1 - \kappa\epsilon)r_s^2 + \kappa\epsilon U_s^2 r_s^2, \tag{12}$$

The rate of the replication of individuals playing the strategy s in the first subpopulation is given by ϵU_s^1, and in the second subpopulation by $\kappa\epsilon U_s^2$. The parameter κ measures the difference of reproduction speeds in both habitats.

Functions Φ_s, $s \in \{A, B\}$, are defined by

$$\Phi_s = \epsilon\gamma(U_s^1 - U_s^2)[r_s^2\Theta(U_s^1 - U_s^2) + r_s^1\Theta(U_s^2 - U_s^1)], \qquad (13)$$

where Θ is the Heaviside's function,

$$\Theta(x) = \begin{cases} 1, & x \geq 0; \\ 0, & x < 0 \end{cases} \qquad (14)$$

and γ is the migration rate.

Functions Φ_s describe changes of the numbers of the individuals playing strategy s in the relevant habitat due to migration. Φ_s will be referred to as the migration of individuals (who play the strategy s) between two habitats.

Thus, if for example $U_A^1 > U_A^2$, then there is a migration of individuals with the strategy A from the second habitat to the first one:

$$\Phi_A = \delta\gamma r_A^2(U_A^1 - U_A^2), \qquad (15)$$

and since then necessarily $U_B^1 < U_B^2$ [note that $U_A^1 - U_A^2 = (a - b)(x - y)$ and $U_B^1 - U_B^2 = (c - d)(x - y)$], there is a migration flow of individuals with strategy B from the first habitat to the second one:

$$\Phi_B = \epsilon\gamma r_B^1(t)(U_B^1 - U_B^2). \qquad (16)$$

In this case, the migration flow Φ_A describes the increase of the number of individuals which play the strategy A in the first subpopulation due to migration of the individuals playing A in the second subpopulation. This increase is assumed to be proportional to the number of individuals playing A in the second subpopulation and the difference of payoffs of this strategy in both subpopulations. The constant of proportionality is ϵ times the migration rate γ.

The case $\gamma = 0$ corresponds to two separate populations which do not communicate and evolve independently. Our model reduces then to the standard discrete-time replicator dynamics. In this case, the total number of players who use a given strategy changes only due to the increase or decrease of the strategy fitness, as described by functions defined in (11-12).

In the absence of the replication, there is a conservation of the number of individuals playing each strategy in the whole population. This corresponds to our model assumption that individuals can not change their strategies but only habitats in which they live.

For $U_A^1 > U_A^2$ we obtain from (7-10) equations for $r_i(t)$ and $r(t)$:

$$r_1(t+\epsilon) = (1 - \epsilon)r_1(t) + \delta r_1(t)\left[\frac{r_A^1 U_A^1 + r_B^1 U_B^1}{r_1} + \gamma\frac{r_A^2(U_A^1 - U_A^2) + r_B^1(U_B^1 - U_B^2)}{r_1}\right], \qquad (17)$$

$$r_2(t+\epsilon)=(1-\kappa\epsilon)r_2(t)+\delta r_2(t)[\kappa\frac{r_A^2U_A^2+r_B^2U_B^2}{r_2}+\gamma\frac{r_A^2(U_A^2-U_A^1)+r_B^1(U_B^2-U_B^1)}{r_2}],$$
(18)

$$r(t+\delta)=(1-\epsilon)r_1(t)+(1-\kappa\delta)r_2(t)$$
$$+\delta r(t)[\alpha(\frac{r_A^1}{r_1}U_A^1+\frac{r_B^1}{r_1}U_B^1)+(1-\alpha)\kappa(\frac{r_A^2}{r_2}U_A^2+\frac{r_B^2}{r_2}U_B^2)],$$
(19)

where all functions in square brackets depend on t.

Now, like in the derivation of the standard replicator dynamics, we consider frequencies of individuals playing the relevant strategies in both habitats. Thus, we focus on the temporal evolution of the frequencies, x and y, and the relative size of the first subpopulation, α. We divide (7) by (17), (9) by (18), and (17) by (19). Performing the limit $\epsilon \to 0$ we obtain the following differential equations:

$$\frac{dx}{dt}=x[(1-x)(U_A^1-U_B^1)+\gamma[(\frac{y(1-\alpha)}{x\alpha}-\frac{y(1-\alpha)}{\alpha})(U_A^1-U_A^2)-(1-x)(U_B^1-U_B^2)]],$$
(20)

$$\frac{dy}{dt}=y[\kappa(1-y)(U_A^2-U_B^2)+\gamma[(1-y)(U_A^2-U_A^1)-\frac{(1-x)\alpha}{1-\alpha}(U_B^2-U_B^1)]],$$
(21)

$$\frac{d\alpha}{dt}=\alpha(1-\alpha)[xU_A^1+(1-x)U_B^1-(yU_A^2+(1-y)U_B^2)]$$

$$+\alpha\gamma[\frac{y(1-\alpha)}{\alpha}(U_A^1-U_A^2)+(1-x)(U_B^1-U_B^2)]$$
(22)

$$+\alpha(1-\alpha)(\kappa-1)(1-yU_A^2-(1-y)U_B^2).$$

Similar equations are derived for the case $U_A^1 < U_A^2$ (since our model is symmetric with respect to the permutation of the subpopulations, it is enough to renumerate the relevant indices and redefine the parameter κ).

Assume first that $U_A^1(0) > U_A^2(0)$, which is equivalent to $x(0) > y(0)$. It follows from (7-10) that a fraction of A-players from the subpopulation 2 will migrate to the subpopulation 1 and a fraction of B-players will migrate in the opposite direction. This will cause x to increase and y to decrease. However, if $x(0) < x^*$ and $y(0) < x^*$, then $U_A^1 < U_B^1$ and $U_A^2 < U_B^2$, therefore B-players will have more offspring than A-players. This has the opposite effect on the relative number of A-players in the first subpopulation than migration. If $x(0) < y(0)$, then migration takes place in the reverse directions.

The outcome of the competition between migration and replication depends, for a given payoff matrix, on the relation between $x(0) - y(0)$, γ and κ. We are interested in formulating sufficient conditions for the parameters of the model, for which most individuals of the whole population will play in the long run the efficient strategy A. We prove the following theorem [55].

Theorem 5 *If*

$$\gamma[x(0) - y(0)] > max[\frac{d-b}{d-c}, \frac{\kappa(a-c)}{a-b}],$$

then $x(t) \rightarrow_{t\rightarrow\infty} 1$ *and* $y(t) \rightarrow_{t\rightarrow\infty} 0$.
   If $\kappa < (a-1)/(d-1)$, *then* $\alpha(t) \rightarrow_{t\rightarrow\infty} 1$.
   If

$$\gamma[y(0) - x(0)] > max[\frac{\kappa(d-b)}{d-c}, \frac{a-c}{a-b}],$$

then $x(t) \rightarrow_{t\rightarrow\infty} 0$ *and* $y(t) \rightarrow_{t\rightarrow\infty} 1$.
   If $\kappa > (d-1)/(a-1)$, *then* $\alpha(t) \rightarrow_{t\rightarrow\infty} 0$.

Proof:
   Assume first that $x(0) > y(0)$. From (20-21) we get the following differential inequalities:

$$\frac{dx}{dt} > x(1-x)[U_A^1 - U_B^1) + \gamma(U_B^2 - U_B^1)], \tag{23}$$

$$\frac{dy}{dt} < y(1-y)[\kappa(U_A^2 - U_B^2) + \gamma(U_A^2 - U_A^1)], \tag{24}$$

Using explicit expressions for U_s^i we get

$$\frac{dx}{dt} > x(1-x)[(a-c+d-b)x + b - d + \gamma(d-c)(x-y)], \tag{25}$$

$$\frac{dy}{dt} < y(1-y)[\kappa[(a-c+d-b)y + b - d] - \gamma(a-b)(x-y)], \tag{26}$$

   We note that if $\gamma(d-c)(x(0) - y(0)) > d - b$ then $\gamma(d-c)(x(0)-y(0)) + b - d + (a-c+d-b)x(0) > 0$, i.e. $dx/dt(0) > 0$.
   Analogously, if $\gamma(a-b)(x(0)-y(0)) > \kappa(a-c)$, then $\gamma(a-b)(x(0)-y(0)) > \kappa[(a-c+d-b)+b-d] > \kappa[(a-c+d-b)y(o)+b-d]$, therefore $dy/dt(0) < 0$. Thus, combining both conditions we conclude that $x(t) - y(t)$ is an increasing function so $x(t) > y(t)$ for all $t \geq 0$, hence we may use (20-22) all the time. We get that $x(t) \rightarrow_{t\rightarrow\infty} 1$ and $y(t) \rightarrow_{t\rightarrow\infty} 0$, and the first part of the thesis follows. Now from (22) it follows that if $a - d + (\kappa - 1)(1 - d) > 0$, i.e. $\kappa < (a-1)/(d-1)$, then $\alpha(t) \rightarrow_{t\rightarrow\infty} 1$.
   The second part of Theorem 5, corresponding to initial conditions $y(0) > x(0)$, can be proved analogously, starting from eqs. (7-10) written for the case $U_A^1(0) < U_A^2(0)$ and their continuous counterparts. We omit details.
   The above conditions for κ mean that the population consisting of just A-players replicates faster (exponentially in $(a - 1)t$) than the one consisting of just B-players (exponentially in $(d-1)\kappa t$). The same results would follow if the coefficients of the payoff matrix of the game played in one habitat would differ from those in the second habitat by an additive constant.

We showed that introduction of the mechanism of attraction by the habitat with a higher expected payoff in the standard replicator dynamics helps the whole population to reach the state in which in the long run most individuals play the efficient strategy.

More precisely, we proved that for a given rate of migration, if the fractions of individuals playing the efficient strategy in both habitats are not too close to each other, then the habitat with a higher fraction of such players overcomes the other one in the long run. The fraction of individuals playing the efficient strategy tends to unity in this habitat and consequently in the whole population. Alternatively, we may say that the bigger the rate of migration is, larger is the basin of attraction of the efficient equilibrium. In particular, we showed that for a large range of parameters of our dynamics, even if the initial conditions in both habitats are in the basin of attraction of the risk-dominant equilibrium (with respect to the standard replication dynamics without migration), in the long run most individuals play the efficient strategy.

6 Replicator Dynamics with Time Delay

Here we consider two-player games with two strategies, two pure non- symmetric Nash equilibria, and a unique symmetric mixed one, that is $a < c$ and $d < b$ in a general payoff matrix given in the beginning of Chapter 3. Let us recall that the Hawk-Dove game is of such type.

Recently Tao and Wang [91] investigated the effect of a time delay on the stability of the mixed equilibrium in the replicator dynamics. They showed that it is asymptotically stable if a time delay is small. For sufficiently large delays it becomes unstable.

We construct two models of discrete-time replicator dynamics with a time delay [2]. In the social-type model, players imitate opponents taking into account average payoffs of games played some units of time ago. In the biological-type model, new players are born from parents who played in the past. We show that in the first type of dynamics, the unique symmetric mixed Nash equilibrium is asymptotically stable for small time delays and becomes unstable for large ones when the population oscillates around its stationary state. In the second type of dynamics, however, the Nash equilibrium is asymptotically stable for any time delay. Our proofs are elementary, they do not rely on the general theory of delay differential and difference equations.

6.1 Social-Type Time Delay

Here we assume that individuals at time t replicate due to average payoffs obtained by their strategies at time $t - \tau$ for some delay $\tau > 0$ (see also a discussion after (32)). As in the standard replicator dynamics, we assume that during the small time interval ϵ, only an ϵ fraction of the population takes part in pairwise competitions, that is plays games. Let $r_i(t)$, $i = A, B$, be the number of individuals playing at the time t the strategy A and B respectively, $r(t) = r_A(t) + r_B(t)$ the total number of players and $x(t) = \frac{r_1(t)}{r(t)}$ a fraction of the population playing A.

We propose the following equations:

$$r_i(t + \epsilon) = (1 - \epsilon)r_i(t) + \epsilon r_i(t)U_i(t - \tau); \quad i = A, B. \tag{27}$$

Then for the total number of players we get

$$r(t + \epsilon) = (1 - \epsilon)r(t) + \epsilon r(t)\bar{U}_o(t - \tau), \tag{28}$$

where $\bar{U}_o(t - \tau) = x(t)U_A(t - \tau) + (1 - x(t))U_B(t - \tau)$.

We divide (27) by (28) and obtain an equation for the frequency of the strategy A,

$$x(t + \epsilon) - x(t) = \epsilon \frac{x(t)[U_A(t - \tau) - \bar{U}_o(t - \tau)]}{1 - \epsilon + \epsilon\bar{U}_o(t - \tau)} \tag{29}$$

and after some rearrangements we get

$$x(t + \epsilon) - x(t) = -\epsilon x(t)(1 - x(t))[x(t - \tau) - x^*]\frac{\delta}{1 - \epsilon + \epsilon\bar{U}_o(t - \tau)}, \tag{30}$$

where $x^* = (d - b)/(d - b + a - c)$ is the unique mixed Nash equilibrium of the game.

Now the corresponding replicator dynamics in the continuous time reads

$$\frac{dx(t)}{dt} = x(t)[U_A(t - \tau) - \bar{U}_o(t - \tau)] \tag{31}$$

and can also be written as

$$\frac{dx(t)}{dt} = x(t)(1 - x(t))[U_A(t - \tau) - U_B(t - \tau)] = -\delta x(t)(1 - x(t))(x(t - \tau) - x^*). \tag{32}$$

The first equation in (32) can be also interpreted as follows. Assume that randomly chosen players imitate randomly chosen opponents. Then the probability that a player who played A would imitate the opponent who played B at time t is exactly $x(t)(1 - x(t))$. The intensity of imitation depends on the delayed information about the difference of corresponding payoffs at time $t - \tau$. We will therefore say that such models have a social-type time delay.

Equations (31-32) are exactly the time-delay replicator dynamics proposed and analyzed by Tao and Wang [91]. They showed that if $\tau < c - a + b - d\pi/2(c - a)(b - d)$, then the mixed Nash equilibrium, x^*, is asymptotically stable. When τ increases beyond the bifurcation value $c - a + b - d\pi/2(c - a)(b - d)$, x^* becomes unstable. We have the following theorem [2].

Theorem 6 x^* *is asymptotically stable in the dynamics (30) if τ is sufficiently small and unstable for large enough τ.*

Proof: We will assume that τ is a multiple of ϵ, $\tau = m\epsilon$ for some natural number m. Observe first that if $x(t - \tau) < x^*$, then $x(t + \epsilon) > x(t)$, and if $x(t - \tau) > x^*$, then $x(t + \epsilon) < x(t)$. Let us assume first that there is t' such that $x(t'), x(t' - \epsilon), x(t' - 2\epsilon), \ldots, x(t' - \tau) < x^*$. Then $x(t)$, $t \geq t'$ increases up to the moment t_1 for which $x(t_1 - \tau) > x^*$. If such t_1 does not exist then $x(t) \rightarrow_{t\to\infty} x^*$ and the theorem is proved. Now we have $x^* < x(t_1 - \tau) < x(t_1 - \tau + \epsilon) < \ldots < x(t_1)$ and $x(t_1 + \epsilon) < x(t_1)$ so t_1 is a turning point. Now $x(t)$ decreases up to the moment t_2 for which $x(t_2 - \tau) < x^*$. Again, if such t_2 does not exist, then the theorem follows. Therefore let us assume that there is an infinite sequence, t_i, of such turning points. Let $\eta_i = |x(t_i) - x^*|$. We will show that $\eta_i \rightarrow_{i\to\infty} 0$.

For $t \in \{t_i, t_i + \epsilon, \ldots, t_{i+1} - 1\}$ we have the following bound for $x(t + \epsilon) - x(t)$:

$$|x(t + \epsilon) - x(t)| < \frac{1}{4}\eta_i \frac{\epsilon\delta}{1 - \epsilon + \epsilon\bar{U}_o(t - \tau)}. \tag{33}$$

This means that

$$\eta_{i+1} < (m + 1)\epsilon K \eta_i, \tag{34}$$

where K is the maximal possible value of $\frac{\delta}{4(1-\epsilon+\epsilon\bar{U}_o(t-\tau))}$. We get that if

$$\tau < \frac{1}{K} - \epsilon, \tag{35}$$

then $\eta_i \rightarrow_{i\to\infty} 0$ so $x(t)$ converges to x^*.

Now if for every t, $|x(t + \epsilon) - x^*| < \max_{k\in\{0,1,\ldots,m\}} |x(t - k\epsilon) - x^*|$, then $x(t)$ converges to x^*. Therefore assume that there is t'' such that $|x(t'' + \epsilon) - x^*| \geq \max_{k\in\{0,1,\ldots,m\}} |x(t'' - k\epsilon) - x^*|$. If τ satisfies (35), then it follows that $x(t + \epsilon), \ldots, x(t + \epsilon + \tau)$ are all on the same side of x^* and the first part of the proof can be applied. We showed that $x(t)$ converges to x^* for any initial conditions different from 0 and 1 hence x^* is globally asymptotically stable.

Now we will show that x^* is unstable for any large enough τ.

Let $\gamma > 0$ be arbitrarily small and consider a following perturbation of the stationary point x^*: $x(t) = x^*, t \leq 0$ and $x(\epsilon) = x^* + \gamma$. It folows from (30) that $x(k\epsilon) = x(\epsilon)$ for $k = 1, \ldots, m + 1$. Let $K' = \min_{x\in[x^*-\gamma, x^*+\gamma]} \frac{x(1-x)\delta}{4(1-\epsilon+\epsilon\bar{U}_o(t-\tau))}$. If $\frac{m}{2}\epsilon K'\gamma > 2\gamma$, that is $\tau > \frac{4}{K'}$, then it follows from (30) that after $m/2$ steps (we assume without loss of generality that m is even) $x((m + 1 + m/2)\epsilon) < x^* - \gamma$. In fact we have $x((2m + 1)\epsilon) < \ldots < x((m + 1)\epsilon)$ and at least $m/2$ of $x's$ in this sequence are smaller than $x^* - \gamma$. Let $\bar{t} > (2m + 1)\epsilon$ be the smallest t such that $x(t) > x^* - \gamma$. Then we have $x(\bar{t} - m\epsilon), \ldots, x(\bar{t} - \epsilon) < x^* - \gamma < x(\bar{t})$ hence after $m/2$ steps, $x(t)$ crosses $x^* + \gamma$ and the situation repeats itself. We showed that if

$$\tau > \frac{4}{K'}, \tag{36}$$

then there exists an infinite sequence, $\tilde{t}_i$, such that $|x(\tilde{t}_i) - x^*| > \gamma$ and therefore x^* is unstable. Moreover, $x(t)$ oscillates around x^*.

6.2 Biological-Type Time Delay

Here we assume that individuals born at time $t - \tau$ are able to take part in contests when they become mature at time t or equivalently they are born τ units of time after their parents played and received payoffs. We propose the following equations:

$$r_i(t + \epsilon) = (1 - \epsilon)r_i(t) + \epsilon r_i(t - \tau)U_i(t - \tau); \quad i = A, B. \tag{37}$$

Then the equation for the total number of players reads

$$r(t + \epsilon) = (1 - \epsilon)r(t) + \epsilon r(t)[\tfrac{x(t)r_A(t-\tau)}{r_A(t)}U_A(t - \tau)$$

$$+ \tfrac{(1-x(t))r_B(t-\tau)}{r_B(t)}U_B(t - \tau)]. \tag{38}$$

We divide (37) by (38) and obtain an equation for the frequency of the first strategy,

$$x(t + \epsilon) - x(t) = \epsilon \frac{x(t - \tau)U_A(t - \tau) - x(t)\bar{U}(t - \tau)}{(1 - \epsilon)\frac{r(t)}{r(t-\tau)} + \epsilon\bar{U}(t - \tau)}, \tag{39}$$

where $\bar{U}(t - \tau) = x(t - \tau)U_A(t - \tau) + (1 - x(t - \tau))U_B(t - \tau)$.

We proved in [2] the following

Theorem 7 x^* *is asymptotically stable in the dynamics (39) for any value of the time delay* τ.

We begin by showing our result in the following simple example. The payoff matrix is given by $U = \begin{pmatrix} 0 & 1 \\ 1 & 0 \end{pmatrix}$ hence $x^* = \frac{1}{2}$ is the mixed Nash equilibrium which is asymptotically stable in the replicator dynamics without the time delay. The equation (39) now reads

$$x(t + \epsilon) - x(t) = \epsilon \frac{x(t - \tau)(1 - x(t - \tau)) - 2x(t)x(t - \tau)(1 - x(t - \tau))}{(1 - \epsilon)\frac{r(t)}{r(t-\tau)} + 2\epsilon x(t - \tau)(1 - x(t - \tau))}. \tag{40}$$

After simple algebra we get

$$x(t + \epsilon) - \frac{1}{2} + \frac{1}{2} - x(t) = \epsilon(1 - 2x(t))\frac{x(t - \tau)(1 - x(t - \tau))}{(1 - \epsilon)\frac{r(t)}{r(t-\tau)} + 2\epsilon x(t - \tau)(1 - x(t - \tau))}, \tag{41}$$

$$x(t + \epsilon) - \frac{1}{2} = (x(t) - \frac{1}{2})\frac{1}{1 + \frac{\epsilon r(t-\tau)}{(1-\epsilon)r(t)}2x(t - \tau)(1 - x(t - \tau))}$$

hence

$$|x(t + \epsilon) - \frac{1}{2}| < |x(t) - \frac{1}{2}|. \tag{42}$$

It follows that x^* is globally asymptotically stable.

Now we present the proof for the general payoff matrix with a unique symmetric mixed Nash equilibrium.

Proof of Theorem 7:

Let $c_t = \frac{x(t)U_A(t)}{U(t)}$. Observe that if $x(t) < x^*$, then $c_t > x(t)$, if $x(t) > x^*$, then $c_t < x(t)$, and if $x(t) = x^*$, then $c_t = x^*$. We can write (39) as

$$x(t + \epsilon) - x(t) =$$

$$\epsilon \frac{x(t - \tau)U_A(t - \tau) - c_{t-\tau}\bar{U}(t - \tau) + c_{t-\tau}\bar{U}(t - \tau) - x(t)\bar{U}(t - \tau)}{(1 - \epsilon)\frac{p(t)}{p(t-\tau)} + \epsilon\bar{U}(t - \tau)} \tag{43}$$

and after some rearrangements we obtain

$$x(t + \epsilon) - c_{t-\tau} = (x(t) - c_{t-\tau})\frac{1}{1 + \frac{\epsilon p(t-\tau)}{(1-\epsilon)p(t)}\bar{U}(t - \tau)}. \tag{44}$$

We get that at time $t + \epsilon$, x is closer to $c_{t-\tau}$ than at time t and it is on the same side of $c_{t-\tau}$. We will show that c is an increasing or a constant function of x. Let us calculate the derivative of c with respect to x.

$$c' = \frac{f(x)}{(xU_A + (1 - x)U_B)^2}, \tag{45}$$

where

$$f(x) = (ac + bd - 2ad)x^2 + 2d(a - b)x + bd. \tag{46}$$

A simple analysis shows that $f > 0$ on $(0, 1)$ or $f = 0$ on $(0, 1)$ (in the case of $a = d = 0$). Hence $c(x)$ is either an increasing or a constant function of x. In the latter case, $\forall_x c(x) = x^*$, as it happens in our example, and the theorem follows.

We will now show that

$$|x(t + \tau + \epsilon) - x^*| < \max\{|x(t) - x^*|, |x(t + \tau) - x^*|\} \tag{47}$$

hence $x(t)$ converges to x^* for any initial conditions different from 0 and 1 so x^* is globally asymptotically stable.

If $x(t) < x^*$ and $x(t + \tau) < x^*$, then $x(t) < c_t \leq x^*$ and also $x(t + \tau) < c_{t+\tau} \leq x^*$.

From (44) we obtain

$$\begin{cases} x(t + \tau) < x(t + \tau + \epsilon) < c_t & if \ x(t + \tau) < c_t \\ x(t) < x(t + \tau + \epsilon) = c_t & if \ x(t + \tau) = c_t \\ x(t) < c_t < x(t + \tau + \epsilon) < x(t + \tau) & if \ x(t + \tau) > c_t \end{cases}$$

hence (47) holds.

If $x(t) > x^*$ and $x(t+\tau) < x^*$, then $x(t+\tau) < x^* < c_t < x(t)$ and either $x(t+\tau) < x(t+\tau+\epsilon) < x^*$ or $x^* < x(t+\tau+\epsilon) < c_t$ which means that (47) holds.

The cases of $x(t) > x^*$, $x(t+\tau) > x^*$ and $x(t) < x^*$, $x(t+\tau) < x^*$ can be treated analogously. We showed that (47) holds.

7 Stochastic Dynamics of Finite Populations

In the next two chapters we will discuss various stochastic dynamics of populations with a fixed number of players interacting in discrete moments of time. We will analyze symmetric two-player games with two or three strategies and multiple Nash equilibria. We will address the problem of equilibrium selection - which strategy will be played in the long run with a high frequency.

Our populations are characterized either by numbers of individuals playing respective strategies in well-mixed populations or by a complete profile - assignment of strategies to players in spatial games. Let Ω be a state space of our system. For non-spatial games with two strategies, $\Omega = \{0, 1, \ldots, n\}$, where n is the number of players or $\Omega = 2^\Lambda$ for spatial games with players located on the finite subset Λ of $\mathbf{Z}, \mathbf{Z}^2$, or any other infinite graph, and interacting with their neighbours. In well-mixed populations, in discrete moments of times, some individuals switch to a strategy with a higher mean payoff. In spatial games, players choose strategies which are best responses, i.e. ones which maximize the sum of the payoffs obtained from individual games. The above rules define deterministic dynamics with some stochastic part corresponding to a random matching of players or a random choice of players who may revise their strategies. We call this mutation-free or noise-free dynamics. It is a Markov chain with a state space Ω and a transition matrix P^0. We are especially interested in absorbing states, i.e. rest points of our mutation-free dynamics. Now, with a small probability, ϵ, players may mutate or make mistakes of not chosing the best reply. The presence of mutatation allows the system to make a transition from any state to any other state with a positive probability in some finite number of steps or to stay indefinitively at any state for an arbitrarily long time. This makes our Markov chains with a transition matrix P^ϵ ergodic ones. They have therefore unique stationary measures. To describe the long-run behavior of stochastic dynamics of finite populations, Foster and Young [22] introduced a concept of stochastic stability. A state of the system is **stochastically stable** if it has a positive probability in the stationary measure of the corresponding Markov chain in the zero-noise limit, that is the zero probability of mistakes or the zero-mutation level. It means that along almost any time trajectory the frequency of visiting this state converges to a positive value given by the stationary measure. Let μ^ϵ be the stationary measure of our Markov chain.

Definition 2 $X \in \Omega$ *is* **stochastically stable** *if* $\lim_{\epsilon \to 0} \mu^\epsilon(X) > 0$.

It is a fundamental problem to find stochastically stable states for any stochastic dynamics of interest. We will use the following tree representation of stationary measures of Markov chains proposed by Freidlin and Wentzell [23, 104], see also [82]. Let (Ω, P^ϵ) be an ergodic Markov chain with a state space Ω, transition probabilities given by the transition matrix $P^\epsilon : \Omega \times \Omega \to [0,1]$, where $P^\epsilon(Y, Y')$ is a conditional probability that the system will be in the state $Y' \in \Omega$ at the time $t + 1$, if it was in the state $Y \in \Omega$ at the time t, and a unique stationary measure, μ^ϵ, also called a stationary state. A stationary state is an eigenvector of P^ϵ corresponding to the eigenvalue 1, i.e. a solution of a system of linear equations,

$$\mu^\epsilon P^\epsilon = \mu^\epsilon, \qquad (48)$$

where μ^ϵ is a row wector $[\mu_1^\epsilon, \ldots, \mu_{|\Omega|}^\epsilon]$. After specific rearrangements one can arrive at an expression for the stationary state which involves only positive terms. This will be very useful in describing the asymptotic behaviour of stationary states.

For $X \in \Omega$, let an X-tree be a directed graph on Ω such that from every $Y \neq X$ there is a unique path to X and there are no outcoming edges out of X. Denote by $T(X)$ the set of all X-trees and let

$$q^\epsilon(X) = \sum_{d \in T(X)} \prod_{(Y,Y') \in d} P^\epsilon(Y, Y'), \qquad (49)$$

where the product is with respect to all edges of d.

We have that

$$\mu^\epsilon(X) = \frac{q^\epsilon(X)}{\sum_{Y \in \Omega} q^\epsilon(Y)} \qquad (50)$$

for all $X \in \Omega$.

We assume that our noise-free dynamics, i.e. in the case of $\epsilon = 0$, has at least one absorbing state and there are no absorbing sets (recurrent classes) consisting of more than one state. It then follows from (50) that only absorbing states can be stochastically stable.

Let us begin with the case of two absorbing states, X and Y. Consider a dynamics in which $P^\epsilon(Z, W)$ for all $Z, W \in \Omega$, is of order ϵ^m, where m is the number of mistakes involved to pass from Z to W. The noise-free limit of μ^ϵ in the form (50) has a 0/0 character. Let m_{XY} be a minimal number of mistakes needed to make a transition from the state X to Y and m_{YX} the minimal number of mistakes to evolve from Y to X. Then $q^\epsilon(X)$ is of the order $\epsilon^{m(YX)}$ and $q^\epsilon(Y)$ is of the order $\epsilon^{m(XY)}$. If for example $m_{YX} < m_{XY}$, then $\lim_{\epsilon \to 0} \mu^\epsilon(X) = 1$ hence X is stochastically stable.

In general, to study the zero-noise limit of the stationary measure, it is enough to consider paths between absorbing states. More precisely, we construct X-trees with absorbing states X^k, $k = 1, \ldots, l$ as vertices; the family of such X-trees is denoted by $\tilde{T}(X)$. Let

$$q_m(X) = max_{d \in \tilde{T}(X)} \prod_{(Y,Y') \in d} \tilde{P}(Y, Y'), \tag{51}$$

where $\tilde{P}(Y, Y') = max \prod_{(W,W')} P(W, W')$, where the product is taken along any path joining Y with Y' and the maximum is taken with respect to all such paths. Now we may observe that if $lim_{\epsilon \to 0} q_m(X^i)/q_m(X^k) = 0$, for every $i = 1, \ldots, l$, $i \neq k$, then X^k is stochastically stable. Therefore we have to compare trees with the biggest products in (51); such trees are called maximal.

The above characterisation of the stationary measure was used to find stochastically stable states in non-spatial [40, 45, 75, 97, 102, 103] and spatial games [17, 18]. We will use it below in our examples.

In many cases, there exists a state X such that $lim_{\epsilon \to 0} \mu^\epsilon(X) = 1$ in the zero-noise limit. Then we say that X was selected in the zero-noise limit of a given stochastic dynamics. However, for any low but fixed mutation level, when the number of players is very large, the frequency of visiting any single state can be arbitrarily low. It is an ensemble of states that can have a probability close to one in the stationary measure. The concept of the ensemble stability is discussed in Chapter 9.

8 Stochastic Dynamics of Well-Mixed Populations

Here we will discuss stochastic dynamics of well-mixed populations of players interacting in discrete moments of time. We will analyze two-player games with two strategies and two pure Nash equilibria. The efficient strategy (also called payoff dominant) when played by the whole population results in its highest possible payoff (fitness). The risk-dominant one is played by individuals averse to risk. The strategy is risk dominant if it has a higher expected payoff against a player playing both strategies with equal probabilities [29]. We will address the problem of equilibrium selection - which strategy will be played in the long run with a high frequency.

We will review two models of adaptive dynamics of a population with a fixed number of individuals. In both of them, the selection part of the dynamics ensures that if the mean payoff of a given strategy is bigger than the mean payoff of the other one, then the number of individuals playing the given strategy increases. In the first model, introduced by Kandori, Mailath, and Rob [40], one assumes (as in the standard replicator dynamics) that individuals receive average payoffs with respect to all possible opponents - they play against the average strategy. In the second model, introduced by Robson and Vega-Redondo [75], at any moment of time, individuals play only one or few games with randomly chosen opponents. In both models, players may mutate with a small probability, hence the population may move against a selection pressure. Kandori, Mailath, and Rob showed that in their model, the risk-dominant strategy is stochastically stable - if the mutation level is small enough we observe it in the long run with the frequency close to one [40]. In the model of Robson and Vega-Redondo, the efficient strategy is stochastically stable [75, 97]. It is one of very few models in which an efficient strategy is stochastically stable in the presence of a risk-dominant one. The population evolves in the long run to a state with the maximal fitness.

The main goal of this chapter is to investigate the effect of the number of players on the long-run behaviour of the Robson-Vega-Redondo model [54]. We will discuss a sequential dynamics and the one where each individual enjoys each period a revision opportunity with the same probability. We will show that for any arbitrarily low but a fixed level of mutations, if the number of players is sufficiently large, then a risk-dominant strategy is played in the long run with a frequency closed to one - a stochastically stable efficient strategy is observed with a very low frequency. It means that when the number of players increases, the population undergoes a transition between an efficient payoff-dominant equilibrium and a risk-dominant one. We will also show that for some range of payoff parameters, stochastic stability itself depends on the number of players. If the number of players is below certain value (which may be arbitrarily large), then a risk-dominant strategy is stochastically stable. Only if the number of players is large enough, an efficient strategy becomes stochastically stable as proved by Robson and Vega-Redondo.

Combining the above results we see that for a low but fixed noise level, the population undergoes twice a transition between its two equilibria as the number of individuals increases [57]. In addition, for a sufficiently large number of individuals, the population undergoes another equilibrium transition when the noise decreases.

Let us formally introduce our models. We will consider a finite population of n individuals who have at their disposal one of two strategies: A and B. At every discrete moment of time, $t = 1, 2, \ldots$ individuals are randomly paired (we assume that n is even) to play a two-player symmetric game with payoffs given by the following matrix:

$$U = \begin{array}{c@{\quad}c@{\quad}c} & A & B \\ A & a & b \\ B & c & d, \end{array}$$

where $a > c, d > b, a > d$, and $a + b < c + d$ so (A, A) is an efficient Nash equilibrium and (B, B) is a risk-dominant one.

At the time t, the state of our population is described by the number of individuals, z_t, playing A. Formally, by the state space we mean the set

$$\Omega = \{z, 0 \leq z \leq n\}.$$

Now we will describe the dynamics of our system. It consists of two components: selection and mutation. The selection mechanism ensures that if the mean payoff of a given strategy, $\pi_i(z_t), i = A, B$, at the time t is bigger than the mean payoff of the other one, then the number of individuals playing the given strategy increases in $t + 1$. In their paper, Kandori, Mailath, and Rob [40] write

$$\pi_A(z_t) = \frac{a(z_t - 1) + b(n - z_t)}{n - 1}, \tag{52}$$

$$\pi_B(z_t) = \frac{cz_t + d(n - z_t - 1)}{n - 1},$$

provided $0 < z_t < n$.

It means that in every time step, players are paired infnitely many times to play the game or equivalently, each player plays with every other player and his payoff is the sum of corresponding payoffs. This model may be therefore considered as an analog of replicator dynamics for populations with a fixed numbers of players.

The selection dynamics is formalized in the following way:

$$z_{t+1} > z_t \; \; if \; \; \pi_A(z_t) > \pi_B(z_t), \tag{53}$$

$$z_{t+1} < z_t \; \; if \; \; \pi_A(z_t) < \pi_B(z_t),$$

$$z_{t+1} = z_t \; \; if \; \; \pi_A(z_t) = \pi_B(z_t),$$

$$z_{t+1} = z_t \; \; if \; \; z_t = 0 \; \; or \; \; z_t = n.$$

Now mutations are added. Players may switch to new strategies with the probability ϵ. It is easy to see that for any two states of the population, there is a positive probability of the transition between them in some finite number of time steps. We have therefore obtained an ergodic Markov chain with $n+1$ states and a unique stationary measure which we denote by μ_n^ϵ. Kandori, Mailath, and Rob proved that the risk-dominant strategy B is stochastically stable [40].

Theorem 8 $\lim_{\epsilon \to 0} \mu_n^\epsilon(0) = 1$

This means that in the long run, in the limit of no mutations, all players play B.

The general set up in the Robson-Vega-Redondo model [75] is the same. However, individuals are paired only once at every time step and play only one game before a selection process takes place. Let p_t denote the random variable which describes the number of cross-pairings, i.e. the number of pairs of matched individuals playing different strategies at the time t. Let us notice that p_t depends on z_t. For a given realization of p_t and z_t, mean payoffs obtained by each strategy are as follows:

$$\tilde{\pi}_A(z_t, p_t) = \frac{a(z_t - p_t) + bp_t}{z_t}, \tag{54}$$

$$\tilde{\pi}_B(z_t, p_t) = \frac{cp_t + d(n - z_t - p_t)}{n - z_t},$$

provided $0 < z_t < n$. Robson and Vega-Redondo showed that the payoff-dominant strategy is stochastically stable [75].

Theorem 9 $\lim_{\epsilon \to 0} \mu_n^\epsilon(n) = 1$

We will outline their proof.

First of all, one can show that there exists k such that if n is large enough and $z_t \geq k$, then there is a positive probability (a certain realization of p_t) that after a finite number of steps of the mutation-free selection dynamics, all players will play A. Likewise, if $z_t < k$ (for any $k \geq 1$), then if the number of players is large enough, then after a finite number of steps of the mutation-free selection dynamics all players will play B. In other words, $z = 0$ and $z = n$ are the only absorbing states of the mutation-free dynamics. Moreover, if n is large enough, then if $z_t \geq n - k$, then the mean payoff obtained by A is always (for any realization of p_t) bigger than the mean payoff obtained by B (in the worst case all B-players play with A-players). Therefore the size of the basin of attraction of the state $z = 0$ is at most $n - k - 1$ and that of $z = n$ is at least $n - k$. Observe that mutation-free dynamics is not deterministic (p_t describes the random matching) and therefore basins of attraction may overlap. It follows that the system needs at least $k+1$ mutations to evolve from $z = n$ to $z = 0$ and at most k mutations to evolve from $z = 0$ to $z = n$. Now using the tree representation of stationary states, Robson and Vega-Redondo finish the proof and show that the efficient strategy is stochastically stable.

However, as outlined above, their proof requires the number of players to be sufficiently large. We will now show that a risk-dominant strategy is stochastically stable if the number of players is below certain value which can be arbitrarily large.

Theorem 10 *If* $n < \frac{2a-c-b}{a-c}$, *then the risk-dominant strategy* B *is stochastically stable in the case of random matching of players.*

Proof: If the population consists of only one B-player and $n - 1$ A-players and if $c > [a(n-2)+b]/(n-1)$, that is $n < (2a-c-b)/(a-c)$, then $\tilde{\pi}_B > \tilde{\pi}_A$. It means that one needs only one mutation to evolve from $z = n$ to $z = 0$. It is easy to see that two mutations are necessary to evolve from $z = 0$ to $z = n$.

To see stochastically stable states, we need to take the limit of no mutations. We will now examine the long-run behavior of the Robson-Vega-Redondo model for a fixed level of mutations in the limit of the infinite number of players.

Now we will analyze the extreme case of the selection rule (53) - a sequential dynamics where in one time unit only one player can change his strategy. Although our dynamics is discrete in time, it captures the essential features of continuous-time models in which every player has an exponentially distributed waiting time to a moment of a revision opportunity. Probability that two or more players revise their strategies at the same time is therefore equal to zero - this is an example of a birth and death process.

The number of A-players in the population may increase by one in $t + 1$, if a B-player is chosen in t which happens with the probability $(n - z_t)/n$. Analogously, the number of B-players in the population may increase by one in $t + 1$, if an A-player is chosen in t which happens with the probability $(z_t)/n$.

The player who has a revision opportunity chooses in $t+1$ with the probability $1 - \epsilon$ the strategy with a higher average payoff in t and the other one with the probability ϵ. Let

$$r(k) = P(\tilde{\pi}_A(z_t, p_t) > \tilde{\pi}_B(z_t, p_t)) \ and \ l(k) = P(\tilde{\pi}_A(z_t, p_t) < \tilde{\pi}_B(z_t, p_t)).$$

The sequential dynamics is described by the following transition probabilities:

if $z_t = 0$, then $z_{t+1} = 1$ with the probability ϵ and $z_{t+1} = 0$ with the probability $1 - \epsilon$,

if $z_t = n$, then $z_{t+1} = n - 1$ with the probability ϵ and $z_{t+1} = n$ with the probability $1 - \epsilon$,

if $z_t \neq 0, n$, then $z_{t+1} = z_t + 1$ with the probability

$$r(k)\frac{n - z_t}{n}(1 - \epsilon) + (1 - r(k))\frac{n - z_t}{n}\epsilon$$

and $z_{t+1} = z_t - 1$ with the probability

$$l(k)\frac{z_t}{n}(1 - \epsilon) + (1 - l(k))\frac{z_t}{n}\epsilon.$$

In the dynamics intermediate between the parallel (where all individuals can revise their strategies at the same time) and the sequential one, each individual has a revision opportunity with the same probability $\tau < 1$ during the time interval of the lenght 1. For a fixed ϵ and an arbitrarily large but fixed n, we consider the limit of the continuous time, $\tau \to 0$, and show that the limiting behaviour is already obtained for a sufficiently small τ, namely $\tau < \epsilon/n^3$.

For an interesting discussion on the importance of the order of taking different limits ($\tau \to 0, n \to \infty$, and $\epsilon \to 0$) in evolutionary models (especially in the Aspiration and Imitation model) see Samuelson [78].

In the intermediate dynamics, instead of $(n - z_t)/n$ and z_t/n probabilities we have more involved combinatorial factors. In order to get rid of these inconvenient factors, we will enlarge the state space of the population. The state space Ω' is the set of all configurations of players, that is all possible assignments of strategies to individual players. Therefore, a state $z_t = k$ in Ω

consists of $\binom{n}{k}$ states in Ω'. Observe that the sequential dynamics on Ω' is not anymore a birth and death process. However, we are able to treat both dynamics in the same framework.

We showed in [54] that for any arbitrarily low but fixed level of mutation, if the number of players is large enough, then in the long run only a small fraction of the population plays the payoff-dominant strategy. Smaller the mutation level is, fewer players use the payoff-dominant strategy.

The following two theorems were proven in [54].

Theorem 11 *In the sequential dynamics, for any $\delta > 0$ and $\beta > 0$ there exist $\epsilon(\delta, \beta)$ and $n(\epsilon)$ such that for any $n > n(\epsilon)$*

$$\mu_n^\epsilon(z \leq \beta n) > 1 - \delta.$$

Theorem 12 *In the intermediate dynamics, for any $\delta > 0$ and $\beta > 0$ there exist $\epsilon(\delta, \beta)$ and $n(\epsilon)$ such that for any $n > n(\epsilon)$ and $\tau < \frac{\epsilon}{n^3}$*

$$\mu_n^\epsilon(z \leq \beta n) > 1 - \delta.$$

We can combine Theorems 9, 10, and 12 and obtain [57].

Theorem 13 *In the intermediate dynamics, for any $\delta > 0$ and $\beta > 0$ there exists $\epsilon(\delta, \beta)$ such that, for all $\epsilon < \epsilon(\delta, \beta)$, there exist $n_1 < n_2 < n_3(\epsilon) < n_4(\epsilon)$ such that*

if $n < n_1 = \frac{2a-c-b}{a-c}$, then $\mu_n^\epsilon(z = 0) > 1 - \delta$,
if $n_2 < n < n_3(\epsilon)$, then $\mu_n^\epsilon(z = n) > 1 - \delta$,
if $n > n_4(\epsilon)$ and $\tau < \epsilon/n^3$, then $\mu_n^\epsilon(z \leq \beta n) > 1 - \delta$.

Small τ means that our dynamics is close to the sequential one. We have that $n_3(\epsilon), n_4(\epsilon), n_3(\epsilon) - n_2$, and $n_4(\epsilon) - n_3(\epsilon) \to \infty$ when $\epsilon \to 0$.

It follows from Theorem 13 that the population of players undergoes several **equilibrium transitions**. First of all, for a fixed noise level, when the number of players increases, the population switches from a B-equilibrium, where most of the individuals play the strategy B, to an A-equilibrium and then back to B one. We know that if $n > n_2$, then $z = n$ is stochastically stable. Therefore, for any fixed number of players, $n > n_4(\epsilon)$, when the noise level decreases, the population undergoes a transition from a B-equilibrium to A one. We see that in order to study the long-run behaviour of stochastic population dynamics, we should estimate the relevant parameters to be sure what limiting procedures are appropriate in specific examples.

Let us note that the above theorems concern an ensemble of states, not an individual one. In the limit of the infinite number of players, that is the infinite number of states, every single state has zero probability in the stationary state. It is an ensemble of states that might be stable [51, 53]. The concept of ensemble stability will be discussed in Chapter 9.

9 Spatial Games with Local Interactions

9.1 Nash Configurations and Stochastic Dynamics

In spatial games, players are located on vertices of certain graphs and they interact only with their neighbours; see for example [5, 7, 17, 18, 30–34, 41, 44, 62–64, 85, 86, 88, 103] and a recent review paper [89] and references therein.

Let Λ be a finite subset of the simple lattice $\mathbf{Z}^d$. Every site of Λ is occupied by one player who has at his disposal one of m different pure strategies. Let S be the set of strategies, then $\Omega_\Lambda = S^\Lambda$ is the space of all possible configurations of players, that is all possible assignments of strategies to individual players. For every $i \in \Lambda$, X_i is a strategy of the i–th player in the configuration $X \in \Omega_\Lambda$ and X_{-i} denotes strategies of all remaining players; X therefore can be represented as the pair (X_i, X_{-i}). Every player interacts only with his nearest neighbours and his payoff is the sum of the payoffs resulting from individual plays. We assume that he has to use the same strategy for all neighbours. Let N_i denote the neighbourhood of the i–th player. For the nearest-neighbour interaction we have $N_i = \{j; |j - i| = 1\}$, where $|i - j|$ is the distance between i and j. For $X \in \Omega_\Lambda$ we denote by $\nu_i(X)$ the payoff of the i–th player in the configuration X:

$$\nu_i(X) = \sum_{j \in N_i} U(X_i, X_j), \qquad (55)$$

where U is a $m \times m$ matrix of payoffs of a two-player symmetric game with m pure strategies.

Definition 3 $X \in \Omega_\Lambda$ *is a* **Nash configuration** *if for every* $i \in \Lambda$ *and* $Y_i \in S$,

$$\nu_i(X_i, X_{-i}) \geq \nu_i(Y_i, X_{-i})$$

Here we will discuss only coordination games, where there are m pure symmetric Nash equilibria and therefore m homogeneous Nash configurations, where all players play the same strategy.

In the Stag-hunt game in Example 1, we have two homogeneous Nash configurations, X^{St} and X^H, where all individuals play St or H respectively.

We describe now the sequential deterministic dynamics of the **best-response rule**. Namely, at each discrete moment of time $t = 1, 2, \ldots$, a randomly chosen player may update his strategy. He simply adopts the strategy, X_i^{t+1}, which gives him the maximal total payoff $\nu_i(X_i^{t+1}, X_{-i}^t)$ for given X_{-i}^t, a configuration of strategies of remaining players at the time t.

Now we allow players to make mistakes, that is they may not choose best responses. We will discuss two types of such stochastic dynamics. In the first one, the so-called **perturbed best response**, a player follows the best-response rule with probability $1 - \epsilon$ (in case of more than one best-response strategy he chooses randomly one of them) and with probability ϵ he makes a mistake and chooses randomly one of the remaining strategies. The probability of mistakes (or the noise level) is state-independent here.

In the so called **log-linear dynamics** [5, 6], the probability of chosing by the i-th player the strategy X_i^{t+1} at the time $t + 1$ decreases with the loss of the payoff and is given by the following conditional probability:

$$p_i^\epsilon(X_i^{t+1}|X_{-i}^t) = \frac{e^{\frac{1}{\epsilon}\nu_i(X_i^{t+1},X_{-i}^t)}}{\sum_{Y_i \in S} e^{\frac{1}{\epsilon}\nu_i(Y_i,X_{-i}^t)}}, \tag{56}$$

Let us observe that if $\epsilon \to 0$, p_i^ϵ converges pointwise to the best-response rule. Both stochastic dynamics are examples of ergodic Markov chains with $|S^\Lambda|$ states. Therefore they have unique stationary states denoted by μ_Λ^ϵ.

Stationary states of the log-linear dynamics can be explicitly constructed for the so-called potential games. A game is called a **potential game** if its payoff matrix can be changed to a symmetric one by adding payoffs to its columns [49]. As we know, such a payoff transformation does not change strategic character of the game, in particular it does not change the set of its Nash equilibria. More formally, we have the following definition.

Definition 4 *A two-player symmetric game with a payoff matrix U is a* **potential game** *if there exists a symmetric matrix V, called a potential of the game, such that for any three strategies $A, B, C \in S$*

$$U(A,C) - U(B,C) = V(A,C) - V(B,C). \tag{57}$$

It is easy to see that every game with two strategies has a potential V with $V(A, A) = a - c$, $V(B, B) = d - b$, and $V(A, B) = V(B, A) = 0$. It follows that an equilibrium is risk-dominant if and only if it has a bigger potential.

For players on a lattice, for any $X \in \Omega_\Lambda$,

$$V(X) = \sum_{(i,j) \subset \Lambda} V(X_i, X_j) \tag{58}$$

is then the potential of the configuration X.

For the sequential log-linear dynamics of potential games, one can explicitely construct stationary measures [103].

We begin by the following general definition concerning a Markov chain with a state space Ω and a transition matrix P.

Definition 5 *A measure μ on Ω satisfies a* **detailed balance condition** *if*

$$\mu(X)P_{XY} = \mu(Y)P_{YX}$$

for every $X, Y \in \Omega$

Lemma

If μ satisfies the detailed balance condition then it is a stationary measure

Proof:

$$\sum_{X \in \Omega} \mu(X) P_{XY} = \sum_{X \in \Omega} \mu(Y) P_{YX} = \mu(Y)$$

The following theorem is due Peyton Young [103]. We will present here his proof.

Theorem 14 *The stationary measure of the sequential log-linear dynamics in a game with the potential V is given by*

$$\mu_\Lambda^\epsilon(X) = \frac{e^{\frac{1}{\epsilon} V(X)}}{\sum_{Z \in \Omega_\Lambda} e^{\frac{1}{\epsilon} V(Z)}}. \tag{59}$$

Proof:

We will show that μ_Λ^ϵ in (59) satisfies the detailed balance condition. Let us notice that in the sequential dynamics, $P_{XY} = 0$ unless $X = Y$ or Y differs fom X at one lattice site only, say $i \in \Lambda$.

Let

$$\lambda = \frac{1}{|\Lambda|} \frac{1}{\sum_{Z \in \Omega_\Lambda} e^{\frac{1}{\epsilon} V(Z)}} \frac{1}{\sum_{Z_i \in S} e^{\frac{1}{\epsilon} \sum_{j \in N_i} U(Z_i, X_j)}}$$

Then

$$\mu_\Lambda^\epsilon(X) P_{XY} = \lambda e^{\frac{1}{\epsilon}(\sum_{(h,k) \subset \Lambda} V(X_h, X_k) + \sum_{j \in N_i} U(Y_i, X_j))}$$

$$= \lambda e^{\frac{1}{\epsilon}(\sum_{(h,k) \subset \Lambda} V(X_h, X_k) + \sum_{j \in N_i} (U(X_i, X_j) - V(X_i, X_j) + V(Y_i, X_j)))}$$

$$= \lambda e^{\frac{1}{\epsilon}(\sum_{(h,k) \subset \Lambda} V(Y_h, Y_k) + \sum_{j \in N_i} U(X_i, X_j))} = \mu_\Lambda^\epsilon(Y) P_{YX}.$$

We may now explicitly perform the limit $\epsilon \to 0$ in (59). In the Stag-hunt game, X^H has a bigger potential than X^{St} so $\lim_{\epsilon \to 0} \mu_\Lambda^\epsilon(X^H) = 1$ hence X^H is stochastically stable (we also say that H is stochastically stable).

The concept of a Nash configuration in spatial games is very similar to the concept of a ground-state configuration in lattice-gas models of interacting particles. We will discuss similarities and differences between these two systems of interacting entities in the next section.

9.2 Ground States and Nash Configurations

We will present here one of the basic models of interacting particles. In classical lattice-gas models, particles occupy lattice sites and interact only with their neighbours. The fundamental concept is that of a ground-state configuration. It can be formulated conveniently in the limit of an infinite lattice (the infinite number of particles). Let us assume that every site of the $\mathbf{Z}^d$

lattice can be occupied by one of m different particles. An infinite-lattice configuration is an assignment of particles to lattice sites, i.e. an element of $\Omega = \{1,\ldots,m\}^{\mathbf{Z}^d}$. If $X \in \Omega$ and $i \in \mathbf{Z}^d$, then we denote by X_i a restriction of X to i. We will assume here that only nearest-neighbour particles interact. The energy of their interaction is given by a symmetric $m \times m$ matrix V. An element $V(A,B)$ is the interaction energy of two nearest-neighbour particles of the type A and B. The total energy of a system in the configuration X in a finite region $\Lambda \subset \mathbf{Z}^d$ can be then written as

$$H_\Lambda(X) = \sum_{(i,j)\subset\Lambda} V(X_i, X_j). \tag{60}$$

Y is a **local excitation** of X, $Y \sim X$, $Y, X \in \Omega$, if there exists a finite $\Lambda \subset \mathbf{Z}^d$ such that $X = Y$ outside Λ.

For $Y \sim X$, the **relative energy** is defined by

$$H(Y,X) = \sum_{(i,j)\in\mathbf{Z}^d} (V(Y_i, Y_j) - V(X_i, X_j)), \tag{61}$$

where the summation is with respect to pairs of nearest neighbours on $\mathbf{Z}^d$. Observe that this is the finite sum; the energy difference between Y and X is equal to zero outside some finite Λ.

Definition 6 $X \in \Omega$ *is a* **ground-state configuration** *of V if*

$$H(Y,X) \geq 0 \quad for \quad any \quad Y \sim X.$$

That is, we cannot lower the energy of a ground-state configuration by changing it locally.

The energy density $e(X)$ of a configuration X is

$$e(X) = \liminf_{\Lambda \to \mathbf{Z}^2} \frac{H_\Lambda(X)}{|\Lambda|}, \tag{62}$$

where $|\Lambda|$ is the number of lattice sites in Λ.

It can be shown that any ground-state configuration has the minimal energy density [84]. It means that local conditions present in the definition of a ground-state configuration force the global minimization of the energy density.

We see that the concept of a ground-state configuration is very similar to that of a Nash configuration. We have to identify particles with agents, types of particles with strategies and instead of minimizing interaction energies we should maximize payoffs. There are however profound differences. First of all, ground-state configurations can be defined only for symmetric matrices; an interaction energy is assigned to a pair of particles, payoffs are assigned to individual players and may be different for each of them. Ground-state configurations are stable with respect to all local changes, Nash configurations are stable only with respect to one-player changes. It means that for the same symmetric matrix U, there may exist a configuration which is a Nash configuration but not a ground-state configuration for the inter-action matrix $-U$. The simplest example is given by the following matrix:

Example 5

$$
U = \quad
\begin{array}{c|cc}
 & A & B \\
\hline
A & 2 & 0 \\
B & 0 & 1 \\
\end{array}
$$

(A, A) and (B, B) are Nash configurations for a system consisting of two players but only (A, A) is a ground-state configuration for $V = -U$. We may therefore consider the concept of a ground-state configuration as a refinement of a Nash equilibrium.

For any classical lattice-gas model there exists at least one ground-state configuration. This can be seen in the following way. We start with an arbitrary configuration. If it cannot be changed locally to decrease its energy it is already a ground-state configuration. Otherwise we may change it locally and decrease the energy of the system. If our system is finite, then after a finite number of steps we arrive at a ground-state configuration; at every step we decrease the energy of the system and for every finite system its possible energies form a finite set. For an infinite system, we have to proceed ad infinitum converging to a ground-state configuration (this follows from the compactness of Ω in the product of discrete topologies). Game models are different. It may happen that a game with a nonsymmetric payoff matrix may not posess a Nash configuration. The classical example is that of the Rock-Scissors-Paper game. One may show that this game dos not have any Nash configurations on $\mathbf{Z}$ and $\mathbf{Z}^2$ but many Nash configurations on the triangular lattice.

In short, ground-state configurations minimize the total energy of a parti-cle system, Nash configurations do not necessarily maximize the total payoff of a population.

Ground-state configuration is an equilibrium concept for systems of in-teracting particles at zero temperature. For positive temperatures, we must take into account fluctuations caused by thermal motions of particles. Equi-librium behaviour of the system results then from the competition between its energy V and entropy S (which measures the number of configurations corresponding to a macroscopic state), i.e. the minimization of its free energy $F = V - TS$, where T is the temperature of the system - a measure of thermal motions. At the zero temperature, $T = 0$, the minimization of the free energy reduces to the minimization of the energy. This zero-temperature limit looks very similar to the zero-noise limit present in the definition of the stochastic stability. Equilibrium behaviour of a system of interacting particles can be described by specifying probabilities of occurence for all particle configurations. More formally, it is described by a Gibbs state (see [26] and references therein).

We construct it in the following way. Let Λ be a finite subset of $\mathbf{Z}^d$ and ρ_Λ^T the following probability mass function on $\Omega_\Lambda = (1, \ldots, m)^\Lambda$:

$$\rho_\Lambda^T(X) = (1/Z_\Lambda^T) \exp(-H_\Lambda(X)/T), \tag{63}$$

for every $X \in \Omega_\Lambda$, where

$$Z_\Lambda^T = \sum_{X \in \Omega_\Lambda} \exp(-H_\Lambda(X)/T) \tag{64}$$

is a normalizing factor.

We define a **Gibbs state** ρ^T as a limit of ρ_Λ^T as $\Lambda \to \mathbf{Z}^d$. One can prove that a limit of a translation-invariant Gibbs state for a given interaction as $T \to 0$ is a measure supported by ground-state configurations. One of the fundamental problems of statistical mechanics is a characterization of low-temperature Gibbs states for given interactions between particles.

Let us observe that the finite-volume Gibbs state in (63) is equal to stationary state μ_Λ^ϵ in (59) if we identify T with ϵ and $V \to -V$.

9.3 Ensemble Stability

The concept of stochastic stability involves individual configurations of players. In the zero-noise limit, a stationary state is usually concentrated on one or at most few configurations. However, for a low but fixed noise and for a sufficiently large number of players, the probability of any individual configuration of players is practically zero. The stationary measure, however, may be highly concentrated on an ensemble consisting of one Nash configuration and its small perturbations, i.e. configurations where most players use the same strategy. Such configurations have relatively high probability in the stationary measure. We call such configurations ensemble stable. Let μ_Λ^ϵ be a stationary measure.

Definition 7 $X \in \Omega_\Lambda$ is γ-**ensemble stable** if $\mu_\Lambda^\epsilon(Y \in \Omega_\Lambda; Y_i \neq X_i) < \gamma$ for any $i \in \Lambda$ if $\Lambda \supset \Lambda(\gamma)$ for some $\Lambda(\gamma)$.

Definition 8 $X \in \Omega_\Lambda$ is **low-noise ensemble stable** if for every $\gamma > 0$ there exists $\epsilon(\gamma)$ such that if $\epsilon < \epsilon(\gamma)$, then X is γ-ensemble stable.

If X is γ-ensemble stable with γ close to zero, then the ensemble consisting of X and configurations which are different from X at most few sites has the probability close to one in the stationary measure. It does not follow, however, that X is necessarily low-noise ensemble or stochastically stable as it happens in examples presented below [51].

Example 6

Players are located on a finite subset Λ of $\mathbf{Z}^2$ (with periodic boundary conditions) and interact with their four nearest neighbours. They have at their disposal three pure strategies: $A, B,$ and C. The payoffs are given by the following symmetric matrix:

$$
U = \quad
\begin{array}{c|ccc}
 & A & B & C \\
A & 1.5 & 0 & 1 \\
B & 0 & 2 & 1 \\
C & 1 & 1 & 2 \\
\end{array}
$$

Our game has three Nash equilibria: $(A, A), (B, B),$ and (C, C), and the corresponding spatial game has three homogeneous Nash configurations: $X^A, X^B,$ and X^C, where all individuals are assigned the same strategy. Let us notice that X^B and X^C have the maximal payoff in every finite volume and therefore they are ground-state configurations for $-U$ and X^A is not.

The unique stationary measure of the log-linear dynamics (56) is given by (59) with $U = V$ which is a finite-volume Gibbs state (63) with V replaced by $-U$ and T by ϵ. We have

$$
\sum_{(i,j)\subset\Lambda} U(X_i^k, X_j^k) - \sum_{(i,j)\in\Lambda} U(Y_i, Y_j) > 0,
$$

for every $Y \neq X^B$ and X^C, $k = B, C$, and

$$
\sum_{(i,j)\subset\Lambda} U(X_i^B, X_j^B) = \sum_{(i,j)\subset\Lambda} U(X_i^C, X_j^C).
$$

It follows that $\lim_{\epsilon\to0} \mu_\Lambda^\epsilon(X^k) = 1/2$, for $k = B, C$ so X^B and X^C are stochastically stable. Let us investigate the long-run behaviour of our system for large Λ, that is for a large number of players.

Observe that

$$
\lim_{\Lambda\to\mathbf{Z}^2} \mu_\Lambda^\epsilon(X) = 0
$$

for every $X \in \Omega = S^{\mathbf{Z}^2}$.

Therefore, for a large Λ we may only observe, with reasonably positive frequencies, ensembles of configurations and not particular configurations. We will be interested in ensembles which consist of a Nash configuration and its small perturbations, that is configurations, where most players use the same strategy. We perform first the limit $\Lambda \to \mathbf{Z}^2$ and obtain an infinite-volume Gibbs state in the temperature $T = \epsilon$,

$$\mu^\epsilon = \lim_{\Lambda \to \mathbf{Z}^2} \mu^\epsilon_\Lambda. \tag{65}$$

In order to investigate the stationary state of our example, we will apply a technique developed by Bricmont and Slawny [8,9]. They studied low-temperature stability of the so-called dominant ground-state configurations. It follows from their results that

$$\mu^\epsilon(X_i = C) > 1 - \delta(\epsilon) \tag{66}$$

for any $i \in \mathbf{Z}^2$ and $\delta(\epsilon) \to 0$ as $\epsilon \to 0$ [51].

The following theorem is a simple consequence of (66).

Theorem 15 X^C *is low-noise ensemble stable.*

We see that for any low but fixed ϵ, if the number of players is large enough, then in the long run, almost all players use C strategy. On the other hand, if for any fixed number of players, ϵ is lowered substantially, then B and C appear with frequencies close to $1/2$.

Let us sketch briefly the reason of such a behavior. While it is true that both X^B and X^C have the same potential which is the half of the payoff of the whole system (it plays the role of the total energy of a system of interacting particles), the X^C Nash configuration has more lowest-cost excitations. Namely, one player can change its strategy and switch to either A or B and the potential will decrease by 4 units. Players in the X^B Nash configuration have only one possibility, that is to switch to C; switching to A decreases the potential by 8. Now, the probability of the occurrence of any configuration in the Gibbs state (which is the stationary state of our stochastic dynamics) depends on the potential in an exponential way. One can prove that the probability of the ensemble consisting of the X^C Nash configuration and configurations which are different from it at few sites only is much bigger than the probability of the analogous X^B-ensemble. It follows from the fact that the X^C-ensemble has many more configurations than the X^B-ensemble. On the other hand, configurations which are outside X^B and X^C-ensembles appear with exponentially small probabilities. It means that for large enough systems (and small but not extremely small ϵ) we observe in the stationary state the X^C Nash configuration with perhaps few different strategies. The above argument was made into a rigorous proof for an infinite system of the closely related lattice-gas model (the Blume-Capel model) of interacting particles by Bricmont and Slawny in [8].

In the above example, X^B and X^C have the same total payoff but X^C has more lowest-cost excitations and therefore it is low-noise ensemble stable. We will now discuss the situation, where X^C has a smaller total payoff but nevertheless in the long run C is played with a frequency close to 1 if the noise

level is low but not extremely low. We will consider a family of games with
the following payoff matrix:

Example 7

$$
U = \begin{array}{c c c c}
 & A & B & C \\
A & 1.5 & 0 & 1 \\
B & 0 & 2+\alpha & 1 \\
C & 1 & 1 & 2,
\end{array}
$$

where $\alpha > 0$ so B is both payoff and pairwise risk-dominant.

We are interested in the long-run behavior of our system for small positive
α and low ϵ. One may modify the proof of Theorem 15 and obtain the following
theorem [51].

Theorem 16 *For every $\gamma > 0$, there exist $\alpha(\gamma)$ and $\epsilon(\gamma)$ such that for every
$0 < \alpha < \alpha(\gamma)$, there exists $\epsilon(\alpha)$ such that for $\epsilon(\alpha) < \epsilon < \epsilon(\gamma)$, X^C is
γ-ensemble stable, and for $0 < \epsilon < \epsilon(\alpha)$, X^B is γ-ensemble stable.*

Observe that for $\alpha = 0$, both X^B and X^C are stochastically stable (they
appear with the frequency $1/2$ in the limit of zero noise) but X^C is low-noise
ensemble stable. For small $\alpha > 0$, X^B is both stochastically (it appears with
the frequency 1 in the limit of zero noise) and low-noise ensemble stable.
However, for an intermediate noise $\epsilon(\alpha) < \epsilon < \epsilon(\gamma)$, if the number of players
is large enough, then in the long run, almost all players use the strategy C
(X^C is ensemble stable). If we lower ϵ below $\epsilon(\alpha)$, then almost all players start
to use the strategy B. $\epsilon = \epsilon(\alpha)$ is the line of the first-order phase transition.
In the thermodynamic limit, there exist two Gibbs states (equilibrium states)
on this line. We may say that at $\epsilon = \epsilon(\alpha)$, the population of players undergoes
a sharp **equilibrium transition** from C to B-behaviour.

9.4 Stochastic Stability in Non-Potential Games

Let us now consider non-potential games with three strategies and three symmetric Nash equilibria: (A, A), (B, B), and (C, C). Stationary measures of such
games cannot be explicitly constructed. To find stochastically stable states
we will use here the tree representation of stationary measures described in
Chapter 7. We will discuss some interesting examples.

Example 8

Players are located on a finite subset of the one-dimensional lattice $\mathbf{Z}$ and
interact with their nearest neighbours only. Denote by n the number of players.
For simplicity we will assume periodic boundary conditions, that is we will
identify the $n + 1$-th player with the first one. In other words, the players are
located on the circle.

The payoffs are given by the following matrix:

$$U = \begin{array}{c c c c} & \text{A} & \text{B} & \text{C} \\ \text{A} & 1+\alpha & 0 & 1.5 \\ \text{B} & 0 & 2 & 0 \\ \text{C} & 0 & 0 & 3 \end{array}$$

with $0 < \alpha \le 0.5$.

As before, we have three homogeneous Nash configurations: X^A, X^B, and X^C. The log-linear and perturbed best-response dynamics for this game were discussed in [52].

Let us note that X^A, X^B, and X^C are the only absorbing states of the noise-free dynamics. We begin with a stochastic dynamics with a state-independent noise. Let us consider first the case of $\alpha < 0.5$.

Theorem 17 *If $0 < \alpha < 0.5$, then X^C is stochastically stable in the perturbed best-response dynamics.*

Proof: It is easy to see that $q_m(X^C)$ is of the order ϵ^2, $q_m(X^B)$ is of the order $\epsilon^{\frac{n}{2}+1}$, and $q_m(X^A)$ is of the order ϵ^{n+2}.

Let us now consider the log-linear rule.

Theorem 18 *If $n < 2 + 1/(0.5 - \alpha)$, then X^B is stochastically stable and if $n > 2+1/(0.5-\alpha)$, then X^C is stochastically stable in the log-linear dynamics.*

Proof: The following are maximal A-tree, B-tree, and C-tree:

$$B \to C \to A, \quad C \to A \to B, \quad A \to B \to C,$$

where the probability of $A \to B$ is equal to

$$\frac{1}{1+1+e^{\frac{1}{\epsilon}(2+2\alpha)}} \Big(\frac{1}{1+e^{-\frac{2}{\epsilon}}+e^{\frac{1}{\epsilon}(-1+\alpha)}}\Big)^{n-2} \frac{1}{1+e^{-\frac{4}{\epsilon}}+e^{-\frac{4}{\epsilon}}}, \qquad (67)$$

the probability of $B \to C$ is equal to

$$\frac{1}{1+1+e^{\frac{4}{\epsilon}}} \Big(\frac{1}{1+e^{-\frac{1}{\epsilon}}+e^{-\frac{1.5}{\epsilon}}}\Big)^{n-2} \frac{1}{1+e^{-\frac{6}{\epsilon}}+e^{-\frac{3}{\epsilon}}}, \qquad (68)$$

and the probability of $C \to A$ is equal to

$$\frac{1}{1+e^{-\frac{3}{\epsilon}}+e^{\frac{3}{\epsilon}}} \Big(\frac{1}{1+e^{-\frac{1}{\epsilon}(2.5+\alpha)}+e^{\frac{1}{\epsilon}(0.5-\alpha)}}\Big)^{n-2} \frac{1}{1+e^{-\frac{2}{\epsilon}(1+\alpha)}+e^{-\frac{2}{\epsilon}(1+\alpha)}}, \qquad (69)$$

Let us observe that

$$P_{B\to C\to A} = O(e^{-\frac{1}{\epsilon}(7+(0.5-\alpha)(n-2))}), \qquad (70)$$

$$P_{C\to A\to B} = O(e^{-\frac{1}{\epsilon}(5+2\alpha+(0.5-\alpha)(n-2))}), \qquad (71)$$

$$P_{A \to B \to C} = O(e^{-\frac{1}{\epsilon}(6+2\alpha)}), \tag{72}$$

where $\lim_{x \to 0} O(x)/x = 1$.

Now if $n < 2 + 1/(0.5 - \alpha)$, then

$$\lim_{\epsilon \to 0} \frac{q_m(X^C)}{q_m(X^B)} = \lim_{\epsilon \to 0} \frac{P_{A \to B \to C}}{P_{C \to A \to B}} = 0 \tag{73}$$

which finishes the proof.

It follows that for a small enough n, X^B is stochastically stable and for a large enough n, X^C is stochastically stable. We see that adding two players to the population may change the stochastic stability of Nash configurations. Let us also notice that the strategy C is globally risk dominant. Nevertheless, it is not stochastically stable in the log-linear dynamics for a sufficiently small number of players.

Let us now discuss the case of $\alpha = 0.5$ [52].

Theorem 19 *If $\alpha = 0.5$, then X^B is stochastically stable for any n in the log-linear dynamics.*

Proof:

$$\lim_{\epsilon \to 0} \frac{q_m(X^C)}{q_m(X^B)} = \lim_{\epsilon \to 0} \frac{e^{-\frac{4}{\epsilon}} e^{-\frac{3}{\epsilon}}}{(1/2)^{n-2} e^{-\frac{3}{\epsilon}} e^{-\frac{3}{\epsilon}}} = 0.$$

X^B is stochastically stable which means that for any fixed number of players, if the noise is sufficiently small, then in the long run we observe B players with an arbitrarily high frequency. However, we conjecture that for any low but fixed noise, if the number of players is big enough, the stationary measure is concentrated on the X^C-ensemble. We expect that X^C is ensemble stable because its lowest-cost excitations occur with a probability of the order $e^{-\frac{3}{\epsilon}}$ and those from X^B with a probability of the order $e^{-\frac{4}{\epsilon}}$. We observe this phenomenon in Monte-Carlo simulations.

Example 9

Players are located on a finite subset Λ of $\mathbf{Z}$ (with periodic boundary conditions) and interact with their two nearest neighbours. They have at their disposal three pure strategies: A, B, and C. The payoffs are given by the following matrix [51]:

$$U = \begin{array}{c|ccc} & A & B & C \\ \hline A & 3 & 0 & 2 \\ B & 2 & 2 & 0 \\ C & 0 & 0 & 3 \end{array}$$

Our game has three Nash equilibria: $(A, A), (B, B)$, and (C, C). Let us note that in pairwise comparisons, B risk dominates A, C dominates B and A dominates C. The corresponding spatial game has three homogeneous Nash configurations: X^A, X^B, and X^C. They are the only absorbing states of the noise-free best-response dynamics.

Theorem 20 X^B *is stochastically stable.*

Proof: The following are maximal A-tree, B-tree, and C-tree:

$$B \to C \to A, \quad C \to A \to B, \quad A \to B \to C.$$

Let us observe that

$$P_{B \to C \to A} = O(e^{-\frac{6}{\epsilon}}), \tag{74}$$

$$P_{C \to A \to B} = O(e^{-\frac{4}{\epsilon}}), \tag{75}$$

$$P_{A \to B \to C} = O(e^{-\frac{6}{\epsilon}}), \tag{76}$$

The theorem follows from the tree characterization of stationary measures.

X^B is stochastically stable because it is much more probable (for low ϵ) to escape from X^A and X^C than from X^B. The relative payoffs of Nash configurations are not relevant here (in fact X^B has the smallest payoff). Let us recall Example 7 of a potential game, where an ensemble-stable configuration has more lowest-cost excitations. It is easier to escape from an ensemble-stable configuration than from other Nash configurations.

Stochatic stability concerns single configurations in the zero-noise limit; ensemble stability concerns families of configurations in the limit of the infinite number of players. It is very important to investigate and compare these two concepts of stability in nonpotential games.

Non-potential spatial games cannot be directly presented as systems of interacting particles. They constitute a large family of interacting objects not thoroughly studied so far by methods statistical physics. Some partial results concerning stochastic stability of Nash equilibria in non-potential spatial games were obtained in [5, 17, 18, 52, 53].

One may wish to say that A risk dominates the other two strategies if it risk dominates them in pairwise comparisons. In Example 9, B dominates A, C dominates B, and finally A dominates C. But even if we do not have such a cyclic relation of dominance, a strategy which is pairwise risk-dominant may not be stochastically stable as in the case of Example 8. A more relevant notion seems to be that of a global risk dominance [45]. We say that A is globally risk dominant if it is a best response to a mixed strategy which assigns probability $1/2$ to A. It was shown in [17, 18] that a global risk-dominant strategy is stochastically stable in some spatial games with local interactions.

A different criterion for stochastic stability was developed by Blume [5]. He showed (using techniques of statistical mechanics) that in a game with m strategies A_i and m symmetric Nash equilibria (A_k, A_k), $k = 1, \dots, m$, A_1 is stochastically stable if

$$\min_{k>1}(U(A_1, A_1) - U(A_k, A_k)) > \max_{k>1}(U(A_k, A_k) - U(A_1, A_k)). \tag{77}$$

We may observe that if A_1 satisfies the above condition, then it is pairwise risk dominant.

9.5 Dominated Strategies

We say that a pure strategy is **strictly dominated** by another (pure or mixed) strategy if it gives a player a lower payoff than the other one regardless of strategies chosen by his opponents.

Definition 9 $k \in S$ is strictly dominated by $y \in \Delta$ if $U_i(k, w_{-i}) < U_i(y, w_{-i})$ for every $w \in \Delta^I$.

Let us see that a strategy can be strictly dominated by a mixed strategy without being strictly dominated by any pure strategy in its support.

Example 10

$$U = \begin{array}{c c c c} & A & B & C \\ A & 5 & 1 & 3 \\ B & 2 & 2 & 2 \\ C & 1 & 5 & 3 \end{array}$$

B is strictly dominated by a mixed strategy assigning the probability $1/2$ both to A and C but is strictly dominated neither by A nor by C.

It is easy to see that strictly dominated pure strategies cannot be present in the support of any Nash equilibrium.

In the replicator dynamics (16), all strictly dominated pure strategies are wiped out in the long run if all strategies are initially present [1, 77].

Theorem 21 *If a pure strategy k is strictly dominated, then $\xi_k(t, x^0) \to_{t \to \infty} 0$ for any $x^0 \in interior(\Delta)$.*

Strictly dominated strategies should not be used by rational players and consequently we might think that their presence should not have any impact on the long-run behaviour of the population. We will show that in the best-reply dynamics, if we allow players to make mistakes, this may not be necessarily true. Let us consider the following game with a strictly dominated strategy and two symmetric Nash equilibria [51].

Example 11

$$U = \begin{array}{c c c c} & A & B & C \\ A & 0 & 0.1 & 1 \\ B & 0.1 & 2 + \alpha & 1.1 \\ C & 1.1 & 1.1 & 2, \end{array}$$

where $\alpha > 0$.
We see that strategy A is strictly dominated by both B and C, hence X^A is not a Nash configuration. X^B and X^C are both Nash configurations but only X^B is a ground-state configuration for $-U$. In the absence of A, B is both payoff and risk-dominant and therefore is stochastically stable and low-noise ensemble stable. Adding the strategy A does not change dominance relations;

B is still payoff and pairwise risk dominant. However, Example 11 fulfills all the assumptions of Theorem 16 and we get that X^C is γ-ensemble stable at intermediate noise levels. The mere presence of a strictly dominated strategy A changes the long-run behaviour of the population.

Similar results were discussed by Myatt and Wallace [58]. In their games, at every discrete moment of time, one of the players leaves the population and is replaced by another one who plays the best response. The new player calculates his best response with respect to his own payoff matrix which is the matrix of a common average payoff modified by a realization of some random variable with the zero mean. The noise does not appear in the game as a result of players' mistakes but is the effect of their idiosyncratic preferences. The authors then show that the presence of a strictly dominated strategy may change the stochastic stability of Nash equilibria. However, the reason for such a behavior is different in their and in our models. In our model, it is relatively easy to get out of X^C and this makes X^C ensemble stable. Mayatt and Wallace introduce a strictly dominated strategy in such a way that it is relatively easy to make a transition to it from a risk and payoff-dominant equilibrium and then with a high probability the population moves to a second Nash configuration which results in its stochastic stability.

This is exactly a mechanism present in Examples 8 and 9.

10 Review of Other Results

We discussed the long-run behaviour of populations of interacting individuals playing games. We have considered deterministic replicator dynamics and stochastic dynamics of finite populations.

In spatial games, individuals are located on vertices of certain graphs and they interact only with their neighbours.

In this paper, we considered only simple graphs - finite subsets of the regular $\mathbf{Z}$ or $\mathbf{Z}^2$ lattice. Recently there appeared many interesting results of evolutionary dynamics on random graphs, Barabasi-Albert free-scale graphs, and small-world networks [3, 79–81, 87, 89, 90, 98, 99]. Especially the Prisoner's Dilemma was studied on such graphs and it was shown that their heterogeneity favors the cooperation in the population [79–81, 89].

In well-mixed populations, individuals are randomly matched to play a game. The deterministic selection part of the dynamics ensures that if the mean payoff of a given strategy is bigger than the mean payoff of the other one, then the number of individuals playing the given strategy increases. In discrete moments of time, individuals produce offspring proportional to their payoffs. The total number of individuals is then scaled back to the previous value so the population size is constant. Individuals may mutate so the population may move against a selection pressure. This is an example of a stochastic frequency-dependent Wright-Fisher process [12, 19–21, 105].

There are also other stochastic dynamics of finite populations. The most important one is the Moran process [12, 19, 50]. In this dynamics, at any time step a single individual is chosen for reproduction with the probability proportional to his payoff, and then his offspring replaces the random chosen individual. It was showed recently that in the limits of the infinite population, the Moran process results in the replicator dynamics [94, 95].

The stochastic dynamics of finite populations has been extensively studied recently [31,38,43,65,68–70,92]. The notion of an evolutionarily stable strategy for finite populations was introduced [15,61,65,92,96,101]. One of the important quantity to calculate is the fixation probability of a given strategy. It is defined as the probability that a strategy introduced into a population by a single player will take over the whole population. Recently, Nowak et. al. [65] have formulated the following weak selection 1/3 law. In two-player games with two strategies, selection favors the strategy A replacing B if the fraction of A-players in the population for which the average payoff for the strategy A is equal to the average payoff of the strategy B if is smaller than 1/3, i.e. the mixed Nash equilibrium for this game is smaller than 1/3. The 1/3 law was proven to hold both in the Moran [65,92] and the Fisher-Wright process [38].

In this review we discussed only two-player games. Multi-player games were studied recently in [10, 11, 28, 39, 42, 53, 73, 74].

We have not discussed at all population genetics in the context of game theory. We refer to [12, 36] for results and references.

Acknowledgments: These lecture notes are based on the short course given in the Stefan Banach International Mathematical Center in the framework of the CIME Summer School "From a Microscopic to a Macroscopic Description of Complex Systems" which was held in Będlewo, Poland, 4-9 September 2006.

I would like to thank the Banach Center for a financial support to participate in this School and the Ministry of Science and Higher Education for a financial support under the grant N201 023 31/2069.

References

1. Akin, E.: Domination or equilibrium. Mathematical Biosciences, **50**, 239–250 (1980)
2. Alboszta, J., Miękisz, J.: Stability of evolutionarily stable strategies in discrete replicator dynamics with time delay. J. Theor. Biol., **231**, 175–179 (2004)
3. Antal, T., Redner, S., Sood, V.: Evolutionary dynamics on degree-heterogeneous graphs. Phys. Rev. Lett., **96**, 188104 (2006)
4. Axelrod, R.: The Evolution of Cooperation. Basic Books, New York (1984)
5. Blume, L.E.: The statistical mechanics of strategic interaction. Games Econ. Behav., **5**, 387–424 (1993)
6. Blume, L.E.: How noise matters. Games Econ. Behav., **44**, 251–271 (2003)
7. Brauchli, K., Killingback, T., Doebeli, M.: Evolution of cooperation in spatially structured populations. J. Theor. Biol., **200**, 405–417 (1999)

8. Bricmont, J., Slawny, J.: First order phase transitions and perturbation theory. In: Statistical Mechanics and Field Theory: Mathematical Aspects. Lecture Notes in Physics, **257**, Springer-Verlag (1986)
9. Bricmont, J., Slawny, J.: Phase transitions in systems with a finite number of dominant ground states. J. Stat. Phys., **54**, 89–161 (1989)
10. Broom, M., Cannings, C., Vickers, G.T.: Multi-player matrix games. Bull. Math. Biology, **59**, 931–952 (1997)
11. Bukowski, M., Miękisz, J.: Evolutionary and asymptotic stability in symmetric multi-player games. Int. J. Game Theory, **33**, 41–54 (2004)
12. Bürger, R.: The Mathematical Theory of Selection, Recombination, and Mutation. Wiley (2000)
13. Condensed Matter and Quantitative Biology archives at xxx.lanl.gov
14. Cressman, R.: Evolutionary Dynamics and Extensive Form Games. MIT Press, Cambridge, USA (2003)
15. Dostálková, I., Kindlmann, P.: Evolutionarily stable strategies for stochastic processes. Theor. Popul. Biol., **65** 205–210 (2004)
16. Econophysics bulletin at www.unifr.ch/econophysics
17. Ellison, G.: Learning, local interaction, and coordination. Econometrica, **61**, 1047–1071 (1993)
18. Ellison, G. Basins of attraction, long-run stochastic stability, and the speed of step-by-step evolution. Review of Economic Studies, **67**, 17–45 (2000)
19. Ewens, W.J.: Mathematical Population Genetics. Springer (2004)
20. Fisher, R. A.: On the dominance ratio. Proc. Roy. Soc. Edinb., **42**, 321–341 (1922)
21. Fisher, R. A.: The Genetical Theory of Natural Selection. Clarendon Press, Oxford (1930)
22. Foster, D., Young, P.H.: Stochastic evolutionary game dynamics. Theor. Popul. Biol., **38**, 219–232 (1990)
23. Freidlin, M.I, Wentzell, A.D.: Random Perturbations of Dynamical Systems. Springer Verlag, New York (1984)
24. Fudenberg, D., Tirole, J.: Game Theory. MIT Press, Cambridge, USA (1991)
25. Fudenberg, D., Levine, D.K.: The Theory of Learning in Games. MIT Press, Cambridge, USA (1998)
26. Georgii, H. O.: Gibbs Measures and Phase Transitions. Walter de Gruyter, Berlin (1988)
27. Gintis, H.: Game Theory Evolving. Princeton University Press, Princeton (2000)
28. Gulyas, L., Płatkowski, T.: On evolutionary 3-person Prisoner's Dilemma games on 2d lattice. Lecture Notes in Computer Science, **3305**, 831–840 (2004)
29. Harsányi, J., Selten, R.: A General Theory of Equilibrium Selection in Games. MIT Press, Cambridge, USA (1988)
30. Hauert, C.: Effects of space in 2x2 games. Int. J. Bifurcat. Chaos, **12**, 1531–1548 (2002).
31. Hauert, C., Doebeli, M.: Spatial structure often inhibits the evolution of cooperation in the snowdrift game. Nature, **428**, 643–646 (2004)
32. Hauert, C.: Models of cooperation based on the Prisoner's Dilemma and the Snowdrift game. Ecology Letters, **8**, 748–766 (2005)
33. Hauert, C.: Spatial effects in social dilemmas. J. Theor. Biol., **240**, 627–636 (2006)

34. Herz, A.V.M.: Collective phenomena in spatially extended evolutionary games. J. Theor. Biol., **169**, 65–87 (1994)
35. Hofbauer, J., Schuster, P., Sigmund, K.: A note on evolutionarily stable strategies and game dynamics. J. Theor. Biol., **81**, 609–612 (1979)
36. Hofbauer, J., Sigmund, K.: Evolutionary Games and Population Dynamics. Cambridge University Press, Cambridge (1998)
37. Hofbauer, J., Sigmund, K.: Evolutionary game dynamics. Bulletin AMS, **40**, 479–519 (2003)
38. Imhof, L.A., Nowak, M.A.: Evolutionary game dynamics in a Wright-Fisher process. J. Math. Biol., **52**, 667–681 (2006)
39. Kamiński, D., Miękisz, J., Zaborowski, M.: Stochastic stability in three-player games. Bull. Math. Biol., **67**, 1195–1205 (2005)
40. Kandori, M., Mailath, G.J, Rob, R.: Learning, mutation, and long-run equilibria in games. Econometrica, **61**, 29–56 (1993)
41. Killingback, T., Doebeli, M.: Spatial evolutionary game theory: hawks and doves revisited. Proc. R. Soc. London, **263**, 1135–1144 (1996)
42. Kim, Y.: Equilibrium selection in n-person coordination games. Games Econ. Behav., **15**, 203–227 (1996)
43. Lieberman, E., Hauert, C., Nowak, M.A.: Evolutionary dynamics on graphs. Nature, **433**, 312–316 (2005)
44. Lindgren, K., Nordahl, M.G.: Evolutionary dynamics of spatial games. Physica D, **75**, 292–309 (1994)
45. Maruta, T.: On the relationship between risk-dominance and stochastic stability. Games Econ. Behav., **19**, 221–234 (1977)
46. Maynard Smith, J., Price, G.R.: The logic of animal conflicts. Nature **246**, 15–18 (1973)
47. Maynard Smith, J.: The theory of games and the evolution of animal conflicts. J. Theor. Biol., **47**, 209–221 (1974)
48. Maynard Smith, J.: Evolution and the Theory of Games. Cambridge University Press, Cambridge (1982)
49. Monderer, D., Shapley, L.S.: Potential games. Games Econ. Behav., **14**, 124–143 (1996)
50. Moran, P.A.P.: The Statistical Processes of Evolutionary Theory. Clarendon, Oxford (1962)
51. Miękisz, J.: Statistical mechanics of spatial evolutionary games. Phys. A: Math. Gen., **37**, 9891–9906 (2004)
52. Miękisz, J.: Stochastic stability in spatial games. J. Stat. Phys., **117**, 99–110 (2004)
53. Miękisz, J.: Stochastic stability in spatial three-player games. Physica A, **343**, 175–184 (2004)
54. Miękisz, J.: Equilibrium selection in evolutionary games with random matching of players. J. Theor. Biol., **232**, 47–53 (2005)
55. Miękisz, J., Płatkowski, T.: Population dynamics with a stable efficient equilibrium. J. Theor. Biol., **237**, 363–368 (2005)
56. Miękisz, J.: Long-run behavior of games with many players. Markov Processes and Relat. Fields, **11**, 371–388 (2005)
57. Miękisz, J.: Equilibrium transitions in finite populations of players. In: Game Theory and Mathematical Economics. Banach Center Publications, **71**, 237–242 (2006)

58. Myatt, D.P., Wallace, C.: A multinomial probit model of stochastic evolution. J. Econ. Theory, **113**, 286–301 (2003)
59. Nash, J.: Equilibrium points in n-person games. Proc. Nat. Ac. Sc., **36**, 48–49 (1950)
60. Nash, J.: Non-cooperative games. Ann. Math., **54**, 287–295 (1951)
61. Neill, D.B.: Evolutionary stability for large populations. J. Theor. Biol., **227**, 397–401 (2004)
62. Nowak, M.A., May, R.M.: Evolutionary games and spatial chaos. Nature, **359**, 826–829 (1992)
63. Nowak, M.A., May, R.M.: The spatial dilemmas of evolution. Int. J. Bifurcat. Chaos, **3**, 35–78 (1993)
64. Nowak, M.A., Bonhoeffer, S., May, R.M.: More spatial games. Int. J. Bifurcat. Chaos, **4**, 33–56 (1994)
65. Nowak, M.A., Sasaki, A., Taylor, C., Fudenberg, D.: Emergence of cooperation and evolutionary stability in finite populations. Nature, **428**, 646–650 (2004)
66. Nowak, M.A., Sigmund, K.: Evolutionary dynamics of biological games. Science, **303**, 793–799 (2004)
67. Nowak, M.A.: Evolutionary Dynamics. Harvard University Press, Cambridge, USA (2006)
68. Ohtsuki, H., Nowak, M.A.: Evolutionary games on cycles. Proc. R. Soc. B., **273**, 2249–2256 (2006)
69. Ohtsuki, H., Nowak, M.A.: The replicator equation on graphs. J. Theor. Biol., **243**, 86–97 (2006)
70. Ohtsuki, H., Hauert, C., Lieberman, E., Nowak, M.A.: A simple rule for the evolution of cooperation on graphs and social networks. Nature, **441**, 502–505 (2006)
71. Ordershook, P.: Game Theory and Political Theory. Cambridge University Press (1986)
72. Poundstone, W.: Prisoner's Dilemma. Doubleday (1992)
73. Płatkowski, T.: Evolution of populations playing mixed multi-player games. Mathematical and Computer Modelling, **39**, 981–989 (2004)
74. Płatkowski, T., Stachowska, J.: ESS in multiplayer mixed games. Applied Mathematics and Computations, **167**, 592–606 (2005)
75. Robson, A., Vega-Redondo, F.: Efficient equilibrium selection in evolutionary games with random matching. J. Econ. Theory **70**, 65–92 (1996)
76. Russell, B.: Common Sense and Nuclear Warfare. Allen and Unwin Ltd., London (1959)
77. Samuelson, L., Zhang, J.: Evolutionary stability in asymmetric games. J. Econ. Theory, **57**, 363–391 (1992)
78. Samuelson, L.: Evolutionary Games and Equilibrium Selection. MIT Press, Cambridge, USA (1997)
79. Santos, F.C., Pacheco, J.M.: Scale-free networks provide a unifying framework for the emergence of cooperation. Phys. Rev. Lett., **95**, 098104 (2005)
80. Santos, F.C., Pacheco, J.M., Lenaerts, T.: Evolutionary dynamics of social dilemmas in structured heterogeneous populations. Proc. Natl. Acad. Sci. USA, **103**, 3490–3494 (2006)
81. Santos, F.C., Rodrigues, J.F., Pacheco, J.M.: Graph topology plays a determinant role in the evolution of cooperation. Proc. R. Soc. B, **273**, 51–55 (2006)
82. Shubert, B.: A flow-graph formula for the stationary distribution of a Markov chain. Trans. Systems Man. Cybernet., **5**, 565–566 (1975)

83. Sigmund, K.: Games of Life - Explorations in Ecology, Evolution and Behaviour. Oxford University Press (1993)
84. Sinai, Y.G.: Theory of Phase Transitions: Rigorous Results. Pergamon, Oxford (1982)
85. Szabo, G., Toke, C.: Evolutionary prisoner's dilemma game on a square lattice. Phys. Rev. E, **58**, 69–73 (1998)
86. Szabo, G., Antal, T., Szabo, P., Droz, M.: Spatial evolutionary prisoner's dilemma game with three strategies and external constraints. Phys. Rev. E, **62**, 1095–1103 (2000)
87. Szabo, G., Szolnoki, A., Izsak, R.: Rock-scissors-paper game on regular small-world networks. J. Phys. A: Math. Gen., **37**, 2599–2609 (2004)
88. Szabo, G., Vukov, J., Szolnoki, A.: Phase diagrams for an evolutionary prisoner's dilemma game on two-dimensional lattices. Phys. Rev. E, **72**, 047107 (2005)
89. Szabo, G., Gabor, F.: Evolutionary games on graphs. arXiv: cond-mat/0607344 (2006)
90. Szolnoki, A., Szabo, G.: Phase transitions for rock-scissors-paper game on different networks. Phys. Rev. E, **70**, 037102 (2004)
91. Tao, Y., Wang, Z.: Effect of time delay and evolutionarily stable strategy. J. Theor. Biol., **187**, 111–116 (1997)
92. Taylor, C., Fudenberg, D., Sasaki, A., Nowak, M.A.: Evolutionary game dynamics in finite populations. Bull. Math. Biol., **66**, 1621–1644 (2004)
93. Taylor, P.D., Jonker, L.B.: Evolutionarily stable strategy and game dynamics. Mathematical Biosciences, **40**, 145–156 (1978)
94. Traulsen, A., Claussen, J.C., Hauert, C.: Coevolutionary dynamics: from finite to infinite populations. Phys. Rev. Lett., **95**, 238701 (2005)
95. Traulsen, A., Claussen, J.C., Hauert, C.: Coevolutionary dynamics in large but finite populations. Phys. Rev. E, **74**, 011901 (2006)
96. Traulsen, A., Pacheco, J., Imhof, L.A.: Stochasticity and evolutionary stability. Phys. Rev. E, **74**, 021905 (2006)
97. Vega-Redondo, F.: Evolution, Games, and Economic Behaviour. Oxford University Press, Oxford (1996)
98. Vukov, J., Szabo, G.: Evolutionary prisoner's dilemma game on hierarchical lattices. Phys. Rev. E, **71**, 71036133 (2005)
99. Vukov, J., Szabo, Szolnoki, A.: Cooperation in the noisy case: Prisoner's dilemma game on two types of regular random graphs. Phys. Rev. E, **74**, 067103 (2006)
100. Weibull, J.: Evolutionary Game Theory. MIT Press, Cambridge, USA (1995)
101. Wild, G., Taylor, P.D.: Fitness and evolutionary stability in game theoretic models of finite populations. Proc. R. Soc. B, **271**, 2345–2349 (2004)
102. Young, P.H.: The evolution of conventions. Econometrica, **61**, 57–84 (1993)
103. Young, H.P.: Individual Strategy and Social Structure: An Evolutionary Theory of Institutions. Princeton University Press, Princeton (1998)
104. Wentzell, A.D., Freidlin, M.I.: On small random perturbations of dynamical systems. Russian Math. Surveys, **25**, 1–55 (1970)
105. Wright, S.: Evolution in Mendelian populations. Genetics, **6**, 97–159 (1931)
106. Zeeman, E.: Dynamics of the evolution of animal conflicts. J. Theor. Biol., **89**, 249–270 (1981)

List of Participants

1. Giuseppina Albano
   pialbano@unisa.it
   Italy

2. Jacek Banasiak (**lecturer**)
   banasiak@ukzn.ac.za
   Durban, South Africa

3. Luisa Arlotti
   Arlotti@uniud.it
   Italy

4. Zhanna Artemichenko
   artemich@imath.kiev.ua
   Ukraine

5. Marine Aubert
   aubert@ipno.in2p3.fr
   France

6. Svetlana Azarina
   azarinas@mail.ru
   Russia

7. Marek Bodnar
   mbodnar@mimuw.edu.pl
   Poland

8. Vincenzo Capasso (**editor, lecturer**)
   Vincenzo.Capasso@mat.unimi.it
   Milano, Italia

9. Mark Chaplain (**lecturer**)
   chaplain@maths.dundee.ac.uk
   Dundee, Scotland

10. Remigius Chidozie Nnadozie
    204504921@ukzn.ac.za
    South Africa

11. Urszula Forys
    urszula@mimuw.edu.pl
    Poland

12. Victor Gerasimenko
    gerasym@imath.kiev.ua
    Ukraine

13. Ana Maria Glavan
    aglavan@math.uc3m.es
    Spain

14. Natalie Kalev-Kronik
    natalie@imbm.org
    Israel

15. Bogdan Kazmierczak
    bkazmier@ippt.gov.pl
    Poland

16. Alena Khmelynitskaya
    avkhmelynitskaya@rambler.ru
    Russia

17. Yuri Kogan
    yuri@imbm.org
    Israel

18. Remigiusz Kowalczyk
    kowalc@mimuw.edu.pl
    Poland

19. Mirosław Lachowicz (**editor, lecturer**)
    `lachowic@mimuw.edu.pl`
    Warszawa, Polska

20. Jean-Yves Le Boudec
    `jean-yves.leboudec@epfl.ch`
    Switzerland

21. Meri Lisi
    `lisi7@unisi.it`
    Italy

22. Gabriela Lorelai Litcanu
    `glitcanu@yahoo.com`
    Poland

23. Thomas Lorenz
    `thomas.lorenz@`
    `iwr.uni-heidelberg.de`
    Germany

24. Anna Marciniak-Czochra
    `anna.marciniak@`
    `iwr.uni-heidelberg.de`
    Germany

25. Barbara Martinucci
    `bmartinucci@unisa.it`
    Italy

26. Jacek Miękisz (**lecturer**)
    `miekisz@mimuw.edu.pl`
    Warszawa, Polska

27. Daniela Morale
    `daniela.morale@mat.unimi.it`
    Italy

28. Cristian Morales-Rodrigo
    `cristianmatematicas@yahoo.com`
    Poland

29. Jochen Mundinger
    `jochen.mundinger@epfl.ch`
    Switzerland

30. Zbigniew Peradzynski
    `zperadz@mimuw.edu.pl`
    Poland

31. Monika Piotrowska
    `monika@mimuw.edu.pl`
    Germany

32. Oleksiy Pliukhin
    `pliukhin@imath.kiev.ua`
    Ukraine

33. Mariya Ptashnyk
    `Mariya.Ptashnyk@`
    `IWR.Uni-Heidelberg.DE`
    Germany

34. Tatiana Ryabukha
    `vyrtum@imath.kiev.ua`
    Ukraine

35. Viacheslav Shtyk
    `shtyk@imath.kiev.ua`
    Ukraine

36. Zuzanna Szymańska
    `mysz@icm.edu.pl`
    Poland

37. Radosław Wieczorek
    `r.wieczorek@impan.gov.pl`
    Poland

38. Eugene Yashagin
    `eyashag@mail.ru`
    Russia

Index

LIST OF C.I.M.E. SEMINARS

Published by C.I.M.E

Published by Ed. Cremonese, Firenze

1966   39. Calculus of variations
        40. Economia matematica
        41. Classi caratteristiche e questioni connesse
        42. Some aspects of diffusion theory

1967   43. Modern questions of celestial mechanics
        44. Numerical analysis of partial differential equations
        45. Geometry of homogeneous bounded domains

1968   46. Controllability and observability
        47. Pseudo-differential operators
        48. Aspects of mathematical logic

1969   49. Potential theory
        50. Non-linear continuum theories in mechanics and physics and their applications
        51. Questions of algebraic varieties

1970   52. Relativistic fluid dynamics
        53. Theory of group representations and Fourier analysis
        54. Functional equations and inequalities
        55. Problems in non-linear analysis

1971   56. Stereodynamics
        57. Constructive aspects of functional analysis (2 vol.)
        58. Categories and commutative algebra

1972   59. Non-linear mechanics
        60. Finite geometric structures and their applications
        61. Geometric measure theory and minimal surfaces

1973   62. Complex analysis
        63. New variational techniques in mathematical physics
        64. Spectral analysis

1974   65. Stability problems
        66. Singularities of analytic spaces
        67. Eigenvalues of non linear problems

1975   68. Theoretical computer sciences
        69. Model theory and applications
        70. Differential operators and manifolds

Published by Ed. Liguori, Napoli

1976   71. Statistical Mechanics
        72. Hyperbolicity
        73. Differential topology

1977   74. Materials with memory
        75. Pseudodifferential operators with applications
        76. Algebraic surfaces

Published by Ed. Liguori, Napoli & Birkhäuser

1978   77. Stochastic differential equations
        78. Dynamical systems

1979   79. Recursion theory and computational complexity
        80. Mathematics of biology

1980   81. Wave propagation
        82. Harmonic analysis and group representations
        83. Matroid theory and its applications

Published by Springer-Verlag

Lecture Notes in Mathematics

For information about earlier volumes
please contact your bookseller or Springer
LNM Online archive: springerlink.com

Vol. 1803: G. Dolzmann, Variational Methods for Crystalline Microstructure – Analysis and Computation (2003)
Vol. 1804: I. Cherednik, Ya. Markov, R. Howe, G. Lusztig, Iwahori-Hecke Algebras and their Representation Theory. Martina Franca, Italy 1999. Editors: V. Baldoni, D. Barbasch (2003)
Vol. 1805: F. Cao, Geometric Curve Evolution and Image Processing (2003)
Vol. 1806: H. Broer, I. Hoveijn. G. Lunther, G. Vegter, Bifurcations in Hamiltonian Systems. Computing Singularities by Gröbner Bases (2003)
Vol. 1807: V. D. Milman, G. Schechtman (Eds.), Geometric Aspects of Functional Analysis. Israel Seminar 2000-2002 (2003)
Vol. 1808: W. Schindler, Measures with Symmetry Properties (2003)
Vol. 1809: O. Steinbach, Stability Estimates for Hybrid Coupled Domain Decomposition Methods (2003)
Vol. 1810: J. Wengenroth, Derived Functors in Functional Analysis (2003)
Vol. 1811: J. Stevens, Deformations of Singularities (2003)
Vol. 1812: L. Ambrosio, K. Deckelnick, G. Dziuk, M. Mimura, V. A. Solonnikov, H. M. Soner, Mathematical Aspects of Evolving Interfaces. Madeira, Funchal, Portugal 2000. Editors: P. Colli, J. F. Rodrigues (2003)
Vol. 1813: L. Ambrosio, L. A. Caffarelli, Y. Brenier, G. Buttazzo, C. Villani, Optimal Transportation and its Applications. Martina Franca, Italy 2001. Editors: L. A. Caffarelli, S. Salsa (2003)
Vol. 1814: P. Bank, F. Baudoin, H. Föllmer, L.C.G. Rogers, M. Soner, N. Touzi, Paris-Princeton Lectures on Mathematical Finance 2002 (2003)
Vol. 1815: A. M. Vershik (Ed.), Asymptotic Combinatorics with Applications to Mathematical Physics. St. Petersburg, Russia 2001 (2003)
Vol. 1816: S. Albeverio, W. Schachermayer, M. Talagrand, Lectures on Probability Theory and Statistics. Ecole d'Eté de Probabilités de Saint-Flour XXX-2000. Editor: P. Bernard (2003)
Vol. 1817: E. Koelink, W. Van Assche (Eds.), Orthogonal Polynomials and Special Functions. Leuven 2002 (2003)
Vol. 1818: M. Bildhauer, Convex Variational Problems with Linear, nearly Linear and/or Anisotropic Growth Conditions (2003)
Vol. 1819: D. Masser, Yu. V. Nesterenko, H. P. Schlickewei, W. M. Schmidt, M. Waldschmidt, Diophantine Approximation. Cetraro, Italy 2000. Editors: F. Amoroso, U. Zannier (2003)
Vol. 1820: F. Hiai, H. Kosaki, Means of Hilbert Space Operators (2003)
Vol. 1821: S. Teufel, Adiabatic Perturbation Theory in Quantum Dynamics (2003)
Vol. 1822: S.-N. Chow, R. Conti, R. Johnson, J. Mallet-Paret, R. Nussbaum, Dynamical Systems. Cetraro, Italy 2000. Editors: J. W. Macki, P. Zecca (2003)
Vol. 1823: A. M. Anile, W. Allegretto, C. Ringhofer, Mathematical Problems in Semiconductor Physics. Cetraro, Italy 1998. Editor: A. M. Anile (2003)
Vol. 1824: J. A. Navarro González, J. B. Sancho de Salas, $\mathscr{C}^\infty$ – Differentiable Spaces (2003)
Vol. 1825: J. H. Bramble, A. Cohen, W. Dahmen, Multiscale Problems and Methods in Numerical Simulations. Martina Franca, Italy 2001. Editor: C. Canuto (2003)
Vol. 1826: K. Dohmen, Improved Bonferroni Inequalities via Abstract Tubes. Inequalities and Identities of Inclusion-Exclusion Type. VIII, 113 p, 2003.

Vol. 1827: K. M. Pilgrim, Combinations of Complex Dynamical Systems. IX, 118 p, 2003.
Vol. 1828: D. J. Green, Gröbner Bases and the Computation of Group Cohomology. XII, 138 p, 2003.
Vol. 1829: E. Altman, B. Gaujal, A. Hordijk, Discrete-Event Control of Stochastic Networks: Multimodularity and Regularity. XIV, 313 p, 2003.
Vol. 1830: M. I. Gil', Operator Functions and Localization of Spectra. XIV, 256 p, 2003.
Vol. 1831: A. Connes, J. Cuntz, E. Guentner, N. Higson, J. E. Kaminker, Noncommutative Geometry, Martina Franca, Italy 2002. Editors: S. Doplicher, L. Longo (2004)
Vol. 1832: J. Azéma, M. Émery, M. Ledoux, M. Yor (Eds.), Séminaire de Probabilités XXXVII (2003)
Vol. 1833: D.-Q. Jiang, M. Qian, M.-P. Qian, Mathematical Theory of Nonequilibrium Steady States. On the Frontier of Probability and Dynamical Systems. IX, 280 p, 2004.
Vol. 1834: Yo. Yomdin, G. Comte, Tame Geometry with Application in Smooth Analysis. VIII, 186 p, 2004.
Vol. 1835: O.T. Izhboldin, B. Kahn, N.A. Karpenko, A. Vishik, Geometric Methods in the Algebraic Theory of Quadratic Forms. Summer School, Lens, 2000. Editor: J.-P. Tignol (2004)
Vol. 1836: C. Năstăsescu, F. Van Oystaeyen, Methods of Graded Rings. XIII, 304 p, 2004.
Vol. 1837: S. Tavaré, O. Zeitouni, Lectures on Probability Theory and Statistics. Ecole d'Eté de Probabilités de Saint-Flour XXXI-2001. Editor: J. Picard (2004)
Vol. 1838: A.J. Ganesh, N.W. O'Connell, D.J. Wischik, Big Queues. XII, 254 p, 2004.
Vol. 1839: R. Gohm, Noncommutative Stationary Processes. VIII, 170 p, 2004.
Vol. 1840: B. Tsirelson, W. Werner, Lectures on Probability Theory and Statistics. Ecole d'Eté de Probabilités de Saint-Flour XXXII-2002. Editor: J. Picard (2004)
Vol. 1841: W. Reichel, Uniqueness Theorems for Variational Problems by the Method of Transformation Groups (2004)
Vol. 1842: T. Johnsen, A. L. Knutsen, K_3 Projective Models in Scrolls (2004)
Vol. 1843: B. Jefferies, Spectral Properties of Noncommuting Operators (2004)
Vol. 1844: K.F. Siburg, The Principle of Least Action in Geometry and Dynamics (2004)
Vol. 1845: Min Ho Lee, Mixed Automorphic Forms, Torus Bundles, and Jacobi Forms (2004)
Vol. 1846: H. Ammari, H. Kang, Reconstruction of Small Inhomogeneities from Boundary Measurements (2004)
Vol. 1847: T.R. Bielecki, T. Björk, M. Jeanblanc, M. Rutkowski, J.A. Scheinkman, W. Xiong, Paris-Princeton Lectures on Mathematical Finance 2003 (2004)
Vol. 1848: M. Abate, J. E. Fornaess, X. Huang, J. P. Rosay, A. Tumanov, Real Methods in Complex and CR Geometry, Martina Franca, Italy 2002. Editors: D. Zaitsev, G. Zampieri (2004)
Vol. 1849: Martin L. Brown, Heegner Modules and Elliptic Curves (2004)
Vol. 1850: V. D. Milman, G. Schechtman (Eds.), Geometric Aspects of Functional Analysis. Israel Seminar 2002-2003 (2004)
Vol. 1851: O. Catoni, Statistical Learning Theory and Stochastic Optimization (2004)
Vol. 1852: A.S. Kechris, B.D. Miller, Topics in Orbit Equivalence (2004)
Vol. 1853: Ch. Favre, M. Jonsson, The Valuative Tree (2004)

Vol. 1854: O. Saeki, Topology of Singular Fibers of Differential Maps (2004)

Vol. 1855: G. Da Prato, P.C. Kunstmann, I. Lasiecka, A. Lunardi, R. Schnaubelt, L. Weis, Functional Analytic Methods for Evolution Equations. Editors: M. Iannelli, R. Nagel, S. Piazzera (2004)

Vol. 1856: K. Back, T.R. Bielecki, C. Hipp, S. Peng, W. Schachermayer, Stochastic Methods in Finance, Bressanone/Brixen, Italy, 2003. Editors: M. Fritelli, W. Runggaldier (2004)

Vol. 1857: M. Émery, M. Ledoux, M. Yor (Eds.), Séminaire de Probabilités XXXVIII (2005)

Vol. 1858: A.S. Cherny, H.-J. Engelbert, Singular Stochastic Differential Equations (2005)

Vol. 1859: E. Letellier, Fourier Transforms of Invariant Functions on Finite Reductive Lie Algebras (2005)

Vol. 1860: A. Borisyuk, G.B. Ermentrout, A. Friedman, D. Terman, Tutorials in Mathematical Biosciences I. Mathematical Neurosciences (2005)

Vol. 1861: G. Benettin, J. Henrard, S. Kuksin, Hamiltonian Dynamics – Theory and Applications, Cetraro, Italy, 1999. Editor: A. Giorgilli (2005)

Vol. 1862: B. Helffer, F. Nier, Hypoelliptic Estimates and Spectral Theory for Fokker-Planck Operators and Witten Laplacians (2005)

Vol. 1863: H. Führ, Abstract Harmonic Analysis of Continuous Wavelet Transforms (2005)

Vol. 1864: K. Efstathiou, Metamorphoses of Hamiltonian Systems with Symmetries (2005)

Vol. 1865: D. Applebaum, B.V. R. Bhat, J. Kustermans, J. M. Lindsay, Quantum Independent Increment Processes I. From Classical Probability to Quantum Stochastic Calculus. Editors: M. Schürmann, U. Franz (2005)

Vol. 1866: O.E. Barndorff-Nielsen, U. Franz, R. Gohm, B. Kümmerer, S. Thorbjønsen, Quantum Independent Increment Processes II. Structure of Quantum Lévy Processes, Classical Probability, and Physics. Editors: M. Schürmann, U. Franz, (2005)

Vol. 1867: J. Sneyd (Ed.), Tutorials in Mathematical Biosciences II. Mathematical Modeling of Calcium Dynamics and Signal Transduction. (2005)

Vol. 1868: J. Jorgenson, S. Lang, $Pos_n(R)$ and Eisenstein Series. (2005)

Vol. 1869: A. Dembo, T. Funaki, Lectures on Probability Theory and Statistics. Ecole d'Eté de Probabilités de Saint-Flour XXXIII-2003. Editor: J. Picard (2005)

Vol. 1870: V.I. Gurariy, W. Lusky, Geometry of Müntz Spaces and Related Questions. (2005)

Vol. 1871: P. Constantin, G. Gallavotti, A.V. Kazhikhov, Y. Meyer, S. Ukai, Mathematical Foundation of Turbulent Viscous Flows, Martina Franca, Italy, 2003. Editors: M. Cannone, T. Miyakawa (2006)

Vol. 1872: A. Friedman (Ed.), Tutorials in Mathematical Biosciences III. Cell Cycle, Proliferation, and Cancer (2006)

Vol. 1873: R. Mansuy, M. Yor, Random Times and Enlargements of Filtrations in a Brownian Setting (2006)

Vol. 1874: M. Yor, M. Émery (Eds.), In Memoriam Paul-André Meyer - Séminaire de Probabilités XXXIX (2006)

Vol. 1875: J. Pitman, Combinatorial Stochastic Processes. Ecole d'Eté de Probabilités de Saint-Flour XXXII-2002. Editor: J. Picard (2006)

Vol. 1876: H. Herrlich, Axiom of Choice (2006)

Vol. 1877: J. Steuding, Value Distributions of L-Functions (2007)

Vol. 1878: R. Cerf, The Wulff Crystal in Ising and Percolation Models, Ecole d'Eté de Probabilités de Saint-Flour XXXIV-2004. Editor: Jean Picard (2006)

Vol. 1879: G. Slade, The Lace Expansion and its Applications, Ecole d'Eté de Probabilités de Saint-Flour XXXIV-2004. Editor: Jean Picard (2006)

Vol. 1880: S. Attal, A. Joye, C.-A. Pillet, Open Quantum Systems I, The Hamiltonian Approach (2006)

Vol. 1881: S. Attal, A. Joye, C.-A. Pillet, Open Quantum Systems II, The Markovian Approach (2006)

Vol. 1882: S. Attal, A. Joye, C.-A. Pillet, Open Quantum Systems III, Recent Developments (2006)

Vol. 1883: W. Van Assche, F. Marcellàn (Eds.), Orthogonal Polynomials and Special Functions, Computation and Application (2006)

Vol. 1884: N. Hayashi, E.I. Kaikina, P.I. Naumkin, I.A. Shishmarev, Asymptotics for Dissipative Nonlinear Equations (2006)

Vol. 1885: A. Telcs, The Art of Random Walks (2006)

Vol. 1886: S. Takamura, Splitting Deformations of Degenerations of Complex Curves (2006)

Vol. 1887: K. Habermann, L. Habermann, Introduction to Symplectic Dirac Operators (2006)

Vol. 1888: J. van der Hoeven, Transseries and Real Differential Algebra (2006)

Vol. 1889: G. Osipenko, Dynamical Systems, Graphs, and Algorithms (2006)

Vol. 1890: M. Bunge, J. Funk, Singular Coverings of Toposes (2006)

Vol. 1891: J.B. Friedlander, D.R. Heath-Brown, H. Iwaniec, J. Kaczorowski, Analytic Number Theory, Cetraro, Italy, 2002. Editors: A. Perelli, C. Viola (2006)

Vol. 1892: A. Baddeley, I. Bárány, R. Schneider, W. Weil, Stochastic Geometry, Martina Franca, Italy, 2004. Editor: W. Weil (2007)

Vol. 1893: H. Hanßmann, Local and Semi-Local Bifurcations in Hamiltonian Dynamical Systems, Results and Examples (2007)

Vol. 1894: C.W. Groetsch, Stable Approximate Evaluation of Unbounded Operators (2007)

Vol. 1895: L. Molnár, Selected Preserver Problems on Algebraic Structures of Linear Operators and on Function Spaces (2007)

Vol. 1896: P. Massart, Concentration Inequalities and Model Selection, Ecole d'Été de Probabilités de Saint-Flour XXXIII-2003. Editor: J. Picard (2007)

Vol. 1897: R. Doney, Fluctuation Theory for Lévy Processes, Ecole d'Été de Probabilités de Saint-Flour XXXV-2005. Editor: J. Picard (2007)

Vol. 1898: H.R. Beyer, Beyond Partial Differential Equations, On linear and Quasi-Linear Abstract Hyperbolic Evolution Equations (2007)

Vol. 1899: Séminaire de Probabilités XL. Editors: C. Donati-Martin, M. Émery, A. Rouault, C. Stricker (2007)

Vol. 1900: E. Bolthausen, A. Bovier (Eds.), Spin Glasses (2007)

Vol. 1901: O. Wittenberg, Intersections de deux quadriques et pinceaux de courbes de genre 1, Intersections of Two Quadrics and Pencils of Curves of Genus 1 (2007)

Vol. 1902: A. Isaev, Lectures on the Automorphism Groups of Kobayashi-Hyperbolic Manifolds (2007)

Vol. 1903: G. Kresin, V. Maz'ya, Sharp Real-Part Theorems (2007)

Vol. 1904: P. Giesl, Construction of Global Lyapunov Functions Using Radial Basis Functions (2007)

Vol. 1905: C. Prévôt, M. Röckner, A Concise Course on Stochastic Partial Differential Equations (2007)

Vol. 1906: T. Schuster, The Method of Approximate Inverse: Theory and Applications (2007)

Vol. 1907: M. Rasmussen, Attractivity and Bifurcation for Nonautonomous Dynamical Systems (2007)

Vol. 1908: T.J. Lyons, M. Caruana, T. Lévy, Differential Equations Driven by Rough Paths, Ecole d'Été de Probabilités de Saint-Flour XXXIV-2004 (2007)

Vol. 1909: H. Akiyoshi, M. Sakuma, M. Wada, Y. Yamashita, Punctured Torus Groups and 2-Bridge Knot Groups (I) (2007)

Vol. 1910: V.D. Milman, G. Schechtman (Eds.), Geometric Aspects of Functional Analysis. Israel Seminar 2004-2005 (2007)

Vol. 1911: A. Bressan, D. Serre, M. Williams, K. Zumbrun, Hyperbolic Systems of Balance Laws. Cetraro, Italy 2003. Editor: P. Marcati (2007)

Vol. 1912: V. Berinde, Iterative Approximation of Fixed Points (2007)

Vol. 1913: J.E. Marsden, G. Misiołek, J.-P. Ortega, M. Perlmutter, T.S. Ratiu, Hamiltonian Reduction by Stages (2007)

Vol. 1914: G. Kutyniok, Affine Density in Wavelet Analysis (2007)

Vol. 1915: T. Bıyıkoğlu, J. Leydold, P.F. Stadler, Laplacian Eigenvectors of Graphs. Perron-Frobenius and Faber-Krahn Type Theorems (2007)

Vol. 1916: C. Villani, F. Rezakhanlou, Entropy Methods for the Boltzmann Equation. Editors: F. Golse, S. Olla (2008)

Vol. 1917: I. Veselić, Existence and Regularity Properties of the Integrated Density of States of Random Schrödinger (2008)

Vol. 1918: B. Roberts, R. Schmidt, Local Newforms for GSp(4) (2007)

Vol. 1919: R.A. Carmona, I. Ekeland, A. Kohatsu-Higa, J.-M. Lasry, P.-L. Lions, H. Pham, E. Taflin, Paris-Princeton Lectures on Mathematical Finance 2004. Editors: R.A. Carmona, E. Çinlar, I. Ekeland, E. Jouini, J.A. Scheinkman, N. Touzi (2007)

Vol. 1920: S.N. Evans, Probability and Real Trees. Ecole d'Été de Probabilités de Saint-Flour XXXV-2005 (2008)

Vol. 1921: J.P. Tian, Evolution Algebras and their Applications (2008)

Vol. 1922: A. Friedman (Ed.), Tutorials in Mathematical BioSciences IV. Evolution and Ecology (2008)

Vol. 1923: J.P.N. Bishwal, Parameter Estimation in Stochastic Differential Equations (2008)

Vol. 1924: M. Wilson, Littlewood-Paley Theory and Exponential-Square Integrability (2008)

Vol. 1925: M. du Sautoy, L. Woodward, Zeta Functions of Groups and Rings (2008)

Vol. 1926: L. Barreira, V. Claudia, Stability of Nonautonomous Differential Equations (2008)

Vol. 1927: L. Ambrosio, L. Caffarelli, M.G. Crandall, L.C. Evans, N. Fusco, Calculus of Variations and Non-Linear Partial Differential Equations. Cetraro, Italy 2005. Editors: B. Dacorogna, P. Marcellini (2008)

Vol. 1928: J. Jonsson, Simplicial Complexes of Graphs (2008)

Vol. 1929: Y. Mishura, Stochastic Calculus for Fractional Brownian Motion and Related Processes (2008)

Vol. 1930: J.M. Urbano, The Method of Intrinsic Scaling. A Systematic Approach to Regularity for Degenerate and Singular PDEs (2008)

Vol. 1931: M. Cowling, E. Frenkel, M. Kashiwara, A. Valette, D.A. Vogan, Jr., N.R. Wallach, Representation Theory and Complex Analysis. Venice, Italy 2004. Editors: E.C. Tarabusi, A. D'Agnolo, M. Picardello (2008)

Vol. 1932: A.A. Agrachev, A.S. Morse, E.D. Sontag, H.J. Sussmann, V.I. Utkin, Nonlinear and Optimal Control Theory. Cetraro, Italy 2004. Editors: P. Nistri, G. Stefani (2008)

Vol. 1933: M. Petkovic, Point Estimation of Root Finding Methods (2008)

Vol. 1934: C. Donati-Martin, M. Émery, A. Rouault, C. Stricker (Eds.), Séminaire de Probabilités XLI (2008)

Vol. 1935: A. Unterberger, Alternative Pseudodifferential Analysis (2008)

Vol. 1936: P. Magal, S. Ruan (Eds.), Structured Population Models in Biology and Epidemiology (2008)

Vol. 1937: G. Capriz, P. Giovine, P.M. Mariano (Eds.), Mathematical Models of Granular Matter (2008)

Vol. 1938: D. Auroux, F. Catanese, M. Manetti, P. Seidel, B. Siebert, I. Smith, G. Tian, Symplectic 4-Manifolds and Algebraic Surfaces. Cetraro, Italy 2003. Editors: F. Catanese, G. Tian (2008)

Vol. 1939: D. Boffi, F. Brezzi, L. Demkowicz, R.G. Durán, R.S. Falk, M. Fortin, Mixed Finite Elements, Compatibility Conditions, and Applications. Cetraro, Italy 2006. Editors: D. Boffi, L. Gastaldi (2008)

Vol. 1940: J. Banasiak, V. Capasso, M.A.J. Chaplain, M. Lachowicz, J. Miękisz, Multiscale Problems in the Life Sciences. From Microscopic to Macroscopic. Będlewo, Poland 2006. Editors: V. Capasso, M. Lachowicz (2008)

Vol. 1941: S.M.J. Haran, Arithmetical Investigations. Representation Theory, Orthogonal Polynomials, and Quantum Interpolations (2008)

Vol. 1942: S. Albeverio, F. Flandoli, Y.G. Sinai, SPDE in Hydrodynamic. Recent Progress and Prospects. Cetraro, Italy 2005. Editors: G. Da Prato, M. Röckner (2008)

Vol. 1943: L.L. Bonilla (Ed.), Inverse Problems and Imaging. Martina Franca, Italy 2002 (2008)

Vol. 1944: A. Di Bartolo, G. Falcone, P. Plaumann, K. Strambach, Algebraic Groups and Lie Groups with Few Factors (2008)

Recent Reprints and New Editions

Vol. 1702: J. Ma, J. Yong, Forward-Backward Stochastic Differential Equations and their Applications. 1999 – Corr. 3rd printing (2007)

Vol. 830: J.A. Green, Polynomial Representations of GL_n, with an Appendix on Schensted Correspondence and Littelmann Paths by K. Erdmann, J.A. Green and M. Schoker 1980 – 2nd corr. and augmented edition (2007)

Vol. 1693: S. Simons, From Hahn-Banach to Monotonicity (Minimax and Monotonicity 1998) – 2nd exp. edition (2008)

Vol. 470: R.E. Bowen, Equilibrium States and the Ergodic Theory of Anosov Diffeomorphisms. With a preface by D. Ruelle. Edited by J.-R. Chazottes. 1975 – 2nd rev. edition (2008)

Vol. 523: S.A. Albeverio, R.J. Høegh-Krohn, S. Mazzucchi, Mathematical Theory of Feynman Path Integral. 1976 – 2nd corr. and enlarged edition (2008)

Lecture Notes in Mathematics Vol. 1940

ISBN 978-3-540-78360-2 © Springer-Verlag Berlin Heidelberg 2009

Vincenzo Capasso
and Mirosław Lachowicz (Eds.)

Multiscale Problems in the Life Sciences

Publisher's Errata

The original version of this chapter unfortunately contained several mistakes. The corrections are given below

page XI, line 25: "Links Between" instead of "Lins Between";

page 201, line 1: "Links Between" instead of "Lins Between";

page 202, Eq. (1): "$\frac{v_{\text{char}}}{c}$" instead of "$frac v_{\text{char}} c$";

page 263, ref. 31: "Modelling aspects of cancer growth" instead of "Modelling tumour growth".